FLORA OF TROPICAL EAST AFRICA

SOLANACEAE

JENNIFER M. EDMONDS, M.A., PH.D., F.L.S.[1,2]

Solanum L. sections *Oliganthes, Melongena* and *Monodolichopus*
by Maria S. Vorontsova M.A., Ph.D., F.L.S.[3]* & Sandra Knapp Ph.D., F.L.S.[3]

Annual or perennial erect or climbing herbs, shrubs or small trees with sympodial growth; vegetative and reproductive parts glabrescent to densely pubescent, with stellate, branched or simple uni- or multi-cellular glandular or eglandular hairs, usually interspersed with spherical four-celled glands. Stems terete or angled, the angles smooth or dentate, with or without straight or curved prickles. Leaves simple to pinnate (pinnatifid), alternate to opposite, with or without prickles on the midrib and veins, usually exstipulate. Inflorescences terminal, leaf-opposed or extra-axillary cymes, simple to much-forked, usually pedunculate, sometimes reduced to a solitary flower. Flowers usually hermaphrodite, rarely unisexual when plants monoecious or more rarely dioecious, actinomorphic or sometimes zygomorphic, rarely bilabiate, mostly 5-merous, occasionally 4- or 6-merous, usually pedicellate, the pedicels sometimes articulated above the base. Calyx (4–)5(–10)-lobed or -dentate, the lobes imbricate or valvate, lower part fused, sometimes cylindrical or tubular, often persistent, sometimes enlarged in fruit. Corolla rotate to campanulate, infundibuliform or tubular, (4–)5(–10)-lobed, rarely subentire; lobes valvate or plicate in bud. Stamens (2–)5(–8), adnate to corolla tube and alternating with the lobes, usually equal, sometimes unequal, often touching but not fused; anthers bithecate, distinct or connivent, dehiscing introrsely by longitudinal slits or by oblique or apical pores, basi-, ventri- or dorsi-fixed; filaments glabrous to pubescent. Ovary superior, 2(4–5)-locular, sessile to shortly stipitate with numerous axile anatropous ovules; style simple, glabrous to partially pubescent; stigma terminal, capitate, longitudinally clavate to bilobed. Fruit a capsule dehiscing by 2–4 valves, or an indehiscent glabrous to pubescent berry, usually 2(–5)-locular, sometimes dry and circumscissile, sometimes with a false septum; sclerotic granules present or absent. Seeds rounded or flat and winged, usually numerous; embryo curved, annular or straight and embedded in copious endosperm; cotyledons semiterete to linear.

About 96 genera with some 2300 species, cosmopolitan with the exception of polar regions. Greatest species concentration and diversity occurs in Central and South America, particularly in the Andean region. Within Africa, the family is relatively poorly represented; only *Discopodium, Lycium, Solanum* and *Withania* contain species which are endemic both to this region and to Africa as a whole. Of these only *Discopodium* and *Withania* are considered to be truly native to tropical Africa (cf. D'Arcy in Hawkes *et al.*(eds), Solanaceae III: 75–137 (1991)). Nevertheless, it is probable that the FTEA region contains the largest number of solanaceous

[1] Faculty of Biology, University of Leeds, Leeds, LS2 9JT & The Herbarium, Royal Botanic Gardens, Kew, Richmond, Surrey, TW9 3AE.
[2] The Bentham-Moxon Trust is gratefully acknowledged for partial financial support for JME during the completion of this treatment.
[3] The Natural History Museum, Cromwell Road, London, SW7 5BD [*now of The Herbarium, Royal Botanic Gardens, Kew]. MSV and SK were supported by the US National Science Foundation DEB-0316614.

1

species found in Africa. Many genera are only represented by one to a few species, which, with the exception of the four genera listed above, were probably introduced either deliberately for their economic, gastronomic, medicinal or ornamental value, or accidentally, especially during the intensive travel and trading that followed the discovery of the Americas by the Europeans. Many of these species originated from the Neotropics, particularly from Central and South America and have successfully established themselves as adventive or casual taxa throughout the FTEA region – often as escapes from cultivation.

Species belonging to the genus *Solanum* exhibit the greatest morphological diversity and taxonomic complexity in Africa as they do elsewhere in the world. Over 70 *Solanum* species occur in the FTEA region; some of these are native to the floral area, others are endemic to more widespread areas within Africa, while some have been introduced, usually from the Neotropics but also from Eurasia.

Solanaceae species are among the most important serving mankind. They include those of considerable economic importance as essential vegetables and fruits (e.g. potatoes, aubergines, tomatoes and peppers etc.); as ornamental plants (e.g. *Petunia*, *Schizanthus* and *Brugmansia*); as drug sources (e.g. anisodamine, anisodine, atropine, scopolamine, hyoscyamine and nicotine); as medicines – particularly by rural communities – and for smoking.

Many authors have reviewed and revised the classification of the Solanaceae since it was first described by de Jussieu in 1789. D'Arcy reviewed the historical treatments of the family [in Solanaceae Newsletter 2: 8–15 (1975) & in Hawkes *et al.* (eds), The Biology and Taxonomy of the Solanaceae: 3–47 (1979)] before revising the infrafamilial classification (1991). Later, Hunziker (Genera of Solanaceae (2001)) comprehensively surveyed the family using morphological, karyological and phytochemical data, and revised and collated its infrafamilial classification before treating and illustrating each category down to generic level. Other authors have tackled the infrageneric classification of a number of genera, with *Solanum* itself attracting the most reviews, including that of Nee for the New World Taxa (in Nee *et al.* (eds), Solanaceae IV: 285–333 (1999)) and Child & Lester (in van den Berg *et al.* (eds), Solanaceae V: 39–60 (2001)). A number of authors have also addressed its phylogeny such as Olmstead *et al.* (in Nee *et al.* (eds) Solanaceae IV: 111–137 (1999)) who used chloroplast DNA sequencing to generate a new infrafamilial classification. Molecular systematic research on the Solanaceae from 1982–2006 was recently summarised by Olmstead & Bohs (in Spooner *et al.* (eds), Solanaceae VI: 255–268 (2007)). They estimated that over 90% of genera and 37% of species had been analysed within that period, with many of the genera having been well-sampled. Systematically, the family is close to the Boraginaceae, Convolvulaceae, Scrophulariaceae and Verbenaceae (cf. Gonçalves in F.Z. 8(4): 1 (2005)), with molecular data derived from cpDNA placing it in a sister group with the Convolvulaceae (Olmstead *et al.* in Ann. Missouri Bot. Gard. 80: 700–722 (1993)).

The family has variously been subdivided into two, three, five (cf. D'Arcy, 1991) or six (Hunziker 2001) subfamilies. However, Olmstead *et al.*'s (1999) molecular analysis suggested that the subfamily Cestroideae was composed of six smaller monophyletic subfamilies, resulting in the recognition of seven solanaceous subfamilies. The family synopsis given below broadly follows D'Arcy's infrafamilial classification of subfamilies and tribes with some the modifications suggested by Olmstead *et al.*'s (1999) molecular results. Olmstead & Bohs (2007) and Olmstead *et al.* in Taxon 57: 1159–1181 (2008) proposed a formal classification, which has not been adopted here, though it broadly agrees with many of the categories used. The supra-generic authorities have been taken from the *Index Nominum Supragenericorum Plantarum Vascularium* website compiled by James L. Reveal. These often differ from those cited by other authors, particularly by Hunziker (2001). The generic authorities and type species have been taken and verified from the *Index Nominum Genericorum* by Farr *et al.* (Regnum Veg. 100, 101 & 102 (1979) and from *Names in Current Use for Extant Plant Genera* by Greuter *et al.* (NCU–3, Regnum Veg. 129 (1993)). An artificial key to all genera represented in the FTEA region is given, even if they are only represented by cultivated plants.

Subfamily 1. **Cestroideae** Burnett

Corolla usually zygomorphic. Stamens often 4 or less, inserted on the upper part of the corolla tube. Fruit a capsule. Seeds variously shaped but never discoid-compressed; embryo usually straight or only slightly curved when the cotyledons incumbent or oblique and broader than the embryo. Mainly occurring in the Americas, but also found in Australia and New Caledonia and some in Namibia. Composed of around eight tribes with 23 genera and approximately 500 species.

Tribe 1. **Cestreae** Dumort.
Shrubs or small trees. Flowers in bracteate, paniculate cymes; pedicels articulated. Calyx 5-toothed. Corolla tubular, occasionally swollen distally; aestivation inflexo-valvate, slightly contorted. Stamens 5, all fertile and usually equal; filaments inserted below or above the middle of the corolla tube, sometimes geniculate or swollen at or above the junction; anthers small, dorsifixed, the thecae separated for lower third. Disc prominent. Fruit a berry or a capsule. Seeds prismatic; embryo straight straight or slightly curved. Composed of three genera, all native to the Americas. Genus 1. **Cestrum**.

Tribe 2. **Francisceae** G. Don
Woody shrubs or trees. Flowers solitary or in terminal cymes. Calyx actinomorphic or slightly zygomorphic, salviform; aestivation imbricate. Stamens 4, didynamous, medifixed, anthers monothecal. Style smooth; stigma bilobed to bifid. Ovary bilocular, multi-ovuled, ovules anatropous. Fruit a coriaceous capsule, 2-valved. Seeds reniform, ellipsoid, ovoid or prismatic, usually angular; cotyledons incumbent, endosperm copius. Monogeneric, confined to C and S America. Genus 2. **Brunfelsia**.

Tribe 3. **Nicotianeae** Dumort.
Usually herbs, occasionally woody. Corolla actinomorphic or zygomorphic; aestivation contorted-conduplicate, imbricate-conduplicate or imbricate. Stamens 4 or 5 (rarely 2), inserted variously in the corolla tube; anthers with independent thecae. Fruit a septicidal many-seeded capsule. Seeds with straight or bent and hemicyclic embryos. Composed of 8 genera, largely found in S America where mainly Andean, but also occurring in N America, Africa and Australia. Genera 3. **Nicotiana** and 4. **Petunia**.

Tribe 4. **Browallieae** Hunz.
Herbs or small shrubs. Flowers in sub-corymbose or racemose cymes or solitary, showy. Calyx campanulate-tubular to tubular, shorter than the corolla. Corolla zygomorphic, salverform to infundibuliform; aestivation imbricate-conduplicate or reciprocative. Stamens 4, didynamous, basifixed; filaments curved; the short anthers with one abortive theca. Style geniculate, tortuous apically, corrugated in distal third; stigmas broad, compressed, 2-horned, the central part 4-lobed. Ovary 2-locular with many ovules. Fruit a capsule, 2-valved, the valves bifid, enveloped by an accrescent papery calyx. Seeds cuboid or elongate-cuboid; embryo slightly curved with copious endosperm. Composed of 2 genera which are sometimes considered synonymous (cf. Mabberley's Plant Book, 3rd ed.; 826 (2008)) and which are found from the southern US to southern S America. Genera 5. **Browallia** and 6. **Streptosolen**.

Tribe 5. **Schwenckieae** Hunz.
Herbs or small shrubs, rarely climbing. Calyx usually campanulate. Corolla zygomorphic, tubular or salver-shaped, aestivation valvate-conduplicate. Stamens 4, didynamous, or composed of 2+2 staminodes, inserted near corolla base; anthers basifixed, opening lengthwise. Ovary 2-locular, multi-ovulate. Fruit a dry capsule. Seeds small, sub-prismatic, angled; embryo straight or slightly curved, small. Composed of 4 genera, mainly found in the Americas. Genus 7. **Schwenckia**.

Subfamily 2. **Schizanthoideae** Hunz.

Herbs with deeply dissected leaves; laminas pinnate or pinnatisect, rarely entire. Inflorescences terminal with resupinate flowers. Calyx 5-partite. Corolla zygomorphic, bilabiate with 5 unequal deeply dissected segments composed of one anterior, two lateral and two posterior 'petals' which fused into a keel; brightly and variously coloured; aestivation imbricate. Androecium 5-partite with two fertile stamens inside posterior keel and three staminodes with vestigial anthers; fertile anthers dorsifixed. Style filiform; stigma discoidal and inconspicuous. Ovary bilocular, many-ovuled. Fruit septicidal capsules, bi-valved. Seeds numerous, compressed. Embryo arcuate, endosperm copious. Hunziker (2001) considered this to be a monotypic small genus originating in S America. Genus 8. **Schizanthus**.

Subfamily 3. **Solanoideae** Kostel.

Corolla usually actinomorphic. Stamens (4–)5(–6), inserted near base of corolla tube. Fruit usually a berry, often with sclerosomes. Seeds compressed, discoidal or reniform; embryo usually coiled, uniform in diameter; cotyledons incumbent, sometimes oblique, or rarely curved outwards apically; usually with abundant endosperm. Composed of 9 tribes, 50 genera and around 1730 species.

Tribe 6. **Datureae** Dumort.
Herbs to small trees. Flowers large, usually solitary though sometimes several aggregated in the axils. Calyx tubular, sometimes circumsessile in fruit. Corolla slightly zygomorphic, infundibuliform to almost tubular; aestivation conduplicate-contorted. Stamens usually 5, occasionally more; filaments inserted variously from base in corolla tube; anthers basifixed, free or coherent, opening longitudinally. Ovary 2–4-locular. Fruit a capsule which dehiscent or woody and indehiscent. Seeds large, flattened, rough or wedge-shaped; embryo straight or circinnate. Composed of 2 genera, originating in the Americas but now found throughout the world. Genera 9. **Datura** and 10. **Brugmansia**.

Tribe 7. **Lycieae** Lowe
Herbs to small trees, usually woody, often spiny. Flowers solitary or clustered among fascicled leaves. Calyx not or only slightly enlarged. Corolla tubular to narrowly campanulate; aestivation imbricate, cochlear or quincuncial. Stamens usually 5, occasionally 4; filaments inserted basally to nearly apically on the corolla tube; anthers dorsifixed, with the thecae divergent below. Ovary 2(–4) locular, few to many-ovulate. Fruit a berry. Seeds discoidal, compressed; embryo strongly curved, flat. Composed of 3 genera which are predominately temperate, often occurring in saline conditions. Genus 11. **Lycium**.

Tribe 8. **Nicandreae** Lowe
Herbs with hollow stems. Flowers large, solitary. Calyx deeply divided, much enlarged after flowering, separated from the corolla by a distinct internode, lobes auriculate at the base. Corolla campanulate, blue; aestivation quincuncial, a little conduplicate-contorted. Stamens 5; filaments geniculate at insertion; anthers basifixed between divergent thecae, opening longitudinally. Ovary irregularly divided by 3–5 false septa. Fruit a berry enclosed by the membranaceous calyx; pericarp thin and translucent. Seeds with a pronounced curved sub-cyclic embryo. Composed of a single American species which has become a ubiquitous ruderal weed. Genus 12. **Nicandra**.

Tribe 9. **Solandreae** Miers
Woody climbers. Flowers solitary, terminal; pedicels not articulated. Calyx usually zygomorphic, 3–5 times shorter than the corolla. Corolla large, slightly zygomorphic, infundibuliform; limb short; aestivation imbricate. Stamens 5, equal,

declinate, attached to the lower half of the corolla tube; anthers basifixed. Ovary 4-locular, multi-ovulate; partly inferior and sunk into the receptacle. Fruit a berry, many-seeded. Seeds discoidal or reniform, compressed; embryo sub-coiled; cotyledons incumbent or slightly oblique; endosperm abundant. Composed of only one genus distributed from the Caribbean into S America. Genus 13. **Solandra**.

Tribe 10. **Solaneae** Dumort.

Herbaceous to woody. Calyx not usually enlarging greatly, sometimes increasing to exceed the fruit at maturity. Corolla rotate, stellate, widely campanulate or slightly infundibuliform; white, yellow, violet or purple; tube very short to long; aestivation valvate, induplicate or plicate. Stamens (4–)5(–6); anthers basifixed, free, connivent or cohering; opening by apical pores or longitudinal slits. Ovary 2 (rarely to many-)-locular, multi-ovulate. Fruit a berry. Embryo strongly curved to sub-annular, terete or flat, sub-peripheral; cotyledons semi-terete. Composed of approximately 30 genera which largely American, but widely distributed elsewhere. Genera 14. **Capsicum**; 15. **Discopodium**; 16. **Iochroma**; 17. **Physalis**; 18. **Solanum** and 19. **Withania**.

KEY TO THE GENERA

1. Stems armed with glabrescent sharp spines on short branches or brachyblasts, spines surrounded by fasciculate clusters of up to 14 leaves from which solitary or paired flowers arise .... 11. *Lycium* (p. 50)

   Stems unarmed or with prickles or bristles, without brachyblasts; leaves usually solitary and alternate or opposite, not fasciculate; flowers varying from extra-axillary to terminal few- to many-flowered, unbranched to branched complex inflorescences, sometimes leaf-opposed 2

2. Fruit a dry to woody 2- to 4-valved capsule; stems always unarmed 3

   Fruit a succulent or dryish indehiscent berry; stems sometimes armed with prickles or bristles 10

3. Leaves pinnatisect, occasionally pinnate with incised or dentate segments; inflorescences always terminal many-flowered cymes; flowers zygomorphic and bilabiate; corollas composed of five unequal segments comprising one anterior, two lateral and two posterior which fused to form a keel .... 8. *Schizanthus* (p. 37)

   Leaves entire or lobed or sinuate-dentate, pinnae absent; flowers solitary or terminal to axillary few- to many-flowered cymes; flowers usually actinomorphic occasionally zygomorphic but corollas always composed of equal segments 4

4. Capsules not enclosed by accrescent calyces; discoid calyx remains forming a frill or collar beneath capsule with sepal lobes spreading or reflexed; valves prominently spiny or tuberculate, rarely smooth; flowers large, to 19 cm long and trumpet-shaped .... 9. *Datura* (p. 39)

   Capsules completely or partially enclosed by enlarged adherent or accrescent calyx lobes, basal calyx rim absent; valves always smooth; flowers less than 10 cm long, infundibuliform, hypocrateriform or salverform 5

10. Corollas always rotate to stellate; anthers connivent around the style, usually prominently visible, dehiscing through apical pores or short apical slits [if dehiscence by internal slits then anthers apically beaked and flowers yellow (*S. lycopersicum,* the tomato)]; inflorescences usually pedunculate . . . . .   18. *Solanum* (p. 77)

Corollas campanulate, infundibuliform, urceolate, salverform to cyathiform, rarely stellate-campanulate (cf. *Capsicum*); anthers not connivent, often included, dehiscing through internal longitudinal slits running their entire length; inflorescences often epedunculate 11

11. Flowers large and conspicuous, corollas > 14 cm long, infundibuliform to trumpet-shaped, with basal tube 5–15 cm long; calyces 8–14 cm long; filaments free for 4–15 cm; anthers 0.9–4 cm long; styles 15–19 cm 12

Flowers smaller, corollas < 5 cm long, campanulate, cyathiform, urceolate or salverform, if infundibuliform tube < 2 cm long; calyces 0.2–3 cm long; filaments free for less than 1 cm; anthers < 0.5 cm long; styles 0.3–3.3 cm long 13

12. Small trees or shrubs; leaves membranaceous to succulent; flowers pendulous, on long slender pedicels, with lobes fused almost to apex and often recurved; anthers linear, 2.1–4 cm long; styles exserted up to 3.5 cm beyond anther; stigmas 3–8 mm long, clasping stylar apices . . . .   10. *Brugmansia* (p. 46)

Woody climbers or epiphytic shrubs; leaves coriaceous; flowers erect on short stout woody pedicels with completely recurved broadly ovate lobes; anthers ellipsoid to ovoid, 0.9–1.3 cm long; styles exserted up to 10 cm beyond anthers; stigmas < 3 mm . . . .   13. *Solandra* (p. 56)

13. Fruiting calyces enlarging, becoming accrescent, chartaceous and conspicuously veined, completely enclosing mature berries 14

Fruiting calyces only slightly enlarging, with the calyx lobes becoming adherent to or apically reflexed from the berry bases; if accrescent then splitting to expose fruits and not chartaceous (occasional *Cestrum* species) 16

14. Woody annual or perennial herbs or shrubs; flowers in axillary fascicles of 2–20 flowers, with deeply triangular recurving corolla lobes; flowering pedicels 0–3 mm long; berries enclosed by urceolate calyces with apically recurved lobes . . . .   19. *Withania* (p. 226)

Annual or perennial herbs; flowers usually solitary, with shallow spreading corolla lobes; flowering pedicels usually > 3 mm long; berries enclosed by ovoid to urceolate calyces with apically connivent or erect lobes 15

15. Annual herbs; flowers blue to mauve, flowering
    pedicels up to 5 cm long; calyx lobes broadly ovate
    with sagittate bases to cordate, enlarging and
    becoming chartaceous to enclose the ripe berries,
    with erect apices . . . .   12. *Nicandra* (p. 53)
    Annual or perennial herbs, flowers green, white or
    yellow, flowering pedicels < 1.2 cm long; calyx
    lobes triangular, fused basally forming tube,
    enlarging to enclose berries, with connivent or
    recurved apices . . . .   17. *Physalis* (p. 69)
16. Flowers campanulate-rotate to -urceolate or stellate,
    solitary or in fascicles, 0.5–1.3 cm long, with tube <
    0.6 cm long; calyces < 0.5 cm long; filaments
    completely glabrous; styles 0.1–0.6 cm 17
    Flowers infundibular, tubular to trumpet-shaped
    rarely slightly urceolate above, usually in terminal
    or subterminal many-flowered cymes, racemes,
    panicles or fascicles, rarely solitary or few-flowered,
    1.8–4 cm long with tube > 1 cm long; calyces tubular
    to campanulate-cupulate, usually > 5 mm long;
    styles 1–3.2 cm 18
17. Annual or perennial herbs or slender short-lived
    shrubs; young stems sparsely pilose to glabrous;
    flowers solitary or paired (occasionally in fours);
    calyces with shallow sepal lobes; disc small and
    inconspicuous; fruiting calyces forming a collar
    around berry (peppers) bases; seeds with thickened
    margins . . . .   14. *Capsicum* (p. 57)
    Shrubs or small trees; young stems densely
    floccose/villous; flowers in few- to many-flowered
    fascicles, rarely solitary; calyces with pronounced
    dentate and often mucronate lobes; disc
    conspicuous, red or orange, fleshy; fruiting calyces
    adherent around berry bases with calyx lobes
    becoming reflexed apically; seeds without
    thickened margins . . . .   15. *Discopodium* (p. 63)
18. Flowers usually blue to purple in floral area; flowering
    pedicels 0.9–2.7 cm; filaments free for 2.5–3.3 cm,
    not enlarged at adnation; anthers elongate,
    basifixed, 3–4.5 mm long; berries with numerous
    small seeds, 1–1.5 mm long, often mixed with
    sclerotic granules . . . .   16. *Iochroma* (p. 67)
    Flowers usually bright red, yellow or orange,
    occasionally white, purple or cream; flowering
    pedicels < 0.6 cm, often absent; filaments free for <
    1 cm, often enlarged at adnation; anthers globose,
    medifixed, 0.5–1.3 mm diameter; berries with few
    (< 17) seeds which 3–5.5 mm long, sclerotic granules
    absent . . . .   1. *Cestrum* (p. 9)

1. CESTRUM

L., Sp. Pl. 1: 191 (1753) & Gen. Pl. ed. 5: 88 (1754); Francey in Candollea 6: 46–398 (1935) & 7: 1–132 (1936); Benítez de Rojas & D'Arcy in Ann. Missouri Bot. Gard. 85: 275–351 (1998); Hunziker in Gen. Solanaceae: 32–35 (2001); Nee in Solanaceae V: 109–136 (2001)

Habrothamnus Endl., Gen.: 667 (1839)

Shrubs, sub-shrubs or small trees, rarely vine-like, evergreen or deciduous. Leaves alternate or opposite, simple, often pseudo-stipulate. Inflorescences axillary or terminal cymes, racemes or panicles, often congested, usually many- but occasionally solitary-flowered, bracts often present; flowers actinomorphic, occasionally zygomorphic, diurnal or nocturnal, often subtended by a small bracteole; pedicels usually erect. Calyx tubular, campanulate or cupulate, with five valvate calyx lobes, persistent and usually enlarged in fruit. Corolla tubular; tube often ampliate above and constricted at the throat, with short, induplicate/valvate lobes, usually spreading after anthesis; often densely pubescent internally. Stamens 5, sub-equal or equal, usually included; filaments variously joined to corolla tube, straight or flexuose, cylindrical or enlarged at point of fusion, sometimes with small appendage near base, glabrous to pubescent where adnate; anthers globose to cordate, medifixed, somewhat versatile. Ovary sometimes shortly stipitate, glabrous, 2-locular, ovules numerous; disc annular or cupulate, small; style filiform, usually exserted; stigma papillate, capitate, bilobed. Fruit a smooth few- to many-seeded berry. Seeds discoidal, sometimes compressed, reniform, elliptic, ovate or prismatic, few to many; testa crustaceous, reticulate-foveate.

Although over 250 species and 650 epithets have been described, Nee (2001) proposed that the correct number is closer to 150. The species are found in warm parts of Mexico, Central America, the Antilles and South America, particularly in Brazil and the Andes from Venezuela to northern Argentina. They have been widely cultivated as ornamentals in many parts of the world, where they are now commonly naturalised. Despite their popularity as garden ornamental and hedging plants, many species are notoriously poisonous, especially to stock; many are used medicinally.

The more important chemical constituents found in *Cestrum* species are summarised in Hunziker (2001). Species belonging to this genus are often called Jessamines. Night-opening flowers are thought to be sphingid (hawkmoth) pollinated, while pollination of red and yellow flowers is by hummingbirds attracted to plants with the sucrose dominant nectars which characterise Cestrums. Their juicy berries are usually bird-dispersed, though some species with foetid foliage and leathery fruits are probably bat-dispersed. Many authors (e.g. Francey, 1935, 1936; Nee, 2001; and Benítez de Rojas & D'Arcy, 1998) have recognized a number of varieties and even forms in most of the species described below. Since none of the names concerned have been encountered on East African material, they have not been included in the synonymy of these species.

The flowers of these species are borne in complex inflorescences which can be compact or lax cymes, racemes, umbels or panicles. The precise structure is extremely difficult to determine from the available herbarium material, and so has not been used diagnostically in the following account.

1. Flowers orange or bright yellow; calyces narrowly tubular, enclosing lower half of corolla tube, lobes subulate apically . . . . 1. *C. aurantiacum*
   Flowers red, pink, purple, white, cream, yellowish-green or pale yellow; calyces broadly tubular, campanulate or cupulate, enclosing less than lower third of corolla tube, lobes acute 2
2. Stems usually ferrugineous; vegetative parts hirsute with long simple or branched hairs; corollas urceolate becoming bulbous above calyx tube before narrowing above . . . . 2. *C. elegans*
   Stems usually light or dark green occasionally brownish-green; vegetative parts pilose with short predominantly simple hairs to glabrescent; corollas infundibuliform or tubular 3
3. Flowers strongly fragrant either diurnally or nocturnally; calyces usually < 4 mm long; filaments fused to upper part of corolla tube, free for < 3 mm 4
   Flowers not strongly fragrant, occasionally with musky odour; calyces 4–6 mm long; filaments fused to middle of corolla tube, free for 6–8 mm 5
4. Flowers with strong nocturnal fragrance; calyx enclosing only base of corolla tube; filaments free for 1–3 mm with basal simple or forked appendage just above adnation; berries white . . . . 3. *C. nocturnum*
   Flowers with strong diurnal fragrance; calyx enclosing lower third to quarter of corolla tube; filaments almost sessile on corolla tube, free for < 0.8 mm without basal appendage; berries purple to blackish . . . . 4. *C. diurnum*
5. Flowers pale yellow to yellowish-green, lobes with densely pilose marginal bands; calyces pilose externally; filaments pilose and slightly enlarged but not geniculate at adnation with corolla tube . . . . 5. *C. parquii*
   Flowers pink, red or purple, lobes without pilose marginal bands; calyces glabrescent externally; filaments glabrous enlarged and geniculate at adnation with the corolla tube . . . . 6. *C.* × *cultum*

1. **Cestrum aurantiacum** *Lindl.* in Edwards' Bot. Reg. 30: Misc. 71 (1844); Dunal in DC., Prodr. 13(1): 603 (1852); Francey in Candollea 6: 102 (1935); Verdcourt & Trump, Common Poisonous Pl. E. Afr.: 161 (1969); U.K.W.F.: 529 (1974); Blundell, Wild Fl. E. Afr.: 188 (1992); Benítez de Rojas & D'Arcy in Ann. Missouri Bot. Gard. 85: 285 (1998); Hepper in Fl. Egypt 6: 135 (1998); Gonçalves in F.Z. 8(4): 11 (2005). Type: raised at the Chiswick Gardens seeds collected by *Skinner* in Chimalapa, Guatemala, *Lindley* s.n. (CGE!, holo.)

Shrubs, sometimes climbing or sprawling, occasionally herbaceous, 1–6.5 m high; stems laxly or densely branched, branches often lenticellate, sparsely pilose to glabrescent with scattered simple or branched eglandular hairs mixed with stalked glands. Leaves pale to dark green with yellowish venation, ovate-lanceolate to lanceolate, 6–11.4 × 2–5.6 cm, bases cuneate, margins entire, apices acute to acuminate, both surfaces usually glabrescent, sometimes coriaceous, pseudostipulate; petioles (1–)2–4 cm long. Inflorescences terminal or lateral lax leafy cymes, racemes or panicles, sometimes subspicate; flowering stems glabrescent to sparsely pilose; flowers often pleasantly fragrant; pedicels 0–1.8 mm long, sparsely to moderately pilose, with apical abscission layer; bracteoles to 11 mm long. Calyx

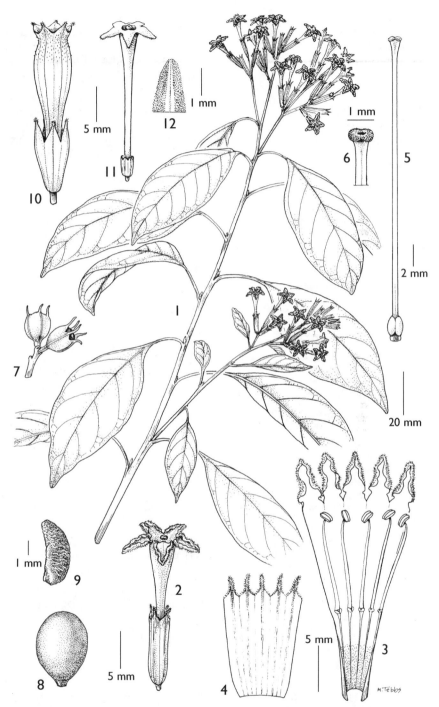

FIG. 1. *CESTRUM AURANTIACUM* — **1**, habit in flower; **2**, flower; **3**, internal view of opened
corolla; **4**, opened calyx; **5**, style and ovary; **6**, stigma; **7**, berries surrounded by accrescent
calyces; **8**, berry with calyx removed; **9**, seed. *C. ELEGANS* — **10**, flower. *C. NOCTURNUM*
— **11**, flower; **12**, underside of corolla lobe. 1–6 from *Ndirito* 13811; 7–9 from *Nyaki* 79;
10 from *Kisena* 17; 11–12 from *Sheil* 1330. Drawn by Margaret Tebbs.

narrowly tubular, 6–10 × 2–3.5 mm, glabrous to sparsely pilose externally, lobes narrowly triangular becoming subulate (ligulate) above, margins tightly inrolled with lobes extending downwards as five dark coloured veins throughout the calyx, acute, usually unequal, 1–4 × 0.2–1.5 mm, usually shortly pilose with ciliate margins in flower, enlarging to 7–8 mm broad, becoming cupulate, accrescent and glabrescent externally in fruit. Corolla usually orange to bright yellow occasionally yellowish-red, unscented, salverform to urceolate, 1.8–2.5 cm long and 0.7–1.2 cm diameter apically, tube 1.1–2 cm long, narrow basally increasing up to 6.5 mm above, glabrescent externally, pilose internally; lobes broadly to narrowly triangular, often apiculate, spreading or reflexed and ovate after anthesis, 3.5–6 × 2–4.5 mm, glabrescent externally apart from densely pilose inrolled marginal bands. Stamens usually equal; filaments free for 5–9 mm, curved apically; anthers yellow to orange, 0.6–1.3 × 0.6–1 mm. Ovary brownish, ovoid, 0.7–1.3 mm long; disc cupulate, 1–1.8 mm broad; style 1–1.9 cm long, usually exserted; stigma 0.3–0.8 mm long. Berry subsessile, white, cream or yellow, ovoid, 5–12 × 5–9 mm, pericarp sometimes splitting apically to expose seeds, partially or completely enclosed by cupulate calyx, which eventually splitting. Seeds 1–5(–8) per berry, dark brown or blackish, ellipsoidal, prismatic, usually curved, 3–5.5 × 2–4 mm, small vestigial or abortive seeds also present. Fig. 1/1–9, p. 11.

KENYA. Kiambu District: Limuru, 7 July 1967, *Ndirito* 13811! & Closeburn, Oct. 1953, *Graham Bell* EAH 10348!; Kericho District: 10 km N of Kericho, 7 Aug. 1969, *Kokwaro* 2070!
TANZANIA. Lushoto District: Lushoto, near road to Giraffe, 1969, *Shabani* 595! & roadside Silviculture Office to town, 30 June 1983, *Kisena* 136!; Mbeya District: base of Pungaluma Hills, edge of Songwe and Pungaluma coffee plantations, 16 Feb. 1990, *Lovett et al.* 4190!
DISTR. **K** 3–5: **T** 2, 3 & 7; native to Mexico, Guatamala, Nicaragua and Costa Rica but now widely cultivated throughout Africa; also in India, Europe and Australia
HAB. Cultivated, also an escape in neglected gardens, old cultivations, roadsides and disturbed montane forest; 850–2600 m
CONSERVATION NOTES. Widespread; least concern (LC)

SYN. *C. warszewiczii* Klotzsch in Allegem. Gartenzeit. 19, 46: 362 (1851). Type: Costa Rica, Cartago-Vulkan, *Warszewicz* 1738 (locality not cited, ?BR)
    C. aurantiacum Lindl. var. *warszewiczii* (Klotzsch) Francey in Candollea 6: 103 (1935)

NOTE. Commonly known as Orange or Yellow Cestrum, this species is widely grown for its showy flowers. It is notoriously toxic to animals and especially poisonous to cattle.
    Two specimens, *Magogo* 206 and *Mwakalinga* 6, are superficially similar to *C. aurantiacum* especially in their calyx morphology. However, the flowers were reportedly blue in the former and pink in the latter (with the berries also being cited as purple). Both were collected from Lushoto (**T** 3) which is notorious for morphological oddities; they are possibly of hybrid origin derived from *C. elegans* and *C. aurantiacum*. The Magogo specimen is distinctly hirsute with hairs exhibiting coloured interstices, has short petioles and the purple-coloured calyx typical of *C. elegans*, while the Mwakalinga specimen is glabrescent and more morphologically similar to *C. aurantiacum*.

2. **Cestrum elegans** (*Neumann*) *Schltdl.* in Linnaea 19: 261 (1847); Dunal in DC., Prodr. 13(1): 600 (1852); Francey in Candollea 6: 123 (1935); Troupin, Fl. Rwanda 3: 362 (1985); Hepper in Fl. Egypt 6: 135 (1998); Gonçalves in F.Z. 8(4): 9 (2005); Friis in Fl. Eth. 5: 105 (2006). Type: Mexico, Oaxaca, *Galeotti* 2619 (?P, holo.)

Evergreen shrubs or small trees, occasionally herbaceous, straggling, 1–4 m high; stems conspicuously hirsute and ferrugineous with long simple or branched hairs, stalked glands usually present throughout. Leaves often scabrid, dark green, ovate-lanceolate to lanceolate, rarely narrowly lanceolate, (4.5–)6–11(–14) × 1.8–4.3(–6.8) cm, bases cuneate, margins entire, apices acute to acuminate, both surfaces conspicuously and softly hirsute, especially on the veins and midribs; petioles (2–)5–10 mm long. Inflorescences dense terminal panicles of spicate racemes or lateral cymes, often nodding; flowering stems densely pilose/villous, often purplish; flowers in compact clusters, sometimes with a faint scent; pedicels densely hirsute, purplish, 1–3 mm long with prominent apical abscission layer; bracts narrowly lanceolate, ± 6 mm long, appearing reddish with stalked glands. Calyx pink, purple or red, campanulate to cupulate, 4–9 × 3–6 mm, lobes equal or unequal, coloured, narrowly triangular, 1.5–4 × 1–3 mm, acute, usually densely pilose with ciliate margins in flower, enlarging slightly and becoming glabrescent and adherent in fruit. Corolla usually red, occasionally pinkish, urceolate, 1.5–2.3 cm long and 5–8 mm diameter apically, narrow basally becoming bulbous above then constricted below lobes, tube 1.5–1.9 cm long, glabrescent below becoming pilose towards the lobes externally; lobes narrowly triangular, spreading after anthesis, 1.5–4 × 1.8–2.5 mm, acute, glabrescent externally apart from densely pilose marginal bands. Stamens usually equal; filaments free for 7–10 mm, curved apically; anthers yellow to orange, 0.8–1.3 × 0.9–1.3 mm. Ovary brownish, globose to ovoid, 1.2–1.4 mm long; disc cupulate, ± 1 mm broad; style 1.3–1.8 cm long, usually exserted; stigma 0.6–1.3 mm diameter. Berry often in dense clusters, glossy, dark red to dark purple, globose to broadly ovoid, 5–13 × 4–9 mm enclosed basally by cupulate enlarged calyx. Seeds 2–17 per berry, dark brown, ellipsoidal, prismatic, (1.8–)3–4.5 × 1.5–2.8 mm, often mixed with small vestigial seeds. Fig. 1/10, p. 11.

KENYA. Naivasha/Kiambu District: Sasamua Dam near South Kinangop, 31 Mar. 1971, *Gilbert* 6330!; Nyeri District: Wandumi School, 16 Apr. 1971, *Mathenge* 762!
TANZANIA. Lushoto District: Silviculture Office, 11 Dec. 1974, *Shunda* 3! & Jaegertal Hotel, 20 May 1969, *Shabani* 430! & Kwembago area, 6 July 1971, *Magogo* 59!
DISTR. **K** 3, 4; **T** 3; also cultivated and becoming naturalised in Congo, Rwanda, Zambia, Malawi, Zimbabwe and South Africa; and in N Africa, Europe, Sri Lanka, Australia and Pacific islands
HAB. Widely cultivated, also an escape in disturbed vegetation; 1350–2600 m
CONSERVATION NOTES. Widespread; least concern (LC)

SYN. *Habrothamnus elegans* Neumann, Ann. Fl. et Pomone: 116 (1844); Hooker in Bot. Reg.: 43 (1844); Fl. des Serres, ser.1, 2, t. 9 (1846).
    H. purpureus Lindl. in Edwards' Bot. Reg.: 30, t. 43 & Misc. No. 19: 12 (1844). Type: described from fresh specimen sent by Van Houtte of Ghent, Belgium, derived from material originally collected in Mexico, *Lindley* 488 (CGE, holo., photo!)
    Cestrum purpureum (Lindl.) Standl. in Contrib. US Nat. Herb. 23: 1280 (1924)

NOTE. The species is a popular ornamental, widely planted for its dense inflorescences of showy red flowers.

3. **Cestrum nocturnum** L., Sp. Pl.: 191 (1753); Dunal in DC., Prodr. 13(1): 631 (1852); Francey in Candollea 7: 67 (1936); U.O.P.Z.: 187 (1949); D'Arcy & Rakotozafy in Fl. Madagascar, Solanaceae: 18 (1994); Hepper in Fl. Egypt 6: 138 (1998); Nee in Solanaceae V: 114 (2001); Gonçalves in F.Z. 8(4): 10 (2005); Friis in Fl. Eth. 5: 105 (2006). Type: "Jamaica, Chilli", *Baeck s.n.*, Herb. Linn. 258.1 (LINN!, lecto.) designated by Deb, J. Econ. Tax. Bot. 1: 36 (1980) [see also Jarvis, Order out of Chaos: 406 (2007)]

Shrubs, sub-shrubs or small trees 1–4 m high, often evergreen; stems light green, much-branched, spreading, occasionally arched; young stems pilose with simple spreading multicellular eglandular hairs, later glabrescent, stalked glands usually present throughout. Leaves light to dark green, oblong-lanceolate, 5.5–9.6(–15) × 2.1–3.4(–5.3) cm, bases cuneate and often decurrent to stem, margins entire, apices acute to acuminate, both surfaces glabrescent but with stalked glands especially on the veins and midribs, membranaceous; pseudostipules usually present; petioles 0.8–1.2(–1.4) cm long. Inflorescences complex leafy axillary and terminal panicles or racemes, often spicate, up to 10 cm long; axes shortly pilose; flowers delicate with strong nocturnal fragrance; bracts ligulate, 1.5–2.5 mm long, glabrescent to shortly pilose, some hairs glandular; pedicels usually erect, 0.5–2.5 mm long, terminating in distinct abscission layer, glabrescent with short stalked glands. Calyx tubular/cupulate, light green, 3–3.5 × 1.5–2 mm, lobes equal or unequal, triangular, 0.5–0.8 × 0.7–1.3 mm, acute, glabrescent with ciliate margins. Corolla white, cream, yellow or green, infundibuliform or tubular, 1.8–2.6 cm long and 6–10 mm diameter apically; tube usually cylindrical, 1.5–2 cm long, glabrous externally and internally; lobes acute or obtuse, narrowly triangular, usually spreading after anthesis, 3–5 × 1.2–3 mm, with densely pilose marginal bands externally, often inrolled (induplicate). Stamens usually equal, with anthers exserted; filaments free for 1–3 mm, curved inwards with sparsely pilose geniculate simple or forked appendage up to 1.3 mm long at point of adnation; anthers brown, 0.6–0.8 × 0.5–0.7 mm. Ovary green, globose or ovoid, 0.7–1.3 mm long, glabrous; disc greenish-white, cupulate, 1–1.3 mm broad; style (1.4–)1.7–2 cm long, usually exserted; stigma 0.7–1.1 mm broad. Berry white, globose to ovoid, 7–10 mm diameter. Seeds 5–8 per berry, ovoid, prismatic, 3.5–4 mm long. Fig. 1/11–12, p. 11.

UGANDA. Bunyoro District: Budongo Forest, Sonso Sawmill, 11 Sept. 1992, *Sheil* 1330!
KENYA. Nairobi, Eastleigh Section 1 off Ganges Road, near St Theresa's Catholic Church, 12 Dec. 1971, *Mwangangi & Kasyoki* 1890!
TANZANIA. Moshi District: Kifura Sawmills, 2 July 1959, *Leone* 14!; Lushoto District: Silviculture Research Station, 1 Mar. 1976, *Shabani* 1097!; Morogoro District: Morogoro, June 1952, *Semsei* 738!
DISTR. U 2; K 4; T 2–3, 6; Z (fide U.O.P.Z.); native to West Indies or Central America, widely introduced and cultivated in the tropics and subtropics
HAB. Occasional escape around villages, sawmills and arboreta; (?0–)1500–1650 m
CONSERVATION NOTES. Widespread; least concern (LC)

NOTE. With its fragrant flowers, this species is often planted for its night-time scent and is known as Queen- or Lady-of-the-Night or Night-blooming Jessamine. The flowers are probably pollinated by night-flying insects attracted to strongly fragrant white flowers. This species is also widely used as a medicinal, hedging and ornamental plant throughout the world, but has been implicated in fatal poisoning of horses.

4. **Cestrum diurnum** *L.*, Sp. Pl.: 191 (1753); Dunal in DC., Prodr. 13(1): 604 (1852); Francey in Candollea 6: 234 (1935); Hepper in Fl. Egypt 6: 138 (1998). Type: Cuba, "Chilli, Havana", Herb. Linn. 258.4 (LINN!, lecto., designated by Howard in Fl. Less. Antill., 6: 274 (1989)) [See also Jarvis, Order out of Chaos: 406 (2007)]

Evergreen shrubs or small trees 2–5 m high; stems with inconspicuous stalked glands throughout. Leaves oblong-lanceolate, 7.2–9.2 × 2.3–3 cm, lower surfaces pilose with simple hairs interspersed with stalked glands; petioles 10–11 mm long. Inflorescences terminal or lateral spicate racemes, cymes, umbels or panicles; pedicels absent or to 1 mm long. Calyx cupulate to campanulate, 2.5–3.5(–4.5) × 1.5–2.5 mm, lobes 0.2–0.8 × 0.5–1 mm, ciliate. Corolla white or creamy-white, diurnally fragrant, tubular to infundibuliform, 1.2–1.6 cm long and 4.5–7 mm diameter apically; tube 10–12 mm long; lobes 1.5–2.1 × 1.5–3 mm, usually spreading after anthesis. Berry purple-black, often glossy, globose to ovoid, 6.5–7 × > 4.5 mm, base enclosed by enlarged cupulate calyces, pericarp splitting apically to expose seeds. Seeds 2 per berry plus several vestigial, light brown, elliptical, 3.1–3.5 × 2 mm.

KENYA. Cultivated: Nairobi, Closeburn, Oct. 1953, *Graham Bell* EA 10351!
DISTR. K 4; possibly native to South America, Florida, Greater Antilles, cultivated fairly widely
HAB. Cultivated; ± 1670 m
CONSERVATION NOTES. Widespread; least concern (LC)

SYN. *C. fastigatum* Jacq., Hort. Schoenbr., 3: 44, t. 330 (1778). Type: Jacq., Hort. Schoenbr., 3: 44, t. 330 (lecto.!)

C. album Dunal in DC., Prodr. 13(1): 605 (1852). Type: Cuba, around Havana "v.s. in h.DC ubi spec. mal." (specimen not found)

C. diurnum L. var. *fastigatum* (Jacq.) Stehle in Fournet, Fl. Illustr. Guad. Mart.: 1281 (1978)

NOTE. This species is commonly known as the Day-Jessamine, reflecting the diurnal opening and potent fragrance of its flowers, though these remain open throughout the day and night.

5. **Cestrum parquii** *L'Hér.*, Stirp. Nov. fasc. 4: 73, t. 36 (1788); Dunal in DC., Prodr. 13(1): 616 (1852); Hiern, Cat. Afr. Pl. Welw. 3: 754 (1898); Francey in Candollea 7: 38 (1936); Verdcourt & Trump, Common Poisonous Pl. E. Afr.: 161 (1969); Fl. Rwanda 3: 362 (1985); D'Arcy & Rakotozafy in Fl. Madagascar, Solanaceae: 19 (1994); Hepper in Fl. Egypt 6: 139 (1998); Gonçalves in F.Z. 8(4): 10 (2005). Type: Chile, Concepcion, *Dombey* s.n. (locality of type not cited, ?P, holo. [NB if holo. not extant, Plate 36 of L'Heritier, Stirp. Nov. fasc., 4 (1788) should be designated as lecto.]

Deciduous erect shrubs 1–3 m high; stems numerous and often suckering, green, terete, lenticellate, glabrescent with inconspicuous stalked glands throughout. Leaves sometimes glossy, dark green, narrowly lanceolate, 8–11 × 2–3.7 cm, bases cuneate, often decurrent to stems, margins entire rarely sinuate, apices acute, both surfaces glabrescent but with stalked glands; stipules ovate to spatulate; petioles 5–9 mm long, glabrescent. Inflorescences congested terminal or lateral panicles; flowering stems shortly pilose/villous with simple, few-celled glandular and eglandular hairs; bracts narrowly lanceolate, ± 3 mm long, shortly pilose, some hairs glandular; pedicels 0.7–4 mm long. Calyx cupulate/tubular, pale green, 4–5.5 mm long, pilose externally, lobes equal or unequal, triangular, 0.8–1 × 0.7–1.2 mm in flower, apex acute, enlarging to 2 × 2.5 mm in fruit, with denser pubescence and apical tufts of hairs. Corolla yellow or yellowish-green, tubular to infundibuliform, (1.2–)1.8–2.2 cm long and 0.7–1.2 cm diameter apically; tube cylindrical or slightly bulbous above, (1.2–)1.4–1.8 cm long, diameter increasing to 3.5–4 mm below lobes, glabrescent externally, shortly pilose internally, with short, acute or obtuse triangular valvate lobes, usually spreading after anthesis when rounded and 2–5 × 1.5–3.5 mm, with densely pilose marginal bands externally. Stamens usually equal, with anthers and stigma visible in corolla throat; filaments free for 6–8 mm, curved apically, enlarged and pilose at adnation; anthers yellow to brown, 0.8–1.1 × 0.8–1 mm. Ovary green sometimes flecked purple, globose or broadly ovoid, 0.9–1.2 mm long, glabrous; disc pale green, cupulate, upper margins crenulate, 0.7–1.3 mm broad; style 14–16 mm long; stigma 1–1.2 mm broad. Berry dark purple to blackish, ovoid, 7–8 × 6–7 mm, surrounded by cupulate calyx with adherent calyx lobes sometimes splitting to expose fruit. Seeds ovoid, prismatic, dark brown, 3.5–4 mm long.

KENYA. Cultivated Nairobi Arboretum, 12 Feb. 1952, *Williams Sangai* 330!; Nyeri District: between Mt Kenya and Highway A2 (± 30 km W), 11 Jan. 1975, *D'Arcy* 7248!
DISTR. **K** 4; native to Chile and Argentina, now a widespread ornamental
HAB. Sometimes occurring as an escape and recorded from open woodland with *Crotalaria*; 1850–3200 m
CONSERVATION NOTES. Widespread; least concern (LC)

NOTE. Though this species is a widespread ornamental it is is well known for its toxicity (especially hepatic toxins) with common names including Green poison berry, Chilean- or green- *Cestrum* as well as Willow-leaved Jessamine. The flowers are reportedly slightly fragrant after dusk, and the leaves can be malodorous.

6. **Cestrum × cultum** *Francey* in Candollea 6: 100 (1935); D'Arcy & Rakotozafy in Fl. Madagascar, Solanaceae: 17 (1994); Gonçalves in F.Z. 8(4): 9 (2005). Type: cultivated in Egypt, Jezireh, Cairo, *Schweinfurth* s.n. (syn., locality not cited); cultivated in garden, Cairo, *Grunow* s.n. (W, syn.); cultivated in Algeria, *Maire* s.n. (syn., locality not cited); Indonesia, Java, *Lörzing* 871 (syn., locality not cited)

Woody herbs or shrubs 2–4 m high; stems much- or sparsely branched, sometimes scrambling, brownish-green; branches often pendent, glabrescent to pilose with spreading, usually simple, hairs. Leaves pale to dark green, narrowly elliptic, 5.2–7 × 1.3–2 cm, bases cuneate often decurrent, margins entire, apices acute, both surfaces glabrescent to pilose with short hairs generally denser on veins where interspersed with stalked glands; stipules narrowly lanceolate; petioles 4–6 mm long, pilose. Inflorescences terminal, panicles or subcorymbose; flowering stems moderately to densely pilose to pilose/villous with spreading coloured hairs as on stems; flowers often with a musky odour, usually sessile and bracteate; bracts linear-lanceolate, 3–5.5 mm long, pilose. Calyx purple, cupulate/tubular, 4–7 × 2–3 mm, glabrescent but with some stalked glands externally, lobes equal or unequal, purple, narrowly triangular, 0.7–1.3 × 1–1.5 mm, acute, with ciliate margins and apical tufts of coloured hairs. Corolla pink, pale- or reddish-purple, infundibuliform, 1.9–2.5 cm long and 8–9 mm diameter apically; tube cylindrical or slightly bulbous above, constricted below lobes, 1.5–2 cm long, diameter increasing to 4–6 mm below lobes, glabrescent externally, densely pilose internally; lobes narrowly or broadly triangular, usually spreading after anthesis when 2–4 × 1.5–5 mm, acute. Stamens usually equal, anthers and stigma visible in corolla throat; filaments free for 6–8 mm, geniculate and enlarged at adnation; anthers yellow to brown, globose, 0.6–1 mm diameter. Ovary broadly ovoid, 1.1–1.3 mm long; disc cupulate, upper margins crenulate, 0.8–1.3 mm broad; style 1.6–1.9 cm long; stigma 0.5–1 mm broad. Berry ? blackish and ovoid. Seeds discoidal, ovoid, prismatic, 4 mm long.

TANZANIA. Lushoto District: S Amiri near Lushoto Roman Church, 14 Apr. 1969, *Ngoundai* 285! & Amani Nursery (cult.), 15 Mar. 1973, *Ruffo* 633! & 5 July 1940, *Greenway* 5966!
DISTR. **T** 3; also cultivated in Zambia, Malawi, South Africa, Madagascar, Egypt and Europe
HAB. Occasional escape from cultivation in parks and gardens; ± 900 m
CONSERVATION NOTES. Widespread; least concern (LC)

NOTE. Known mainly in cultivation, this taxon is thought to be a hybrid between *C. parquii* and *C. elegans*; it is somewhat morphologically intermediate. The precise berry colour of this taxon is uncertain; Francey (1935) stated "bacca ignota", and all other descriptions of it also fail to mention any berry details. Similarly no details of fruit colour were given on the herbarium material seen, perhaps reflecting the general sterility of this hybrid.

2. BRUNFELSIA

L., Sp. Pl. 1: 191 (1753) (*"Brunsfelsia"*) *nom. et orth. conserv.* & Gen. Pl. ed. 5: 87 (1754); Dunal in DC., Prodr. 13(1): 198 (1852); Plowman in Biol. & Tax. Solanaceae: 475–491 (1979) & in Fieldiana Bot., n.s. 39: 1–133 (1998); Hunziker in Gen. Solanaceae: 82–85 (2001)

Franciscea Pohl, Pl. Bras. Ic. 1: 1 (1827); Endl., Gen.1: 676 (1839); Miers, Illus. S. Am. Pl. 2: 73 (1849–1857)
Brunfelsia L. subgen. *Brunfelsiopsis* Urban in N.B.G.B. 1(10): 324 (1897)
Brunfelsiopsis (Urban) Kuntze in Post & Kuntze, Lexicon: 81 (Dec. 1903 "1904")

Evergreen shrubs or small trees. Leaves alternate, simple, hairs usually simple, occasionally branched. Inflorescences usually dense, rarely lax terminal or axillary many-flowered cymes, or reduced to a single flower, subsessile; pedicels erect or pendent, subtended by linear to lanceolate, caducous bracts; flowers slightly zygomophic with bilateral symmetry. Calyx tubular or campanulate, with 5 lobes; usually shorter than corolla tube. Corolla hypocrateriform, basal tube cylindrical straight or slightly curved, dilated above towards five broad recurved lobes with quincuncial-imbricate aestivation, open at mouth which often thickened into fleshy white ring. Stamens four, didynamous, alternating with upper three corolla lobes, usually included, occasionally upper pair exserted, both pairs curved at apex to project anthers towards back of corolla tube; filaments fused to upper quarter of corolla tube, unequal, lower posterior pair shorter than the upper lateral pair, glabrous, becoming thicker and flattened at point of fusion; anthers usually all fertile, rarely upper pair reduced, elongate, medifixed, oblong or reniform, one-celled dehiscing by single slit. Ovary conical to ovoid, bilocular; ovules numerous; disc nectiferous, annular, smooth; style filiform below, erect or curved and broadened beneath the stigma, glabrous, usually included; stigma broadly dilated, shallowly or deeply bilobed or capitate, often surpassed and partially enclosed by upper lateral anthers. Fruit capsular or sub-baccate, fleshy or coriacous, indehiscent or often splitting open by two entire valves, scarcely or partially enclosed by persistent accrescent coriaceous calyx. Seeds deeply reticulate, large, few to many.

± 46 species of shrubs and small trees from South and Central America, many of which are cultivated for their ornamental and medicinal value.

The history and classification of this genus together with its inter-generic and inter-specific relationships and the distiguishing morphological features are extensively dealt with by Plowman (1998). The presence, location, degree and type of indumentum can vary greatly in this genus, even in the same species. Moreover, *Brunfelsia* fruits often appear baccate and have been described as berries in numerous taxonomic descriptions of these species. They are, in fact, capsules which are either indehiscent or split into two smooth or rough valves (cf. Plowman, 1998). Some of the diagnostic features given below have been adapted from Plowman (1998), largely due to the paucity of herbarium material of this genus, not only collected from Africa but also from the Americas.

Plants with simple and branched hairs, often glandular-headed;
    flowers white or greenish fading to yellow; corolla tube 3.5–6 cm
    long; calyx 6–7 mm long; style > 4 cm long; capsule orange
    or yellow . . . .   1. *B. americana*
Plants usually glabrescent (in our area) any hairs present short,
    unbranched and eglandular; flowers purple fading to pale
    violet or white; corolla tube 1.3–2.5 cm long; calyx 1–2.2 cm
    long; style < 2.3 cm long; capsules dark green . . . .   2. *B. uniflora*

1. **Brunfelsia americana** *L.*, Sp. Pl. 1: 191 (1753); DC., Prodr. 10: 200 (1846); U.O.P.Z.: 157 (1949); Plowman in Biol. & Tax. Solanaceae: 475 (1979) & in Fieldiana Bot., n.s. No. 39: 1 (1998); Hepper in Fl. Egypt 6: 161 (1998); Gonçalves in F.Z. 8(4): 5 (2005). Type: "in America meridionali", in Plumier, Nov. Pl. Amer.: 12, t. 22 (1703)!, lecto. designated by Howard in Fl. Lesser Antilles, 6: 270 (1989), as *Brunsfelsia* [icon.] [see also Knapp in Jarvis *et al.*, Regnum Veget., 127: 26 (1993) & Jarvis, Order out of Chaos: 362 (2007)]

Shrubs or small trees 1.5–3(–10) m tall; all vegetative and floral parts pubescent on young parts becoming pilose with eglandular- and glandular-headed branched hairs. Leaves coriaceous, often glossy, light green, elliptic, obovate, ovate to lanceolate, 2.1–5.5(–11) × 1–4(–5) cm, bases cuneate, often decurrent, margins entire, apices acute, mucronate or obtuse, prominently veined beneath, both surfaces glabrescent or pilose, denser beneath and on the veins and midribs; petioles 2–6(–10) mm long, densely pilose to glabrescent. Inflorescences 1-flowered, terminal or axillary, sessile, occasionally with very short peduncles; pedicels erect, 2–7(–15) mm long, pilose, subtended by obovate pilose bracts 1–2.5 mm long. Calyx cupulate or campanulate, 6–7 × 4.3–6 mm, with five acute and often unequal broadly ovate or triangular lobes 0.7–2.5 × 1–2.5 mm. Corolla white or green, fading to yellow, hypocrateriform, sweetly fragrant, tube cylindrical, 3.5–6 cm long and 1.5–3 mm wide basally broadening to 2.5–5 mm above, puberulous externally, terminating in five broadly ovate spreading rounded lobes 0.9–2.4 × 1–2 cm, with crenate margins; overall flower diameter 3–4.3 cm with throat 7–8 mm wide, not constricted. Stamens usually included, occasionally two lateral anthers visible in corolla throat; filaments lower posterior pair free for 7.5–11 mm, upper lateral pair free for 11–13 mm; anthers oblong, medifixed, 2.5–3 × 1.5–3 mm. Ovary ovoid, ± 2.5 × 2 mm, glabrous; disc smooth; style 39–50 mm long, usually included; stigma 1.3–1.8 × 1–1.8 mm, often visible in corolla throat. Fruit an orange or yellow globose capsule, 1.5–2 cm diameter, thick-walled, smooth, splitting from the apex into two fleshy valves, base subtended by persistent calyx with calyx lobes eventually becoming dry and reflexed. Seeds numerous, light to dark brown, obovoid to deltoid, 3–5 × 2–2.5 mm, irregular and angular. Fig. 2/7–9, p. 19.

TANZANIA. Kondoa District: Kondoa Boma, 15 June 1973, *Ruffo* 705!; Morogoro District: Morogoro, Nov. 1955, *Semsei* 2391!
DISTR. **T** 5, 6; originally from the West Indian islands, also found in Zambia, Mozambique and Zimbabwe
HAB. Cultivated, sometimes an escape in gardens and on roadsides; ± 1220 m
CONSERVATION NOTES. Widespread; least concern (LC)

SYN. *B. terminalis* Salisb., Prodr. Stirp.: 109 (1796). Type: Jamaica, *Wright* s.n. (?K, not found)
    B. violacea Lodd, Bot. Cab.: t. 792 (1796). Type: Grown in Paris from material sent from West Indies; Lodd, Bot. Cab., t. 792 (1796)!, lectotype designated here
    B. americana L. var. *pubescens* Griseb., Fl. Brit. W. Ind.: 432 (1864). Type: Jamaica, St Kitts and Antigua, *Wullschlager* s.n. (no type locality given)

NOTE. This is commonly known as 'Lady of the Night' because of its strong night-time fragrance; the flowers are apparently pollinated by night-flying moths.

    2. **Brunfelsia uniflora** (*Pohl*) *D. Don* in Edinb. New Phil. Journ. 7: 85 (1829); Plowman in Biol. & Tax. Solanaceae: 475 (1979) & in Fieldiana Bot. n.s. 39: 113 (1998); Gonçalves in F.Z. 8(4): 5 (2005). Type: Brazil, Rio de Janeiro, *Pohl* [9] (W, holo.; PR, iso., fide Plowman in Fieldiana, Bot., n.s. No. 39: 113 (1998))

    Shrubs 0.5–3 m tall, deciduous, often branched from base; branches many, slender, glabrescent; hairs on all parts short, eglandular. Leaves often glossy, elliptic, obovate or lanceolate, 2.5–8 × 1–4 cm, bases cuneate, often decurrent, margins entire, apices acute to obtuse, membranaceous to coriaceous, both surfaces glabrous to sparsely pilose; petioles 1–5 mm. Inflorescences 1-flowered, terminal, sessile, occasionally with peduncles 1–5 mm long, slender, glabrous to glandular-pubescent, subtended by 1–3 broadly ovate or linear-lanceolate bracts 1–5 mm long; pedicels erect, 1–6 mm. Calyx tubular or tubular-campanulate narrowing towards base, 10–22 × 1.2–2.2 mm basally widening to 5–10 mm above, with five acute and often unequal broadly triangular or obovate lobes 2.5–5.8 × 1.6–4.2 mm, glabrous. Corolla white or purple with a white eye at the throat, usually fading to white or yellow, hypocrateriform, tube cylindrical 1.3–2.5 cm long and 1–2 mm wide, widening around anthers and stigma, terminating in five broadly ovate spreading rounded lobes 1.2–2(–3) cm long and wide; apical flower diameter 1.8–4.8 cm; glabrous or sparsely pubescent externally. Stamens usually included; filaments unequal, lower posterior pair free for 1–3 mm, upper lateral pair free for 1.5–3 mm; anthers orbicular-reniform, medifixed, 0.7–1 mm diameter, lateral pair often visible in corolla throat. Ovary conical to ovoid, 1.8–3 × 1–1.5 mm, glabrous; disc smooth; style 15–23 mm long, usually included; stigma 0.6–0.8 × 0.7–1.5 mm. Fruit a glossy dark green, coriacous, globose or ovoid capsule 7–8 × 8–18 mm, smooth, thin-walled, partially enclosed by accrescent coriaceous persistent calyx, occasionally dehiscent. Seeds numerous, reddish-brown to black, cuboidal to obovoid, 3–5 × 2–3 mm, angulate. Fig. 2/1–6, p. 19.

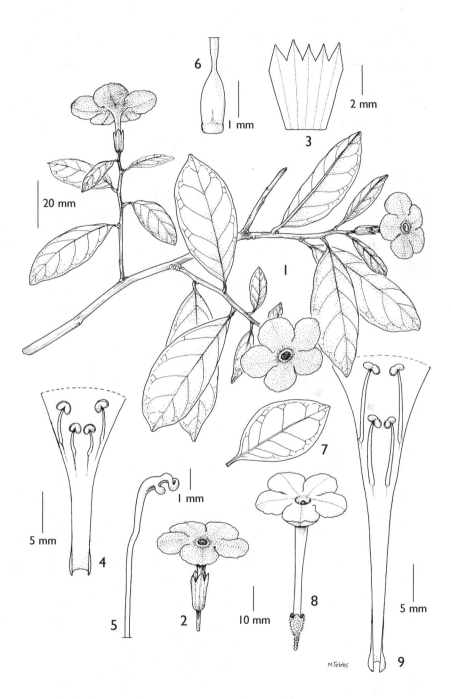

FIG. 2. *BRUNFELSIA UNIFLORA* — **1**, habit in flower; **2**, complete flower; **3**, opened calyx; **4**, opened flower; **5**, style and stigma; **6**, ovary. *B. AMERICANA* — **7**, leaf; **8**, complete flower; **9**, opened flower. 1–2 from *Lye* 23405; 3–6 from *Ngoundai* 405; 7–8 from *Ruffo* 705; 9 from *Semsei* 2391. Drawn by Margaret Tebbs.

UGANDA. Mengo District: Kampala, Makerere University Hill (cult.), 9 Feb. 1999, *Lye*, 23405!
KENYA. Nairobi Arboretum (cult.), 1 May 1953, *Williams Sangai* 547!
TANZANIA. Lushoto District: Amani Bustani, 28 Oct. 1969, *Ngoundai* 405! & Amani Nursery
   (cult.), 16 Mar. 1973, *Ruffo* 601!
DISTR. U 4; **K** 4; **T** 3, 6; originally from South America
HAB. Cultivated and sometimes an escape in gardens and on roadsides; 1050–1200 m
CONSERVATION NOTES. Widespread; least concern (LC)

SYN. *Franciscea uniflora* Pohl, Pl. Bras. 1: 2, t. 1 (1827); Hooker in Bot. Mag. 55: t. 2829 (1828)
   F. hopeana Hook. in Bot. Mag. 55: t. 2829 (1828). Type: from material from Brazil sent by
      Marshall-Beresford and grown by Robert Barclay of Bury Hill: Hooker Bot. Mag. 55: t.
      2829 (1828)! lecto. (designated by Plowman in Fieldiana, Bot., n.s. No. 39: 113 (1998))
   Brunfelsia hopeana (Hook.) Benth. in DC., Prodr. 10: 200 (1846); F.P.U.: 129 (1962);
      Mansfeld, Encycl. Agr. & Hort. Crops: 1856 (2001)
   B. *hopeana* (Hook.) Benth. var. *pubescens* Benth. in DC., Prodr. 10: 200 (1846). Type:
      Trinidad, Bocas Mts, *Lockhart* 197 (K!, holo.; GOET, iso.)

NOTE. This species is particularly variable in leaf size and shape, calyx length, and pubescence
   degree and location. It is often known as 'Yesterday, today, tomorrow' because of the changes
   in flower colour that occur with age. These names were used for plants identified as *B.
   latifolia* [no authority given] in U.O.P.Z. (1949). However, although the inflorescences were
   described "with flowers borne several together", all other features and floral dimensions
   suggest that the plants belong to *B. uniflora*. It is possible that they were mis-identified. The
   flowers are said to be pollinated by day-flying butterflies from several different families.

3. **NICOTIANA**

L., Sp. Pl. 1: 180 (1753) & Gen. Pl. ed. 5: 84 (1754); Goodspeed in Chron. Bot. 16:
1–536 (1954); Horton in Journ. Adelaide Bot. Gard. 3: 1–56 (1981); Japan Tobacco
Inc., Nicotiana Illustrated: 1–293 (1994); Hunziker in Gen. Solanaceae: 49–53 (2001)

Tabacum Chabrey, Omni. Stirp. Sciagr. Icon.: 526, f. 5 (1677)

Annual to perennial herbs, rarely shrubs or trees. Leaves simple, solitary or
alternate, exstipulate, with petioles often winged. Inflorescences terminal many-
flowered bracteate panicles or racemes, rarely solitary-flowered; flowers
actinomorphic or zygomorphic, often opening at night, some nocturnally fragrant,
pedicellate; pedicels erect to recurved. Calyx tubular, campanulate or ovoid, shorter
than the corolla, with five equal or unequal lobes, persistent and usually enlarged in
fruit. Corolla tubular, tube cylindrical below, ampliate and becoming
infundibuliform, salverform, hypocrateriform or urceolate above, with short,
induplicate lobes, usually spreading after anthesis. Stamens 5, equal or subequal
(when either in two groups where 4 are equal with one one shorter or more rarely
longer, or in a 2+2+1 grouping), included or exserted; filaments joined below middle
of corolla tube but at varying levels when the stamens are unequal, cylindrical or
enlarged and geniculate at point of fusion, glabrous to pilose; anthers ovoid,
obovoid, elliptic or globose, sides sometimes parallel, medifixed, visible in throat of
corolla. Ovary usually sessile, glabrous, 2-locular, rarely 4-locular, ovules numerous;
disc prominent, annular or cupulate; style filiform, glabrous, often exserted; stigma
capitate, bilobed, visible in corolla throat. Fruit septicidal and loculicidal capsules,
dehiscing by two smooth, glabrous valves which split into two. Seeds minute,
alveolate-reticulate, angular, elliptic, ovoid, globose or discoid, numerous (up to
400/capsule).

Between 60 and 95 species found in subtropical parts of North and South America, some
Pacific Islands, Australia and SW Africa.

1. **Nicotiana tabacum** *L.*, Sp. Pl. 1: 180 (1753); Dunal in DC., Prodr. 13(1): 557
(1852); A. Rich., Tent. Fl Abyss. 2: 94 (1851); Engl., Hochgebirgsfl. Trop. Afr.: 375
(1892); Hiern, Cat. Afr. Pl. Welw. 3: 754 (1898); C.H. Wright in Fl. Cap. 4(1): 119
(1904) & in F.T.A. 4, 2: 259 (1906); Durand & Durand, Syll. Fl. Congo; 397 (1909);
F.P.N.A. 2: 218 (1947); U.O.P.Z.: 380 (1949); Goodspeed in Chron. Bot. 16: 372
(1954); F.P.U.: 129 (1962); E.P.A. 2: 883 (1963); Heine in F.W.T.A. 2nd ed., 2: 327
(1963); Troupin, Fl. Rwanda 3: 369 (1985); Blundell, Wild Fl. E. Afr.: 189 (1992);
U.K.W.F. 2nd ed.: 244 (1994); D'Arcy & Rakotozafy in Fl. Madagascar, Solanaceae: 31
(1994); Japan Tobacco Inc., Nicotiana Illustrated: 60 & 268 (1994); Hepper in Fl.
Egypt 6: 145 (1998); Mansfeld, Encycl. Ag. & Hort. Crops 4: 1852 (2001); Gonçalves
in F.Z. 8(4): 17 (2005); Thulin in Fl. Somalia 3: 220 (2006); Friis in Fl. Eth. 5: 106
(2006). Type: "in America, nota Europaeis ab 1560", Herb. Linn. 245.1 (LINN!,
lecto. designated by Setchell in Univ. Calif. Publ., Bot., 5: 6 (1912)) [See also Jarvis,
Order out of Chaos: 694 (2007)]

Annual or short-lived perennial herb to 3 m; stems sometimes basally woody, erect,
sparsely branched; all parts conspicuously viscid-villous. Leaves spatulate, lanceolate
or narrowly lanceolate, 8–50 × 2.8–21 cm, upper leaves smaller, bases decurrent,
sessile or with short winged petiole up to 5 cm long, often auriculate, apices acute to
narrowly acuminate, viscid-glandular on both surfaces. Inflorescences terminal much-
branched panicles up to 25 cm long; flowers fragrant; pedicels densely viscid-pilose,
0.5–2 cm long in flower, up to 2.8 cm long in fruit, always erect; bracts linear-
lanceolate to ligulate, densely viscid-pubescent; calyx tubular to narrowly
campanulate, 10–22 × 3–9 mm, viscid-pubescent externally, lobes unequal, narrowly
triangular, 5–11 × 1.5–4 mm, acute to subulate-acuminate, enlarging in fruit. Corolla
white or pink fading to white, often tube greenish and lobes pink to reddish, tubular
below, ampliate becoming infundibuliform above, overall 3.6–5 × 1.1–2.8 cm diameter
apically; tube 2.8–4.2 × 1.5–4 mm wide basally increasing to 5–9 mm below lobes,
viscid-pilose externally, glabrous internally; lobes usually shallow broadly triangular,
1.5–5 × 2.3–10 mm, acuminate, spreading after anthesis. Stamens unequal, the fifth
often shorter; filaments two longer pairs free for 2.4–4.2 cm, shorter free for 2–3.5 cm;
anthers 2–3(–3.6) × 1–2 mm. Ovary elliptic to conical, ?dark brown, 5–8 × 2–5 mm;
disc crenulate, 2–5 mm diameter; style 3–3.8 cm long; stigma 0.7–1.5 × 1–2 mm.
Capsules glabrous, brown, elliptic or ovoid, 1.3–2.2 × 0.8–1.5 cm, with apical
beak, dehiscing by four glabrous smooth valves, wholly or partially covered by
accrescent calyces. Seeds ellipsoid, ovoid, globose, angular or discoid, 0.3–0.8(–1) ×
0.2–0.5(–0.7) mm. Fig. 3/1–8, p. 22.

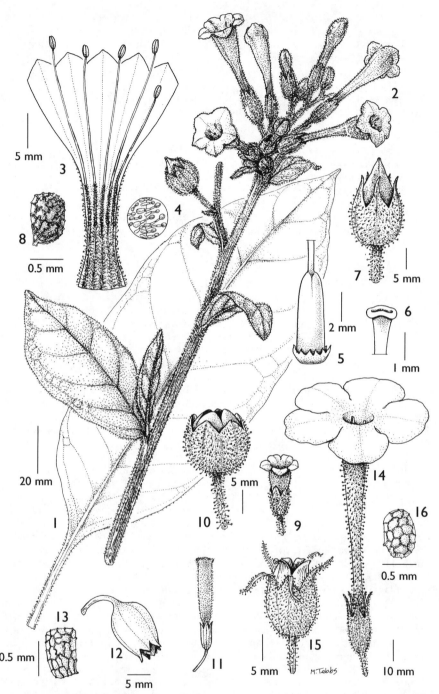

FIG. 3. *NICOTIANA TABACUM* — **1**, leaf; **2**, flowering habit; **3**, opened flower; **4**, stem indument detail; **5**, ovary; **6**, stigma; **7**, capsule within accrescent calyx; **8**, seed. *N. RUSTICA* — **9**, complete flower; **10**, capsule within accrescent calyx. *N. GLAUCA* — **11**, complete flower; **12**, capsule within accrescent calyx; **13**, seed. *N. ALATA* — **14**, complete flower; **15**, calyx lobes exposing capsule; **16**, seed. 1 & 7–8 from *Volkens* 2234; 2–6 from *Mwangangi* 969; 9–10 from *Newbould* 5640; 11–13 from *Mwangangi* 81; 14–16 from *Issa* 103. Drawn by Margaret Tebbs.

KENYA. Northern Frontier District: Ndoto Mountains, valley E of Nguronit Mission Station, 8 June 1979, *Gilbert et al.* 5548!; Nairobi, 23 Feb. 1963, *Njoroge Thairu* 10!; Machakos District: 6 km N of Nunguni at W side of mountain slopes of Kithembe Village, 13 May 1968, *Mwangangi* 969!

TANZANIA. Arusha District: Loitang, Ngurdoto National Park, Ngurdoto Crater, 12 Oct. 1965, *Greenway & Kanuri* 12116!; Kigoma District: Kungwe Mountain, Ntale River, 17 July 1959, *Newbould & Harley* 4451!; Dodoma District: Kazikazi, 5 July 1932, *Burtt* 3681!; Zanzibar, 1927, *Toms* 55!

DISTR. U (see note); **K** 1, 3–6; **T** 2–5; **Z**; probably originated in NW Argentina, naturalised throughout Africa and Indian Ocean islands

HAB. Cultivated, escaped and locally naturalised in woods and dry evergreen forest, bushland, on roadsides and along streams; 800–2600 m

CONSERVATION NOTES. Widespread; least concern (LC)

SYN. *N. fruticosa* L., Syst. Nat. ed. 10, 2: 932 (1759); Sp. Pl. ed 2, 1: 258 (1762). Type: "ad Cap. b. spei, China", Herb. *Linnaeus* 245.2 (LINN!, lecto. designated by Setchell in Univ. Calif. Publ., Bot., 5: 6 (1912)) [See also Jarvis, Order out of Chaos: 693 (2007)]

    N. alba Mill., Gard. Dict., ed. 8., no. 5 (1768). Type: cultivated from seeds sent by *Robert Millar* from Tobago (HS 296.f 61, BM!, lecto. designated here)

    N. angustifolia Mill., Gard. Dict. ed. 8: no. 3 (1768). Based on Plat.185. *Nicotiana major angustifolia*. C.B.P. 170. Type not found.

    N. latissima Mill., Gard. Dict. ed. 8: no. 1 (1768); E.P.A. 2: 883 (1963). Type: Peru, no collector (?BM, holo., not found)

    N. macrophylla Spreng., Index Pl. Hort. Bot. Hal. : 45 (1807). Type: ? Cultivated in Hort. Bot. Halensis (type specimen not cited)

    N. tabacum L. var. *macrophylla* (Spreng.) Schrank in Bot. Zeit.: 260 (1807); E.P.A. 2: 883 (1963)

    Nicotiana mexicana Schltdl. in Ind. Sem. Hort. Hal.: 8 (1840) & in Linnaea 19: 270 (1847). Type: cultivated in Halis Saxonum, ?*Schlechtendal* s.n. (type not cited)

    Tabacum nicotianum Opiz in Bercht. & Opiz, Oekon.-techn. Fl. Böhm, 3: 307 (1841). Basionym: *Nicotiana tabacum* L., type as above.

    Nicotiana pilosa DC., Prodr. 13(1): 559 (1852). Type: ?*Mociño & Sessé*, in hort. Mexico (specimen not cited)

    N. virginica C. Agardh in DC., Prodr. 13(1): 559 (1852), in syn. Based on: Orinoco region of Virginia, USA, specimen not cited

    N. tabacum L. var. *fruticosa* (L.) Hook. in Bot. Mag. 102: 6207 (1876)

    N. tabaca (L.) St. Lager in Ann. Sci. Bot. Lyon, 7: 130 (1880), ?*nom. nud.* [article is a reform of botanical nomenclature]

    N. tabacum L. var. *virginica* (C. Agardh) Comes, Monogr. Nicotiana: 12, t. 1 and 5 (1899)

    N. tabacum L. var. *virginica* (C. Agardh) Anastasia in De Wild., Miss. Laurent: 443 (1907); Durand & Durand, Syll. Fl. Cong.: 398 (1909)

    N. tabacum L. var. *latissima* (Mill.) Chiov. (1936); E.P.A. 2: 883 (1963)

NOTE. Known throughout the world as the common tobacco, *N. tabacum* is widely cultivated for its stimulatory effect, with many cultivars also being grown as ornamentals for their varying flower colours and strong evening fragrances. It is a highly polymorphic species now composed of a large number of cultivar forms used in modern tobaccos. Many more synonyms of this species are given in Goodspeed (1954) and Mansfeld (2001), who both summarise its amphidiploid origin from *N. sylvestris* Speg. & Comes and *N. tomentosiformis* Goodspeed. Plants were cultivated and the leaves already used for chewing and smoking in Mexico, C America, Venezuela, Colombia and Guyana before the Americas were discovered. It is now the most important commercial tobacco species, widely cultivated in the tropics and subtropics and some temperate zones, with the largest cultivation area encompassing China, India, Brazil, the former Soviet Union and the USA. Nicotine content is lower than in *N. rustica*; edible oils have recently been extracted from its seeds (Mansfeld, 2001). Among the many described varieties of this species are nine by Dunal (1852) and two in the Congo by Durand & Durand (1909), though none of these varietal names have been encountered on East African material. Williams (in U.O.P.Z., 1949) listed *N. plumbaginifolia* as occurring in Pemba and Zanzibar. The discriminatory characters given do not accurately reflect those given by Goodspeed (1954) for this species, with many being similar to those exhibited by *N. alata*. It is probable that the specimens were mis-identified.

    Said to be cultivated in Uganda in F.P.U.: 129, but no specimens have been seen.

2. **Nicotiana rustica** *L.*, Sp. Pl. 1: 180 (1753); C.H. Wright in F.T.A. 4, 2: 260 (1906); Dunal in DC., Prodr. 13(1): 563 (1852); Durand & Durand, Syll. Fl. Cong.: 397 (1909); Goodspeed in Chron. Bot. 16: 351 (1954); E.P.A. 2: 883 (1963); Heine in F.W.T.A. 2nd ed., 2: 327 (1963); Blundell, Wild Fl. E. Afr.: 189 (1992); Japan Tobacco Inc., Nicotiana Illustrated: 32 & 266 (1994); Hepper in Fl. Egypt 6: 143 (1998); Gonçalves in F.Z. 8(4): 16 (2005); Friis in Fl. Eth. 5: 106 (2006). Type: "in America, nunc in Europa, Herb. *Linnaeus* 245.3 (LINN!, lecto. designated by Goodspeed in The genus Nicotiana: 351 (1954) [without specimen citation]) & Deb in Journ. Econ. Tax. Bot. 1: 43 (1980). [See also Jarvis, Order out of Chaos: 693 (2007)]

Annual herb to 1.5 m, sometimes becoming shrubby; stems erect, all parts conspicuously viscid-villous, with spreading simple glandular-headed hairs. Leaves often fleshy, viscid, lanceolate, ovate-lanceolate to cordiform, 8–16.5(–26) × 4–9(–22.5) cm, upper leaves often smaller, bases cordate to cuneate, apices obtuse to acute, villous but hairs sparser on upper surfaces; petioles 2–8 cm long. Inflorescences sparsely branched panicles; flowers opening by day; pedicels always erect, 4–5 mm long in flower, 5–10 mm long in fruit, densely glandular-villous. Calyx campanulate to cupulate, 6–10 × 4–8 mm, densely glandular-pilose externally, lobes unequal, broadly triangular, 1.5–4 × 1.3–4.5 mm, acute, enlarging in fruit. Corolla yellow, greenish-yellow to whitish, broadly cylindrical to tubular below, ampliate becoming urceolate above, 1.4–2 × 0.5–1.1 cm diameter apically; tube inflated, 1.2–2 × ± 2 mm wide basally increasing to 8 mm apically below lobes where constricted, pilose externally, glabrous internally; lobes shallow, 1.5–2 × 5–6 mm, acute or slightly apiculate, spreading after anthesis. Stamens unequal with fifth usually longer, occasionally shorter; filaments free for 7–10 mm, with fifth free for 6–7 mm; anthers yellow, cordiform, 1.5–2 × 1.2–1.8 mm. Ovary ovoid to conical, 1.7–2.5 × 1.2–2 mm; disc crenulate, 1.5–2.5 mm diameter; style glabrous, 9–11 mm long; stigma green, 0.6–0.8 × 0.7–1.8 mm. Capsules dark brown, globose, 0.8–1.6 cm in diameter, glabrous, dehiscing by four glabrous smooth valves, subtended by accrescent glabrescent calyx. Seeds ellipsoid, ovoid, obovoid or discoid, 0.7–1.3 × 0.6–0.9 mm. Fig. 3/9 & 10, p. 22.

UGANDA. West Nile District: Midigo, Aringa County (cult.), Sept. 1937, *Eggeling* 3401!; Karamoja District: Pian, near Moruangaberu [Emoruagaberru], 17 July 1958, *Dyson-Hudson* 431! & Moroto, Mar. 1960, *Wilson* 814!

KENYA. Northern Frontier District: Ndoto Mountains, 1 Jan. 1959, *Newbould* 3468!; near Nairobi, Aug. 1903, *Whyte* s.n.!

TANZANIA. Masai District: Lerong, west side of Lemagrut, 1 Feb. 1961, *Newbould* 5640!; Lushoto District: Agroforestry Farm, 13 Sept. 1985, *Kisena* 248! & Silvicultural Agroforestry Plot, 8 Oct. 1996, *Kisena* 1613!

DISTR. U 1; K 1, 4: T 2, 3; originally native to South America (Andes from Ecuador to Bolivia) and Mexico, widely grown for tobacco and snuff by villagers throughout Africa, Europe, Asia, the USA and Australia

HAB. Abandoned villages, shambas and cultivation sites, and on roadsides and rubble; (300–)1050–2350 m

CONSERVATION NOTES. Widespread; least concern (LC)

SYN. *N. rugosa* Mill., Gard. Dict. ed. 8: no. 7 (1768); E.P.A. 2: 883 (1963). Type: no locality given, type not found

　　　N. rustica L. var. *brasilia* Schrank in Bot. Zeit., 6: 264 (1807), as *N. brasila*; E.P.A. 2: 883 (1963). Type: not cited

　　　N. humilis Link, Enum. Hort. Berol: 178 (1821). Type: neither specimen nor locality cited

NOTE. This species is considered to be a cultigen of amphidiploid origin derived from *N. paniculata* L. and *N. undulata* Ruiz & Pavon. It was widely cultivated in the Andes, Mexico, southwestern and eastern USA and eastern Canada, often being known as wild or Aztec tobacco. Though it has become locally naturalised in many countries, it has been largely replaced by *N. tabacum* as the predominant source of tobacco for smoking, chewing and snuff. Deb (Journ. Econ. Tax. Bot, 1: 43 (1980)) described this species as a highly polymorphic cultigen apparently now unknown in the wild state; several varieties have been described, and many cultivars have been developed. It is smoked and cultivated for snuff in Uganda.

3. **Nicotiana glauca** *Graham* in Edinb. New Phil. Journ. 5 : 175 (1828); Hooker in Bot. Mag. 55: t. 2837 (1828); Dunal in DC., Prodr. 13(1): 562 (1852); C.H. Wright in Fl. Cap. 4(1): 120 (1904); Goodspeed in Chron. Bot. 16: 335 (1954); Troupin, Fl. Rwanda 3: 369 (1985); Blundell, Wild Fl. E. Afr.: 189 (1992); U.K.W.F., 2nd ed.: 244 (1994); Japan Tobacco Inc., Nicotiana Illustrated: 2, 32, 264, 266 (1994); Hepper in Fl. Egypt 6: 142 (1998); Mansfeld, Encycl. Ag. & Hort. Crops 4: 1851 (2001); Gonçalves in F.Z. 8(4): 14 (2005); Friis in Fl. Eth. 5: 106 (2006). Type: cultivated Edinburgh from seed collected in Buenos Aires, Argentina, *Smith* s.n. (E, holo., photo!) *fide* Goodspeed in Chron. Bot. 16: 335 (1954) & Purdie *et al.* in Fl. Australia 29: 57 (1982)

Shrub or small tree up to 5(–10) m, quick-growing; stems woody, erect, often densely branched; all stems glabrescent and glaucous. Leaves glaucous, lanceolate, ovate, ovate- lanceolate or elliptic, 3.2–10(–28) × 1.6–5(–18.5) cm, upper leaves often smaller, bases cuneate to cordate, margins entire to sinuate, apices acute, glabrescent; petioles 2.2–5.5(–12) cm long. Inflorescences terminal lax panicles, bracteate; pedicels 4–8 mm long and erect in flower to recurved, 7–9 mm long in fruit when always recurved, glabrescent to sparsely pilose; bracts linear-lanceolate, pilose. Calyx tubular, 8–13 × 3–7 mm, glabrescent externally with ciliate margins, lobes unequal, narrowly triangular, 1–3.5 × 1.1–2.5 mm, acute to acuminate, enlarging in fruit. Corolla yellow to greenish-yellow, tubular below, ampliate beneath lobe lobes, 3–4 cm long and 6–9 mm diameter apically; tube 3–3.6 × ± 3 mm basally increasing to 6–8 mm apically below lobes, densely pilose/villous externally with spreading multicellular glandular- and eglandular-headed hairs, glabrous internally; lobes broadly triangular, 1.5–2.5 × 3.5–5 mm, acute, spreading after anthesis. Stamens unequal with fifth usually slightly shorter; filaments free for 2.1–2.3 cm, with shorter free for ± 1.6 cm; anthers green becoming brown, cordiform, 1.2–2 × 1.2–2 mm. Ovary ovoid to conical, 2–3.5 × 1.7–3.5 mm; disc crenulate, 2–3.5 mm diameter; style 1.8–3.2 cm long; stigma green, 0.6–0.8 × 0.7–1.5 mm. Capsules brown, ellipsoid or ovoid, 0.8–1.2 × 5–8 mm, glabrous, dehiscing by four glabrous smooth valves, subtended by accrescent glabrescent calyx. Seeds ellipsoid, ovoid, cuboidal, rectangular, angular or discoid, 0.3–0.8(–1) × 0.3–0.5(–0.7) mm. Fig. 3/11–13, p. 22.

KENYA. Naivasha District: south of Lake Naivasha, Nov. 1967, *E. Polhill* 71!; Nairobi, Dagoretti Corner near Kenya Science Teachers College, 27 Dec. 1971, *Msafiri & Njunge* 1!; Masai District: between Isenya and Athi River on Kajiado–Nairobi road, 1 Nov. 1978, *Darlington* EA 16303!
DISTR. **K** 3, 4, 6; native to South America, now naturalised in many tropical, subtropical and temperate regions
HAB. Escape from cultivation often found in dry rocky disturbed soils, on steep slopes, waste ground, river- and lake-banks and roadsides, occasionally in grassland, and associated with *Crotalaria* and *Jasminum*; 1600–1900 m
CONSERVATION NOTES. Widespread; least concern (LC)

SYN. *Nicotidendron glauca* (Graham) Griseb. in Goett. Abh. 19: 216 (1874)

NOTE. Often known as the tree tobacco or mustard tree. The species is also grown as a medicinal plant for the production of anabasine, which is a starting material for insecticide (Mansfeld 2001); the leaves are said to be poisonous to cattle, sheep and horses.

4. **Nicotiana alata** *Link & Otto*, Ic. Pl. Rar. 1: 63, t. 32 (1828); Don, Gen. Hist. Dichlam. Pl. 4: 467 (1837); Dunal in DC., Prodr. 13(1): 567 (1852); Goodspeed in Chron. Bot. 16: 393–395 (1954); Japan Tobacco Inc., Nicotiana Illustrated: 86 & 270 (1994); Hepper in Fl. Egypt 6: 148 (1998); Mansfeld, Encycl. Ag. & Hort. Crops 4: 1853 (2001). Type: cultivated in Berlin Botanic Garden from seeds sent from Brazil, *Sellow* s.n. (B†, holo.) *fide* Goodspeed (1954)

Annual to perennial herbs to 1.5 m tall, often rosulate initially; stems erect, sparsely branched mainly from base; all parts densely viscid-pubescent with long simple glandular-headed hairs. Leaves dark green, spatulate, obovate or ovate-lanceolate to elliptic, 7–11 (–26) × 2.6–7.5 cm, bases decurrent into winged petioles which usually auricled and up to 2 cm long, margins undulating, apices acute, densely glandular-pubescent. Inflorescences terminal few-flowered usually lax unbranched racemes, bracteate; flowers nocturnally fragrant; pedicels erect, 0.4–1.7 cm long in flower and 1.2–1.4 cm long in fruit, densely pilose-viscid. Calyx tubular to campanulate, densely viscid-pilose externally, 16–25 × 6–9 mm, with five unequal narrowly triangular dentate or subulate-acicular calyx lobes, 4.4–11 × 1–3 mm, enlarging up to 19 mm long in fruit. Corolla usually white to pale green, 8–10 cm long and 3–5 cm diameter apically; tube slender, cylindrical below, becoming salverform above and dilated below lobes, 6–9.5 × 2 mm diameter basally increasing to 8–10 mm apically, with obovate obtuse and sometimes shortly bilobed lobes 1.8–2.5 × 1.2–2 cm, spreading after anthesis; densely viscid-pubescent externally, glabrous internally. Stamens subequal; filaments free for 2–2.6 cm, the fifth free for 1.5–1.8 cm, cylindrical, glabrous throughout; anthers purple, globose, 2–3 × 1.9–2.5 mm. Ovary ovoid to conical, 4–7.5 × 2–4 mm; disc crenulate, 2–3 mm diameter; style 6.1–7 cm long; stigma 0.7–1 × 2.1–3 mm. Capsules light brown, globose to ovoid, 1.2–1.8 × 0.9–1.4 cm, dehiscing by 2 bifid smooth, glabrous valves which sometimes recurved apically, subtended by enlarged laciniate calyx lobes of enlarged viscid-pilose accrescent calyces. Seeds, brown, ovoid, globose or ellipsoid, 0.5–0.8 × 0.5–0.6 mm. Fig. 3/14–16, p. 22.

KENYA. Nairobi, University of Nairobi Chiromo Campus, 10 Oct. 1981, *Mwangangi* 2008!
TANZANIA. Lushoto District: Mang'ula Village, 10 Nov. 1982, *Kisena* 24! & near Silviculture Office, Jan. 1972, *Issa* 103!; Morogoro District: Morningside, 16 Oct. 1977, *Mhoro* BM 2568!
DISTR. **K** 4; **T** 3, 6; native to SE Brazil, N Uruguay, Argentina and Paraguay, now sparingly cultivated in South Africa, Europe and the US for ornamental uses, and becoming naturalised
HAB. In or near villages, gardens, cattle bomas and rain-forest margins; 1650–1900 m
CONSERVATION NOTES. Widespread; least concern (LC)

SYN. *N. persica* Lindl. in Bot. Reg. 19: t. 1592 (1833). Type: cultivated from seeds sent to RHS from Persia, *Willoch* s.n. (CGE, holo.; photo!)
  N. alata Link & Otto var. *persica* (Lindl.) Comes, Monogr. Nicotiana: 36 (1899)
  N. affinis Moore in Gard. Chron. 16: 141, fig. 31 (1881,II). Type: grown from seed collected in Hyères, France, Fig. 31! in Gard. Chron. 16 (lecto. designated here)

NOTE. Commonly known as the Jasmine- or flowering-tobacco, many horticultural variants of this species have been raised. The main alkoloidal constituent is again nicotine.

4. PETUNIA

Juss. in Ann. Mus. Nat. Hist. Par., 2, 215, t. 47 (1803), *nom. conserv.*; Fries in K. Vetensk.-Acad. Handl. Stockholm 46: 3–72 (1911); Sink, Petunia 9: 3–9 (1984); Hunziker in Gen. Solanaceae: 54 (2001)

Annual or perennial herbs. Leaves usually alternate, occasionally upper ones opposite. Flowers usually solitary, arising from pair of small leaves or bracts; pedicels erect, occasionally deflexed in fruit. Calyx with 5(–6) spreading calyx lobes. Corolla infundibuliform or salver-shaped with lower cylindrical tube flaring above and often terminating in 5 lobes of equal size which become partially reflexed. Stamens usually subequal with the fifth being smaller or larger, included; filaments fused to lower half to third of corolla tube, filiform above becoming thicker below at point of fusion, glabrous; anthers shortly elongate, dorsifixed and somewhat versatile, dehiscing by longitudinal slits, often visible in corolla throat, all fertile. Ovary 2-locular, enclosed basally by annular lobed disc; ovules numerous; style filiform becoming thicker

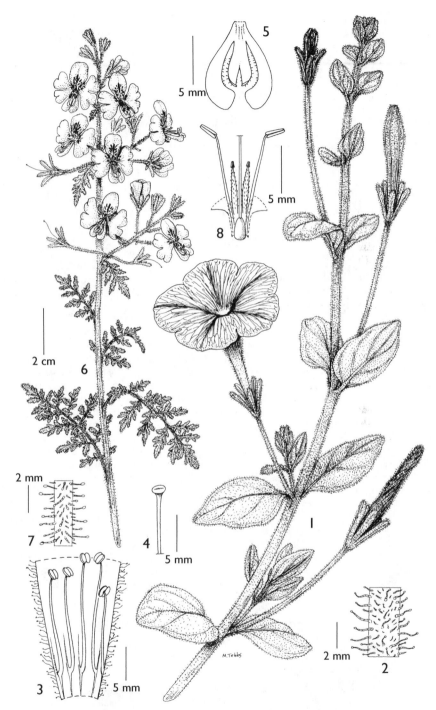

FIG. 4. *PETUNIA* × *HYBRIDA* — **1**, habit in flower; **2**, stem indument; **3**, opened flower showing subequal stamens; **4**, style and stigma; **5**, length section of ovary. *SCHIZANTHUS PINNATUS* — **6**, habit in flower; **7**, stem indument; **8**, floral dissection showing two staminodes, two anthers and gynoecium. 1 & 3–5 from *Mwangangi* 2010; 2 from *Verdcourt* 993; 6–8 from *Verdcourt* 342. Drawn by Margaret Tebbs.

towards the ovary and stigma; stigma bilobed, forked or capitate, often visible in corolla throat. Fruit a smooth, many-seeded, bivalvate, septicidal capsule often dehiscing apically by two shortish slits. Seeds numerous, globular to reniform, tiny, foveolate-reticulate.

Petunia is often considered to be composed of only three herbaceous species, though some authors recognise as many as 40 species while Hunziker's (2001) latest treatment described 34 species. All but *P. parviflora* Juss. inhabit southern South America while this latter species displays a disjunct distribution, growing in the USA, Mexico and Cuba and well as being native in Paraguay, Uruguay and Argentina (Hunziker, 2001). This is another genus which is now widely cultivated throughout the world for its ornamental value.

P. × hybrida (*Hort.*) *Vilm.*, Fl. Pl. Terre, ed. 1: 615 (1863); E. & P. Pf.: 34 (1895); Sink in Petunia: 7 (1984); Hepper in Fl. Egypt 6: 150 (1998); Hunziker in Gen. Solanaceae: 56 (2001); Gonçalves in F.Z. 8(4): 19 (2005); Friis in Fl. Eth. 5: 108 (2006). Type: *Vilmorin* No. 8 "1858", no other specimen or type specimen locality given [N.B. Vilmorin's types were in a private herbarium]

Annual or perennial herbs to 60 cm(–1 m) high, occasionally woody; all vegetative and floral parts conspicuously viscid-pubescent with glandular-headed hairs; stems erect, straggly or decumbent, occasionally trailing, light green to pale brown. Leaves membranaceous or fleshy, light to dark green, elliptic, ovate or ovate-lanceolate, 2.2–5(–7) × 0.8–2.8(–4) cm, bases cuneate, margins entire to sinuate, apices acute to obtuse; sessile or petiolate to 3 mm. Flowers usually solitary, often arising from pair of small ovate leaves; pedicels erect and 1–4.5(–8) cm long in flower, 3–5.5 cm long in fruit when occasionally deflexed, densely glandular-villous. Calyx cupulate at base, 9–18 mm long, with five narrowly ovate to ligulate spreading calyx lobes (5–)8–18 × 2–4 mm, apices acute to obtuse, densely pilose/villous externally. Corolla blue, purple, pink, red, orange, yellow or white, often with distinct venation in contrasting colour, single or double, sometimes sweetly-scented, infundibuliform to salver-shaped, 4–7 cm long, basal tube 2.2–4 cm long, flaring above becoming 3.2–7(–10) cm broad with narrow throat 7–8 mm diameter, often terminating in five broadly ovate lobes which sometimes frilled or crenulate; externally densely glandular-pilose on tube becoming sparser above, glabrous internally. Stamens usually subequal, sometimes didynamous, or with one stamen longer or shorter than remaining four, occasionally equal, included; filaments free for upper 10–20 mm; anthers yellow to light brown (sometimes purplish when flowers purple), 1–1.5 × 1–1.5(–2.5) mm, often visible in corolla throat. Ovary light green, narrowly ovoid to pyriform, glabrescent, 3–6 × 1–2 mm; disc ± 0.7 × 1.7 mm; style light green, 2–3.2 cm long, usually included; stigma light green to brown, 1–2 mm diameter. Capsule greenish to light brown, somewhat chartaceous, ovoid or pyriform, 8–11(–15) × 6–8 mm, glabrous, usually dehiscing by two apical slits, usually enclosed by accrescent calyx lobes 12–16 × 2.5–5 mm. Seeds brown, globular, 0.5–0.7 × 0.5–0.6 mm. Fig. 4/1–5, p. 27

KENYA. Nairobi, University of Nairobi Chiromo Campus, 10 Oct. 1981, *Mwangangi* 2021! & City Park (cult.), 29 July 1953, *Verdcourt* 993!
DISTR. **K** 4; undoubtedly occurs as an escape from cultivation elsewhere in the FTEA region
HAB. Roadsides; ± 1600 m
CONSERVATION NOTES. Widespread; least concern (LC)

NOTE. Usually known as the Common Garden Petunia, this species really constitutes a complex group of hybrids which now form one of the world's most popular groups of bedding plants. Although various *Petunia* species have been cited as its parental species, this taxon is thought to have been derived from hybridisation between *P. integrifolia* (Hook.) Schinz & Thellung and *P. axillaris* (Lam.) Britton (*cf.* Symon, in Journ. Adelaide Bot. Gard. 3: 146 (1981) and Wijsman (in Acta Bot. Neerl., 31: 477–490 (1982)), or between *P. axillaris* and *P. inflata* R.E. Fries (Sink, 1984). It essentially comprises a series of cultivars derived through hybridisation, with the plants exhibiting considerable morphological diversity especially in flower size, form and colour. Many named cultivars are recognised horticulturally, including variants with single and double corollas, those with deeply frilled or crenate edges, and those which are variegated, variously striped, barred or with stellate markings radiating from the throat. Other variants reflect the intended habitat of the marketed cultivars with dwarf, compact, tall or pendulous forms also being common.

5. BROWALLIA

L, Sp. Pl. 1: 631 (1753) & Gen. Pl. ed. 5: 278 (1754) & Syst. Nat. ed. 10: 1118 (1759); Hunziker, Gen. Solanaceae : 86–89 (2001)

Brouvalea Adans., Fam. 2: 211 (1763)

Annual or perennial herbs. Leaves alternate, stipulate. Flowers often solitary, or in few-flowered racemose cymes, zygomorphic, axillary; pedicels deciduous at base. Calyx cylindrical to tubular, rarely campanulate, prominently veined with five calyx lobes. Corolla salverform, usually zygomorphic, tube narrow but swollen apically towards five broad obtuse recurved lobes, open at mouth, hairs usually short, eglandular or simply glandular, glabrescent internally. Stamens four, didynamous, usually included, occasionally with fifth anther or staminode; filaments fused to upper part of corolla tube where thickened, unequal, glabrous below, upper pair flattened and curved usually closing the throat of the corolla tube, lower pair geniculate and densely pilose above; anthers usually all fertile, upper pair sometimes reduced to single cells. Ovary ovoid, glabrous basally, densely pilose above, bilocular; ovules numerous; style filiform below, curved or tortuous and broadened beneath the stigma where corrugated, usually included; stigma broadly dilated, shallowly or deeply bilobed or capitate. Fruit ovoid capsules, smooth; usually splitting open by four entire valves, enclosed by accrescent persistent chartaceous calyx. Seeds cuboid to ovoid, minute, foveolate, numerous.

A small genus with 2–3 (or possibly 6) species, native to North to tropical South America, widely cultivated throughout the world for their ornamental value.

Browallia americana *L.*, Sp. Pl. 2: 631 (1753); Hepper in Fl. Egypt 6: 159 (1998); Gonçalves in F.Z. 8(4): 4 (2005); Friis in Fl. Eth. 5: 109 (2006). Type: Linnaeus' Hort. Cliff.: 319, t. 17 (1738): V Herb. Clifford: 319, *Browallia* 1 (BM!, lecto., designated by Stearn, Introd. Linn. Sp. Pl. (Ray Soc. ed.): 47 (1957) [icon]. [See also Jarvis, Order out of Chaos: 362 (2007)]

Erect annual herbs up to 60 cm (rarely to 1 m); sparsely branched, branches pale green or tinged purple, terete, pilose, eglandular hairs usually intermixed with stalked glands. Leaves membranaceous, light to dark green, elliptic, ovate, ovate-lanceolate or lanceolate, 3–5.4(–8) × 2.2–3(–4) cm, bases cordate to cuneate, laminas often decurrent, margins entire, apices obtuse to acute, sparsely to moderately pilose especially on the veins; petioles 5–20 mm. Flowers solitary or in few-(–10) flowered racemose lax cymes, axillary; rachides and pedicels erect, glabrous to densely viscid with spreading glandular hairs; pedicels 2–9 mm long in flower, (5–)7–18 mm long in fruit. Calyx tubular, 3–6(–10) × 1.5–3 mm, with unequal calyx lobes 1.4–3.3(–5) × 0.7–2.5 mm, generally increasing in fruit, glabrous to densely viscid externally with spreading glandular hairs. Corolla usually blue, often with white or whitish-yellow throat, salverform, often appearing almost two-lipped, 1–2 cm diameter; tube 1–1.8(–2) cm long, swelling to 2.5 mm wide below obtuse occasionally bilobed recurved lobes 5–8 × 3–6.5 mm; throat 1–2(–3) mm diameter; tube and lower lobe surfaces pubescent, upper lobe surfaces and inner surfaces of tube almost glabrous. Stamens with filaments of lower pair free for 1.4–3.5 mm, of upper pair free for 1–2 mm;

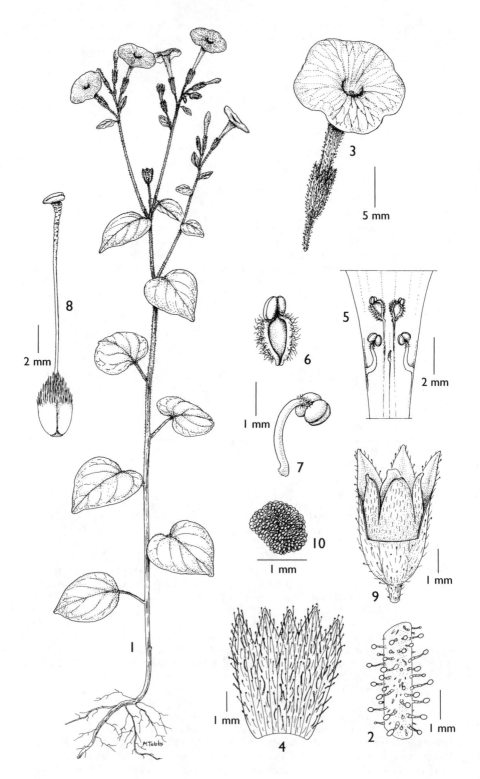

anthers pale yellow, upper pair often reduced to single cell, all curved downwards and 0.5–0.8 × 0.6–1.2 mm. Ovary 2–6 mm long; style 8.5–14 mm long; stigma 0.7 mm high, usually enclosed by the two sets of anthers. Fruit light brown, ovoid capsules, smooth, 4–6 × 3–4.5 mm, usually splitting open by four acute pilose valves 2–2.5 mm broad and long which often recurve apically, completely enclosed by persistent accrescent chartaceous calyx. Seeds brown, angular, cuboidal to rectangular, 0.5–0.8 × 0.3–0.8(–1) mm. Fig. 5, p. 30.

UGANDA. Bunyoro District: Masiudi, Budongo Forest (cult.), Nov. 1931, *Humphreys* 2381!
KENYA. Nairobi, Balmoral Road, Dagoretti Corner, 3 July 1971, *Greenway* 14882! & cultivated City Park, 29 July 1973, *Verdcourt* 994!; Kiambu District: Kiambu, Yara Estate, 10 Sept. 1969, *Hindorf* 809!
TANZANIA. Mbeya District: Mbeya, 11 May 1975, *Hepper & Field* 5504!
DISTR. **U** 2; **K** 4; **T** 7; originally from South America, cultivated and occurring as escape in Nigeria, Cameroon, Malawi and Zambia
HAB. A naturalised escape in gardens, maize-fields, on roadsides, stream banks, in moist sites in grassland; 850–1850 m
CONSERVATION NOTES. Widespread; least concern (LC)

SYN. *B. demissa* L., Syst. Nat. ed. 10: 1118 (1759) & Sp. Pl. ed. 2: 79 (1763); Don, Gen. Hist. Dichlam. Pl. 4: 478 (1837); Dunal in DC., Prodr. 10: 197 (1846); Sims in Bot. Mag. 28: 1136 (1888); Bailey, Man. Cult. Pl.: 880 (1966); U.O.P.Z.: 155 (1949), *nom. illeg.* Type: "in America australi"; lecto. as for *B. americana* L. [See also Jarvis, Order out of Chaos: 362 (2007)]
     B. elata L., Syst. Nat. ed. 10: 1118 (1759); Curtis in Bot. Mag. 1: pl. 34 (1788); Don, Gen. Hist. Dichlam. Pl. 4: 478 (1837); Bailey, Man. Cult. Pl.: 880 (1966). Type: "in Peru", Herb. *Linnaeus* 791.3 (LINN!, lecto. designated by Edmonds in Jarvis, Order out of Chaos: 362 (2007)]
     B. viscosa Kunth, Nov. Gen. Sp. Pl. 2: 373 (1818); Don, Gen. Hist. Dichlam. Pl. 4: 477 (1877); Dunal in DC., Prodr. 10: 197 (1846). Type: Colombia, around Loja and Gonzanama, *Bonpland* s.n. (*fide* D'Arcy in Ann. Missouri Bot. Gard. 60: 578 (1973)). [As far as is known, D'Arcy did not actually mention a type specimen for this species; a specimen in P-HBK labelled n. 3337, *B. viscosa* from Loja should probably be designated as the lectotype (fiche P-HBK56/21!)]

NOTE. The *Hortus Cliffortianus* plate of *B. americana* was first designated as the type of this species by Stearn (1957); this choice therefore preceded that by Deb (J. Econ. Tax. Bot. 1: 34 (1980) who had selected a Linnean sheet of *B. demissa* (LINN 791.2) – an illegitimate replacement name for *B. americana* (C. Jarvis, pers. comm.). A full list of synonymous taxa, many described from South America, and none of which have been encountered on African specimens, is given in D'Arcy (in Ann. Missouri Bot. Gard. 60: 578 (1973)).
    This species is often known as the Jamaican forget-me-not or Bush violet, while cultivars include the pale blue-flowered 'Caerulea', the large-flowered 'Grandiflora' and 'Major', the dwarf-formed 'Nana' and 'Compacta' a variant with a compact habit (*cf.* New RHS Dict. Gard. (1992) & Hepper, (1998)). Although attempts have been made to recognize the considerable variation displayed by *B. americana* specifically, D'Arcy (1973) concluded that it constitutes a single wide-ranging species reflecting the considerable ecologically diverse habitats in which it is found. Looking at a wide range of plants from both South and Central America, he noted the characters showing particular variability included corolla size, pubescence – especially density and the tendency to become viscid, and the extent to which the calyx lobes become expanded and sub-foliar apically during fruit development. Plants with conspicuous glandular hairs are often identified as *B. viscosa*, and D'Arcy conceded that these might represent a distinct taxon. However, the *Hortus Cliffortianus* plate of *B. americana* appears to illustrate prominently hairy young stems, pedicels and calyx suggesting that this generitype was indeed

FIG. 5. *BROWALLIA AMERICANA* — **1**, flowering plant; **2**, stem indument; **3**, complete flower; **4**, opened calyx; **5**, opened flower tube; **6**, reduced upper stamen; **7**, lower stamen; **8**, gynoecium; **9**, accrescent calyx surrounding capsule; **10**, seed. 1–2 & 4–8 from *Hepper & Field* 5504; 3 from *Greenway* 14882; 9–10 from *Hindorf* 809. Drawn by Margaret Tebbs.

viscid. All specimens so far seen from the Flora area belong to this viscid variant, as do those from Zambia, Zimbabwe and Madagascar. The androecial structure of these flowers is complex with the broadly ligulate posterior filaments arching over the anthers and the flattened stigma, effectively sealing the throat of the corolla tube. The straight narrow corolla tubes with their distinctive throat markings facilitate butterfly pollination in this genus.

Most specimens from the FTEA area had been identified as *B. viscosa* Kunth; this is, however synonymous with the extremely variable *B. americana*.

6. STREPTOSOLEN

Miers in Ann. Mag. Nat. Hist., Ser. 2, 5: 208 (1850); Hunziker, Gen. Solanaceae : 90–91 (2001)

Evergreen shrubs, all vegetative and floral parts conspicuously pubescent with eglandular and glandular-headed hairs usually intermixed with stalked glands. Leaves alternate; petioles scabrid. Inflorescences terminal subcorymbose leaf-opposed cymes, usually many-flowered. Calyx tubular or campanulate, somewhat zygomorphic with four or five sometimes unequal teeth; netveined, scabrid externally. Corolla infundibuliform, tubular or trumpet-shaped, basal tube spirally twisted and tapering towards the calyx, terminating in five lobes which partially reflexed together with apical part of the tube. Stamens 4, didynamous, usually included; filaments unequal, two lower posterior fused to basal quarter of corolla tube and two upper lateral fused to apical quarter of the tube, all densely pilose and becoming thicker at point of fusion; anthers unequal, lateral smaller and glabrous, posterior larger and pilose, elongate, dorsifixed. Ovary bilocular, ovules numerous; disc annular, smooth; style filiform below, becoming thicker and rugose for upper half, curved beneath the stigma, glabrous, usually included; stigma broadly dilated, somewhat sagittate, bilobed. Fruit globose or ovoid dehiscent bivalvate capsule, valves often bifid, usually wholly enclosed by accrescent persistent calyx. Seeds numerous, cuboidal to obovoid, reticulate.

This is a monospecific genus found at higher elevations in Peru, Ecuador and Colombia. Although Mabberley (in Mabberley's Plant Book, 3rd ed.: 826 (2008)) considers *Streptosolen* to be a synonym of *Browallia*, the two genera have been retained separately here, as in Hunziker (2001) and all other recent African treatments of the family. Hunziker used habit and a number of floral characters together with pollination vectors to differentiate these genera; they are also reported to differ in their basic chromosome number. Molecular work places these two genera firmly within the same clade (cf. Olmstead *et al.* in Taxon 57: 1167 (2008)) but there is as yet no molecular data on their relationship at the species level. If these two genera prove to be unequivocably conspecific, *Streptosolen* would be synonymised with *Browallia* with the two species *Browallia jamesonii* and *B. americana* occurring in the FTEA area. *Streptosolen* is characterised by colourful tubular orange-red flowers, and is often grown as an ornamental.

Streptosolen jamesonii (*Benth.*) *Miers* in Ann. Mag. Nat. Hist., Ser. 2, 5: 209 (1850); E. & P. Pf.: 37 (1895); T.T.C.L.: 591 (1949); Hunziker, Gen. Solanaceae: 90 (2001); Gonçalves in F.Z. 8(4): 7 (2005); Friis in Fl. Eth. 5: 109 (2006). Type: Ecuador, Loxa, *Hartweg* 818 in herb. Bentham and not Hooker (K!, lecto. designated here) [See notes below]

Evergreen erect or scandent shrubs 1–3(–6.5) m high with much branched scabrid, sparsely pilose main stems; young stems villous. Leaves coriaceous and rugose, dark green, elliptic or ovate, 1.2–4 × 0.6–1.8 cm, bases cuneate, often decurrent, margins entire, apices acute to obtuse, upper and lower surfaces somewhat scabrid, lower surface moderately to densely pilose/villous especially on veins and midribs, upper surface puberulous; petioles (0–)3–12 mm long. Inflorescences terminal subcorymbose leaf-opposed cymes, usually forming many-(–50+) flowered clusters; pedicels usually erect sometimes reflexed, 3–9 mm long,

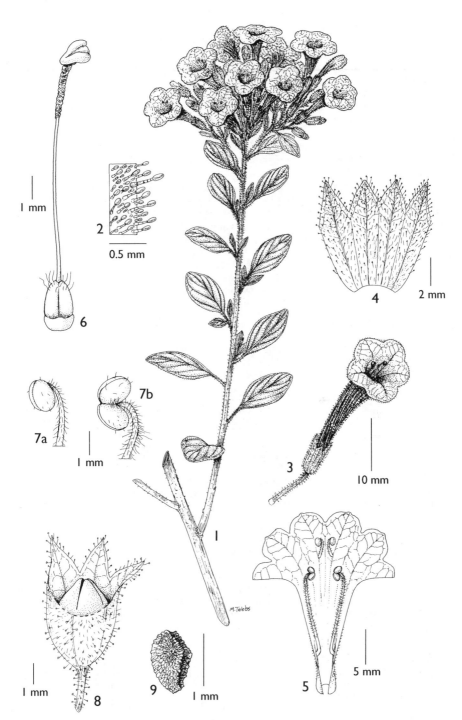

FIG. 6. *STREPTOSOLEN JAMESONII* — **1**, habit in flower; **2**, pedicel indument; **3**, flower; **4**, opened calyx; **5**, opened corolla; **6**, gynoecium; **7a**, anterior stamen; **7a**, posterior stamen; **8**, capsule with accrescent calyx; **9**, seed. 1–7 from *Greenway* 10881; 8–9 from *Ngoundai* 406. Drawn by Margaret Tebbs.

glandular-puberulous. Calyx 4.5–10 mm long, pubescent as pedicels, terminating in four or five short broadly triangular, acute and sometimes unequal teeth 1.5–3 × 1–2.5 mm, densely glandular-puberulous. Corolla red, orange or yellow, infundibuliform, tubular or trumpet-shaped, 2–3 cm long and 1.1–1.8 cm diameter at throat, basal tube spirally twisted and tapering towards the calyx, 1.2–2 cm long, glandular-puberulous, terminating in five broad obtuse lobes 3–8 × 4–10 mm which partially reflexed together with apical part of the tube and densely puberulous with small glands; throat and inner lobe surfaces covered with small glands. Stamens with unequal filaments, two lower free for 1–1.2 cm, two upper free for 1–3.5 mm; anthers unequal, yellow, lateral pair smaller 1–1.3 × 1.2–1.8 mm, with only one theca fully developed, posterior pair larger, 1–2 × 1.6–2.5 mm, sparsely pilose. Ovary ovoid, 1.2–2 × 0.8–1.6 mm, smooth, glabrous below with scattered long filamentous eglandular hairs at apex; disc ± 0.8–2 mm diameter; style 1.4–1.7 cm long, usually included; stigma 0.7–2 mm broad, often visible in corolla throat surrounded by the four anthers. Fruit smooth, light brown, ovoid dehiscent bivalvate capsules, 3.5–4 × 3 mm, valves often bifid and reflexed apically, usually wholly enclosed by accrescent persistent calyx. Seeds brown, cuboid to ovoid, 0.5–1 × 1 mm. Fig. 6, p. 33.

Kenya. Kiambu District: Muguga, 20 July 1963, *Greenway* 10881!; Nairobi, City Council Garden at Ainsworth Hill (cult.), 16 Dec. 1971, *Mwangangi* 1905!
Tanzania. Lushoto District: Amani Bustani, Muheza, 28 Oct. 1969, *Ngoundai* 406! & Lushoto, Boma Road, 10 June 1970, *Mshana* 1!; Iringa District: Mufindi, near F.P.C. House Sao Hill, 15 July 1972, *Ngonyani* 92!
Distr. **K** 4; **T** 3, 7; originally from South America, escaped and naturalised in Ethiopia and Zimbabwe
Hab. Widely cultivated, now occurring as occasional escape in gardens, on roadsides and near rocky pools; 900–2200 m
Conservation notes. Widespread; least concern (LC)

Syn. *Browallia jamesonii* Benth. in DC., Prodr. 10: 197 (1846); J. Smith in Hookers' Bot. Mag.: 771, t. 4605 (1851). Type: "Nova Granada" [Colombia/Ecuador] between Mivir and Naranjal [Naranfus], *Jameson* s.n (K!, syn.); around Loxa, *Hartweg* 818 (K!, syn.), see Note
    Streptosolen benthamii Miers in Ann. Nat. Hist., ser. 2, 5: 210 (1850) & in Illustr. S. Am. Pl. 2: 70 (1849–1857). Based on and types as for *Browallia jamesonii*

Note. The typification of this species is complicated. Bentham in Dunal (1846) cited the two syntypes *Jameson* s.n. and *Hartweg* 818 for his species *Browallia jamesonii*. Miers (1850) later decided that the features exhibited by this species should be recognised generically and using *Browallia jamesonii* as the basionym he described *Streptosolen jamesonii*, citing two specimens from "herb. Hook.", namely *Hartweg* 818 and *Seemann* 872. Since this species was based on Bentham's *Browallia jamesonii*, the second specimen cited by Miers (Sasaranga around Loxam, *Seemann* 872 [in herb. Hooker, (K!)] was not cited by Bentham and is not a syntype. The *Hartweg* 818 specimen is therefore the holotype of *Browallia jamesonii*, and the type on which *Streptosolen jamesonii* was based. However, there are two *Hartweg* 818 specimens in the Kew Herbarium. One is composed of a single shoot as part of a mixed sheet of specimens, and is in Herb. Hooker (dated 1867); this is only annotated with 'Peru' which has been crossed out and relabelled 'Ecuador'. The other is a more complete specimen which is in Herb. Bentham (dated 1854), and is annotated 'A shrub 4–6 ft high. Mountains of Paccha'. Neither specimen is annotated with 'Loxa' as cited by both Bentham (1846) and Miers (1850). Bentham did not specify the herbarium from which his syntype was described, but it is likely to have been his own. Miers probably saw both sheets at Kew, as he used the plant habit description given on the Bentham sheet in his generic description of *Streptosolen* (p. 207). This Bentham sheet of *Hartweg* 818 rather than the Hooker sheet that he cited has therefore been selected as the lectotype of his species *Streptosolen jamesonii*.
    Miers (1850) also described a second species of *Streptosolen*: *S. benthamii* citing the herb. Hooker specimen *Jameson* from "Nova Granada" with the same collection details as that given for the syntype of *Browallia jamesonii*. This specimen is thus both a syntype of *Browallia jamesonii* and the holotype of *Streptosolen benthamii*. There are two Herb. Hooker *Jameson* specimens on different sheets at Kew; both are annotated as *Browallia* and both sheets are

composed of several specimens. The specimens are all conspecific, but only one – an apical fragment which looks as though it was originally part of the larger specimen on the second sheet– is accompanied by a note containing the cited collection details; this has been selected as the syntype of *Browallia jamesonii*.

Apart from their distinctive twisted corollas, plants of this species often exhibit a characteristic pubescence. The pedicels and external calyx and corolla surfaces in particular, are usually densely covered with short spreading few-celled hairs with distinctive brown ovoid glandular heads. This pubescence is evident on all African specimens examined as well as on the *Jameson* specimen of *S. benthamii*. However, it is replaced by dense multicellular and mainly eglandular hairs on the type of the synonymous *Browallia jamesonii* (*Hartweg* 818, K!), while a few other specimens, collected from Peru, Ecuador and Colombia and with slightly larger flowers, also exhibit this atypical pubescence. Further work on this genus, particularly in its native South America, might well show that the variation noted warrants formal taxonomic recognition. The plants are bird-pollinated.

7. SCHWENCKIA

L., Gen. Pl. ed. 6: ("567") 577 (1764), as "*Schwenkia*"; Heine in K.B. 16: 465–469 (1963); Freire de Carvalho in Rodriguesia 44: 307–524 (1978); Hunziker in Gen. Solanaceae: 93 (2001)

Annual or perennial herbs, sometimes shrubby or small trees; hairs simple with eglandular or glandular heads. Leaves usually alternate. Inflorescences many-flowered monochasial racemose or lax paniculate cymes, axillary or terminal subtended by small bracts, pedunculate; flowers solitary or in pairs or triplets, actino- or slightly zygomorphic; pedicels usually erect, occasionally deflexed in fruit. Calyx tubular to campanulate, with 5 calyx lobes, often unequal. Corolla tube narrow and straight, rarely curved, broadening into 5 complex small lobes which often 3- or 2-lobed usually alternating with 5 clavate appendages. Stamens didynamous or reduced to two with two or three staminodes; filaments fused to corolla tube at different levels, filiform or compressed, becoming thicker at point of fusion, glabrous or pilose; fertile anthers ovoid or narrowly ellipsoid, always ventrifixed, dehiscing extrosely, thecae often unequal, staminodes devoid of vestigial anthers. Ovary ovoid, glabrous, 2-locular, ovules numerous; disc annular or cupulate; style filiform, glabrous, usually exserted; stigma capitate or discoidal, inconspicuous. Fruit a smooth, globose to ovoid, many-seeded capsule, bivalvate with entire smooth valves, usually longer than persistent fruiting calyx. Seeds numerous, small, reticulate.

Variously considered to be composed of between 5 and 30 species, Hunziker (2001) considered the genus to consist of around 25 species, which occur from Central America and the Antilles to NE Argentina. The type species *S. americana* L. is the commonest species and now occurs as a widespread weed throughout tropical Africa.

Linnaeus (1764) unintentionally described the genus as *Schwenkia*, while giving the binomial *Schwenckia americana* after the generic description, leading to the frequent mis-spelling of this generic name (cf. Heine, 1963).

Schwenckia americana *L.*, Gen. Pl. ed. 6: 577 [567] (1764); E. & P. Pf.: 37 (1895); Durand & Schinz in Fl. Etat Ind. Congo 1: 208 (1896); Hiern in Cat. Afr. Pl. Welw. 3: 754 (1898); C.H. Wright in F.T.A. 4, 2: 260 (1906); Durand & Durand, Syll. Fl. Cong.: 398 (1909); Heine in F.W.T.A. 2nd ed., 2: 327 (1963) & in K.B. 16: 465–469 (1963); Hunziker in Gen. Solanaceae: 93 (2001); Gonçalves in F.Z. 8(4): 25 (2005). Type: *Schwenkia americana* L., "In Barbyce", Herb. Linn. 31.1 (LINN!, lecto. designated by Knapp in Jarvis *et al.*, Regnum Veg., 127: 86 (1993). [See also Jarvis, Order out of Chaos: 827 (2007)]

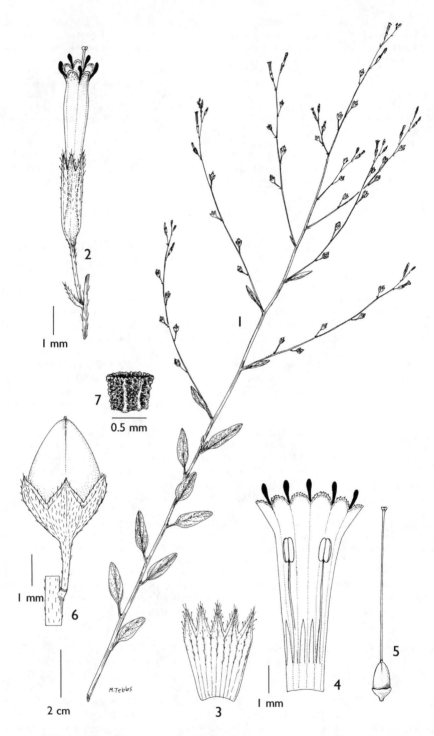

FIG. 7. *SCHWENCKIA AMERICANA* — **1**, habit in flower and fruit; **2**, flower; **3**, opened calyx; **4**, opened flower with two stamens and three staminodes; **5**, gynoecium; **6**, capsule; **7**, seed. 1–5 from *Lye* 2682; 6–7 from *Wood* 585. Drawn by Margaret Tebbs.

Annual or perennial herbs to 50 cm(–1 m) tall; stems erect, occasionally procumbent, usually woody basally, arising from perennial rootstock; all vegetative parts with long spreading and short appressed eglandular or glandular simple hairs. Leaves usually alternate, often light green, elliptic, ovate or ovate-lanceolate, 0.7–2.4(–2.9) × 0.2–0.8(–1) cm and becoming smaller and more sessile apically, bases cuneate, margins entire, apices usually obtuse, occasionally acute, pilose with hairs denser on lower veins; petioles (0–)1–5(–8) mm long. Inflorescences axillary or terminal racemose or paniculate many-flowered cymes, pedunculate with axes glabrous to sparsely pilose; flowers solitary or in pairs, always subtended by small densely pilose bracts; pedicels erect, glabrous to sparsely pilose, 1–3.2 mm long in flower, 2–11(–18) mm long in fruit. Calyx actinomorphic to slightly zygomorphic, 2.5–3.8 mm long in flower and in fruit, with 5 triangular acute and often unequal lobes 0.5–1 × 0.3–0.8 mm, enlarging to 0.7–2.8 × 0.5–1.2 mm in fruit, glabrous to glandular-pilose externally. Corolla usually actinomorphic, whitish, yellow, greenish, purple or violet, tube 5–8 × 0.7–1.4 mm diameter apically, narrow and straight, rarely curved, strongly veined, broadening into 5 lobes 0.5–0.8 × 0.3–0.8 mm, alternating with five clavate appendages arising from corolla tube venation; glabrous externally below, becoming pilose on backs of lobes. Stamens usually included; filaments free for (0.7–)1.2–4 mm; fertile anthers pale yellow, 0.6–1 mm long; staminodes devoid of vestigial anthers. Ovary 0.7–1.5 × 0.6–1.8 mm; disc 0.4 mm high, 0.7–1.1 mm diameter; style 4–6 mm long, enclosed or briefly exserted; stigma 0.1–0.3 mm diameter. Capsules pale brown, globose to ovoid, 3–5 × 2.5–4 mm, smooth and glabrous, apiculate, lower part enclosed by accrescent chartaceous fruiting calyx, fruiting pedicels often ligneous with distinct abscission layer. Seeds verrucose, brown, cuboid, 0.1–0.8 × 0.1–0.8 mm, angulate. Fig. 7, p. 36.

UGANDA. West Nile District: Koboko, Sept. 1940, *Purseglove* P1040!; Busoga District: southern part of Dagusi Island, 16 Jan. 1953, *G.H.S. Wood* 585!; Masaka District: south of Luunga, 31 May 1971, *Katende* 947!
TANZANIA. Kigoma District: Kigoma, Kibirizi, 7 Apr. 1994, *Bidgood & Vollesen* 3057!; Tabora District: 47 km on Tabora–Urumbo road, 13 May 2006, *Bidgood et al.* 5978!
DISTR. U 1–4; T 4; originally from tropical America, now widespread across tropical Africa
HAB. Weed of waste places, cultivation and abandoned fields or plantations, degraded bushland, in secondary forest, secondary woodland, poor quality grassland and in riverine vegetation; may be locally common; 5–1300 m
CONSERVATION NOTES. Widespread; least concern (LC)

SYN. *S. guineense* Schum. & Thonn., Beskr. Guin. Pl.: 8 (1827). Type: none cited - citation reads "Acmindelig paa Vei elber aabne og torre Marker, i alle Aarstider dog maest, Regntiden"
?*S. hirta* Klotzsch in Linnaea 14: 289 (1840); C.H. Wright in F.T.A. 4, 2: 261 (1906). Type: Brazil, Cruz de Casma, Bahia, ?*Schlechtendal* s.n., (?HAL, ?holo., type collector and locality not cited)

NOTE. This species is now widespread throughout tropical Africa, where it was recorded in the early nineteenth century and often occurs as a weed of cultivation. Symon (in *Solanaceae III*: 143 (1991)) considered it to be an adventive in Africa which had arrived through its involuntary transport by man.
Specimens with a conspicuous glandular pubescence are often identified as *S. hirta*.

8. SCHIZANTHUS

Ruiz & Pav., Fl. Peruv. Prodr. 1: 6 (1794) & Fl. Peruv. 1: 13, t. 17 (1798); Dunal in DC. Prodr., 13(1): 202 (1852); Hunziker, Gen. Solanaceae: 379 (2001)

Annual or biennial herbs. Leaves usually alternate, usually pinnatisect, occasionally pinnate, with incised or dentate segments, membranaceous or fleshy. Inflorescences terminal many-flowered monochasial cymes; flowers zygomorphic, 'papilionaceous' in shape; pedicels usually erect, occasionally deflexed in fruit. Calyx somewhat tubular-campanulate, deeply divided into 5 calyx lobes. Corolla with 5 unequal deeply lobed segments or lobes comprising one anterior, two lateral and two posterior which usually fused to form a keel, basal cylindrical tube long or short. Stamens subequal, two lateral fertile protected by the posterior keel, and three staminodes one of which often rudimentary; filaments fused to top of corolla tube, filiform above becoming thicker at point of fusion, pilose at least below; fertile anthers dorsifixed, dehiscing introsely, staminodes with vestigial anthers. Ovary 2-locular, ovules numerous; style filiform, glabrous, usually exserted; stigma capitate or discoidal, inconspicuous. Fruit a smooth, many-seeded, bivalvate capsule, often dehiscing apically by two shortish slits, globose to ovoid, small. Seeds numerous, subspherical to reniform, tiny, foveolate-reticulate.

Between 8 and 20 species, with Hunziker's (*l.c.*) latest treatment recognising 12, all of which are endemic to Chile, although two have spread into neighbouring Argentina.

Schizanthus pinnatus *Ruiz & Pav.*, Fl. Peruv. et Chil. 1: 13, t. 17 (1798); Hepper in Fl. Egypt 6: 155 (1998). Type: Chile, t. 17 in Fl. Peruv. et Chil., 1 (1798)!, fide Hepper loc.cit. [NB. There is a possible lectotype specimen, *Pavon* s.n. in BM!]

Annual or biennial herbs 0.2–1.3 m tall; stems erect, smooth, slender, light green, with single or multiple stems arising from the plant base, sometimes woody basally; vegetative parts viscid-pubescent with long and short spreading hairs, eglandular or glandular. Leaves membranaceous or fleshy, usually pinnatisect, occasionally pinnate or deeply lobed, extremely variable in size from 1.8–5.8(–12) × 0.3–3.8 cm overall, pinnae or segments 0.3–1.2(–2.7) cm × 1–4(–7) mm, usually in 4–9 alternating pairs, linear to ovoid with ± obtuse apices, sessile bases and margins deeply lobed or sinuate, occasionally entire, pubescence denser on veins and lower leaf surfaces, hairs mostly eglandular; petioles absent or to 18 mm long. Inflorescences terminal, many-flowered cymes; flowers zygomorphic, 9–17 mm long; peduncles erect, to 3 cm long; pedicels arising in axil of leaf or small oval bract, erect, occasionally deflexed in fruit, somewhat filiform, 0.7–2.2 cm long in flower, densely viscid, becoming more sparsely pubescent in fruit when 1.8–4.2 cm long. Calyx ± campanulate, 4–9 mm long, calyx lobes ligulate to spatulate with obtuse and sometimes recurved apices, (3.5–)4.5–8 × (0.6–)0.8–2 mm, becoming accrescent around capsules, viscid-pubescent externally. Corolla usually purple or white but varying from blue, pink, red or yellow in cultivars, often with a yellowish throat, cylindrical basal tube shorter than calyx, 2–3 mm long, flaring above into 5 unequal deeply lobed obtuse segments: one anterior which often speckled or striped, two lateral and two posterior, fused to form a keel, pubescent externally, glabrescent internally, 9–17 × 0.8–2.7 mm diameter. Stamens subequal with two fertile and three staminodes one of which often minute; filaments free for 3–7 mm; anthers yellowish or purple-tinged, 1.2–2 mm long; staminodes with filaments free for 1–2 mm, white to cream, 0.3–0.5 mm long. Ovary glabrous, brownish, ovoid, 1.5–2 × 0.6–1.3 mm; style 8–12 mm long, often curved; stigma brownish, 0.2–0.3 mm diameter. Fruit a smooth pale brown capsule, globose to ovoid, (3–)4–6 × (2–)3–5 mm, usually clasped by calyx lobes, often dehiscing apically by two or occasionally by three or four shortish slits. Seeds brown, subspherical, 0.7–1 mm diameter, ridged around deep pits. Fig. 4/6–8, p. 27.

KENYA. Nairobi, City Park (cult.), 29 July 1953, *Verdcourt* 993!
TANZANIA. Lushoto District: W Usambaras, Mkusi, 31 Aug., 1950, *Verdcourt* 342!
DISTR. **K** 4; **T** 3; native to Chile
HAB. Cultivated and naturalised (no habitat details); ± 1700 m
CONSERVATION NOTES. Widespread; least concern (LC)

SYN. *S. porrigens* R. Grah. in Edin. Phil. Journ. 11: 401 (1824); Don, Gen. Hist. Pl. 4: 469 (1837);
    Hooker, Exot. Fl., 2: t. 86 (1824), *nom. nud.*
   S. pinnatifidus Lindl., Bot. Reg. 29: 45 (1843). Type: Chile, Coquimbo, *Bridges* 1355 (Herb
    Lindley, CGE, holo., photo!)
   S. pinnatus Ruiz & Pav. f. *papilionaceus* Vilm., Blumeng., ed. 3, Sieb. & Voss., 1: 773 (1895)
    ?*nom. nud.* Type: not cited

NOTE. This is the most common and variable of the *Schizanthus* species, with many horticultural
forms and varieties being distinguished by stem height and by the flower colour markings.
The flowers are mostly bee-pollinated.

9. **DATURA**

L., Sp. Pl.; 179 (1753) & Gen. Pl. ed. 5: 83 (1754) & Syst. Nat. 2, ed. 10: 932 (1759);
Dunal in DC., Prodr. 13(1): 538 (1852); Safford in Journ. Wash. Acad. Sci. 11(8):
173–189 (1921); Satina & Avery in Chronica Botanica 20: 16–47 (1959); Haegi in
Aust. J. Bot. 24: 415–435 (1976); Hadkins *et al.* in J.L.S. 125: 295–308 (1997);
Persson, Knapp & Blackmore in Solanaceae IV: 171–187 (1999); Hunziker, Gen.
Solanaceae: 149–153 (2001)

Annual or semiperennial herbs and shrubs, often malodorous; main stem often
stout, woody and dichotomously branched, spherical stalked brownish glands
present on all parts. Leaves alternate or opposite. Flowers usually solitary and in
branch forks, always erect, usually fragrant, opening diurnally and remaining open
during anthesis, with prominent venation; pedicels short, elongating during fruiting.
Calyx tubular, usually enclosing the lower half of the corolla tube; calyx lobes five;
base circumsessile, forming a ridged collar which often persistent. Corolla tubular
below becoming funnel- or trumpet-shaped above, sometimes double or triple; tube
long and slender, the lobes fused almost to the apex, with 5 or 10 terminal acuminate
to caudate lobes or teeth. Stamens inserted on lower half of corolla tube and
alternating with lobes, equal, included; filaments glabrous above, slender, filiform,
widening towards point of adnation to corolla tube from where usually sparsely
pilose; anthers oblong, basifixed, free, dehiscing longitudinally, often with long hairs
on dehiscent margins. Ovary superior, conical, softly spinose or tuberculate,
bilocular above, but 4-loculate basally owing to a false septum, ovules numerous,
placentation axile; style long, filiform, glabrous, sometimes exserted beyond anthers
but included; stigma bilobed, clasping the stylar apex. Fruit ovoid or globose 2–4-
celled capsules, dehiscing irregularly or by 2 or 4 valves from the apex; valves usually
spinose or tuberculate, rarely smooth, subtended by the persistent discoid remains of
the calyx which adherent or reflexed, forming a frill or collar beneath capsule;
fruiting pedicels elongated and stout, erect or pendulous. Seeds large, reniform or
discoid, with a thick (suberose) corky testa; funicular caruncle (elaiosome) well-
developed, numerous.

Currently thought to be composed of 11–12 species which have a relatively restricted natural
distribution in semiarid parts of Mexico and the SW USA, extending to Panama in the south
and the Antilles in the east.

All Old World Daturas, collectively known as the Thorn Apples, were introduced from the
Americas during the early years of intensive European colonization of the New World (*cf.*
Symon & Haegi, 1991 in Solanaceae III: 197; Persson *et al.*, 1999). Three naturalised species
are found throughout the East African region, where they are probably also widely cultivated
for their hallucinogenic, medicinal and narcotic effects as well as for their ornamental value.
They all contain the tropane alkaloids scopolamine and hyoscyamine, plus steroidal lactones
of the withanoloid group (*cf.* Hunziker, 2001 for extensive references).

1. Leaves sinuate or with a few shallow obtuse lobes; corolla 11–18 cm; capsules dehiscing irregularly or by 2 to 4 valves which usually covered with equal or unequal tubercles or spines which pubescent throughout; seeds usually greenish/yellow to pale brown 2

Leaves prominently sinuate-dentate with deep acute and often lacerate lobes; corolla 6–9 cm; capsules dehiscing regularly by 4 valves which often reflex, and usually covered with unequal spines which glabrous above and only shortly pilose towards their bases; seeds usually black or dark brown .... 1. *D. stramonium*

2. Plants usually viscid; capsule pendulous, covered with sharp acicular equal spines 4–12 mm long which densely covered with short spreading glandular- and eglandular-headed hairs .... 2. *D. innoxia*

Plants non-viscid; capsule erect or nodding, covered with blunt pyramidal unequal spines (tubercles) 2–6 mm long which covered with short appressed eglandular-headed hairs .... 3. *D. metel*

1. **Datura stramonium** *L.*, Sp. Pl.: 179 (1753) & Syst. Nat. 2, ed. 10: 932 (1759); Dunal in DC., Prodr. 13(1): 540 (1852); Engl., Hochgebirgsfl. Trop. Afr.: 374 (1892); Hiern, Cat. Afr. Pl. Welw. 3: 753 (1898); C.H. Wright in Fl. Cap. 4(1): 118 (1904) & in F.T.A. 4, 2: 257 (1906); Durand & Durand in Syll. Fl. Cong.: 397 (1909); W.F.K.: 90 (1948); Satina & Avery in Chronica Botanica 20: 18 (1959); F.P.U.: 129 (1962); E.P.A. 2: 882 (1963); Heine in F.W.T.A. 2nd. ed.: 326 (1963); Verdc. & Trump, Common Poisonous Pl. E. Afr.: 165 (1969); Troupin, Fl. Rwanda 3: 366 (1985); Blundell, Wild Fl. E. Afr.: 188 (1987); U.K.W.F. 2nd ed.: 244 (1994); Hepper in Fl. Egypt 6: 110 (1998) & in Fl. Egypt 3: 46 (2002); Gonçalves in F.Z. 8(4): 29 (2005); Thulin in Fl. Somalia 3: 219 (2006); Friis in Fl. Eth. 5: 158 (2006). Type: "in America nunc vulgaris per Europam", Herb. *Hort. Clifford* 55, *Datura* 1 (BM!, lecto. designated by D'Arcy in Ann. Missouri Bot. Gard. 60: 624 (1973)). [See Hadkins *et al.* (1997) for problems associated with selection of lectotype & Jarvis, Order out of Chaos: 476 (2007)]

Annual herbs up to 2(–4) m high, occasionally undershrubs, erect or spreading, sometimes unpleasantly aromatic; branches light green, brownish, yellow or purplish, smooth, occasionally hollow, glabrescent to pilose with multicellular hairs. Leaves alternate, often dark green, ovate, ovate-lanceolate, lanceolate or rhomboidal, (5–)8.6–15(–17) × 4.5–13(–15) cm, bases obliquely cuneate to cordate, margins sinuate-dentate with 2–6 deep acute lobes which often lacerate, apices acute to acute/acuminate, glabrescent to moderately pilose; petioles 2.2–7 cm long. Flowers solitary, axillary, erect; pedicels (3–)4–10 mm and erect in flower, elongating to 8–15(–25) mm in fruit. Calyx cylindrical, (26–)30–45(–50) × 6–8 mm, sparsely pilose externally, lobes broadly to narrowly triangular, 3–10 × 1.5–6 mm, apices acute, margins pilose, base circumsessile forming a ridged collar or flange 3.5–8 mm diameter and 1.5–5 mm broad. Corolla greenish, yellow, cream, white, pale or dull purple with prominent veins, tubular below becoming funnel- or trumpet-shaped above, 6–8.2(–9) cm long, glabrescent externally; tube long and slender, with 5 terminal acuminate to caudate teeth or tails 1.3–8 × 0.5–2 mm at base. Stamens occasionally visible in corolla throat; filaments free for 2–3 cm; anthers oblong,

FIG. 8. *DATURA STRAMONIUM* — **1**, leaf; **2**, flower; **3**, opened flower; **4**, ovary; **5**, capsule. *D. INNOXIA* — **6**, leaf; **7**, flower; **8**, glandular stem hairs; **9**, capsule; **10**, capsule spine. *D. METEL* — **11**, flower; **12**, capsule; **13**, capsule spine. 1 from *Richards* 20371; 2 from *Broadhurst Hill* 454; 3–4 from *Purseglove* 3147; 5 from *Gillett* 13739; 6–10 from *Kisena* 1; 11 from *Hucks* 63; 12–13 from *Magogo* 825. Drawn by Margaret Tebbs.

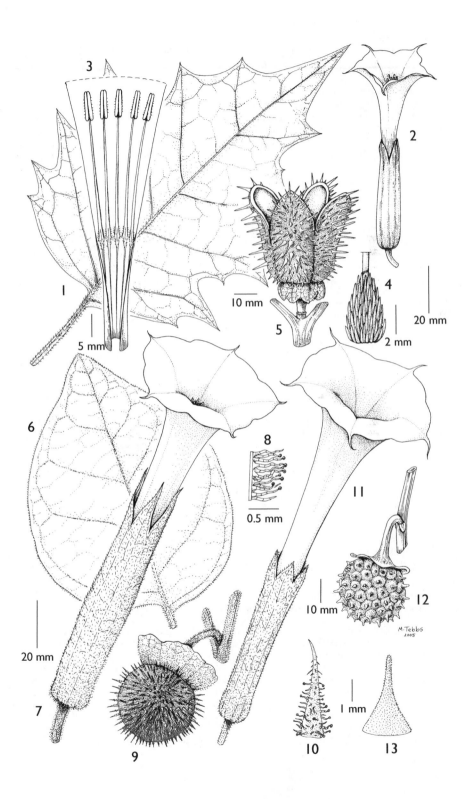

yellow, yellow with purple margins or purplish, (2.5–)4–7 mm long. Ovary dark brown, softly spinose, (2.5–)3–7 × 2.8–6 mm; style dark brown, (3.8–)4.2–6 cm long; stigma clavate, 1.5–3 mm long. Fruit longitudinally ovoid usually spinous capsules, green yellowish or brownish, 2.5–5 × 1.8–4 cm (including spines), dehiscing by 4 valves splitting from the apex and reflexing to expose seeds; valves usually covered with acicular and sharply pointed unequal spines, 7–15 × 0.5–1.8 mm basally, smooth and glabrous but shortly pilose near base, occasionally valves smooth (but not in Flora area); subtended by the persistent remains of the calyx 6–18 × 4–8 mm, usually reflexed to the pedicel; fruiting pedicels elongated, stout, erect. Seeds black or dark-(rarely pale)-brown, reniform, occasionally discoid, 2.5–3.5 × 2.2–3.5 mm, not ridged, minutely and densely punctate, with distinct yellow or whitish elaisome. Fig. 8/1–5, p. 41.

UGANDA. West Nile District: path to Sultan's house at Paida, 20 Aug. 1953, *Chancellor* 187; Teso District: Serere, June 1932, *Chandler* 789!; Kigezi District: Kachwekano Farm, Dec. 1949, *Purseglove* 3147!
KENYA. Northern Frontier District: Moyale, 22 Aug. 1952, *Gillett* 13739!; NE Elgon, July 1954, *Tweedie* 1176!; Masai District: Ol Choro Orogwe Ranch, 3 July 1961, *Glover et al.* 2019!
TANZANIA. Arusha District: Arusha National Park, near Momela Park Headquarters, 7 Apr. 1968, *Greenway & Kanuri* 13438! & Ngurdoto National Park, 7 May 1965, *Richards* 20371!; Mbulu District: Lake Manyara National Park, Endabash River Drift, 5 Mar. 1954, *Greenway & Kanuri* 11315!
DISTR. U 1–4; **K** 1, 3–6; **T** 1–7; probably native to Mexico, now a pantropical weed
HAB. A common weed of disturbed ground, waste places, gardens and plantations, in pastures, arable land, woodland and forest fringes; 550–2400 m
CONSERVATION NOTES. Widespread; least concern (LC)

SYN. *D. tatula* L., Sp. Pl. ed. 2: 256 (1762); Nees von Esenbeck, Trans. Linn. Soc. 17: 75 (1834); Durand & Durand in Syll. Fl. Cong.: 397 (1909). Type: "Habitat...", Herb. Linn. 243.2 (LINN!, lecto. designated by Hadkins *et al.* in Bot. J. Linn. Soc. 125: 305 (1997) [See also Jarvis, Order out of Chaos: 476 (2007)]
   D. inermis Jacq., Hort. Vindob., 3: 44, pl. 82 (1770). Type: *Jussieu* in Hort. Reg. Paris (?P-JUSS, ?holo.) NB. if not extant, Hort. Vindob, 3: pl. 82 should be designated as lectotype
   D. stramonium ß *canescens* Wall. in Roxb., Fl. Indic. 2: 239 (1824). No specimens or catalogue number were cited though the habitat was given as 'Parbutteeya, Muhadeo-Soa. A native of the mountainous parts of Hindoosthan. I have found it on all parts of Nipal which I have visited, both mountains and vallies....' Dunal (1852) later used *D. stramonium* ß *canescens* Wall. as the basionym of *D. wallichii*, citing *Wallich* catal. n. 2637 (G-DC, fiche!, ?holo.; *Wallich* 2637 a! & b!, K-WALL, isosyn.)
   D. stramonium L. var. *tatula* (L.) Dunal in DC., Prodr. 13(1): 540 (1852); De Wildeman, Fl. Etat Ind. Cong. 1: 207 (1896); E.P.A. 2: 883 (1963) [see Satina *et al.* (1959) for discussion of problems associated with this synonymy]
   D. wallichii Dunal in DC., Prodr. 13(1): 539 (1852). Type: see *D. stramonium* ß *canescens* Wall. note
   D. stramonium L. var. *inermis* (Jacq.) Timm. in Pharm. Journ. 68: 571 (1927); E.P.A. 2: 883 (1963)

NOTE. Though typically characterised by spiny capsules, a number of variants with smooth capsules occur and some of these have been formally recognised. Safford (1921) reported this to be a variable character, and that both spiny and smooth capsules could be found on the same plant with the gene for prickles being dominant. As implied by its epithet, *D. inermis* is such a smooth-capsuled variant; all other features of this species described or illustrated on Jacquin's plate imply that it is conspecific with *D. stramonium*. Smooth capsules are not thought to occur in the Floral area. Commonly known as Jimson- or Jamestown- weed, *D. stramonium* has been reported as a contaminant of some 40 different commercial crops in around 100 countries (cf. Hunziker, 2001), with its seeds becoming a toxic contaminant of many seed and grain harvests. All parts of the plant are rich in tropane alkaloids for which species is an important commercial source; they include hyoscyamine, hyoscine and atropine. This species is also the source of the drug stramonium and is now widely used medicinally in the Old World. Its poisonous qualities are well-known and are summarised by Verdcourt & Trump (1969).

Local populations are said to use it to induce drugged sleep, and to make intoxicant beer (**T** 3); it is reported to be lethal to cattle and goats (**K** 1) and is particularly dangerous when the plant occurs as a weed of crops such as wheat and maize when the seed can contaminate the resultant flour (cf. **K** 3). The seeds are also used to drive soldier ants away from native houses (**T** 3), presumably through the beneficial mutualism existing between the seed elaisomes which attract the harvester ants responsible for their dispersal (Hunziker 2001).

With regard to *D. wallichii*, there are 2 specimens labelled 2637 in K-WALL; specimen a) was collected in Nepal and b) from Bihar in northern India. Both seem good examples of *D. stramonium* and are presumably isosyntypes; a third specimen labelled 2639c in the Kew General Herbarium is also a good specimen of this species.

2. **Datura innoxia** *Mill.*, Gard. Dict. ed. 8: no. 5 (1768), as *inoxia*; Satina & Avery in Chronica Botanica 20: 28 (1959); E.P.A. 2: 881 (1963); Heine in F.W.T.A. 2nd ed., 2: 326 (1963); Hepper in Fl. Egypt 6: 115 (1998) & in Fl. Egypt 3: 46 (2002); Gonçalves in F.Z. 8(4): 33 (2005); Thulin in Fl. Somalia 3: 219 (2006); Friis in Fl. Eth. 5: 157 (2006) [as *D. inoxia*]. Type: cultivated Chelsea Physic Garden from seeds collected in Veracruz, *Miller* s.n. (BM!, neo., designated by Barclay in Bot. Mus. Leafl. Harv. Univ. 18: 255 (1959), based on the original listing of the specimen in Phil. Trans. Roy. Soc., 51: 99, no. 1843 (1760)

Annual or semiperennial herbs 0.3–1.5 m tall and 2 m wide, erect or spreading, often malodorous; branches smooth, violet, greyish-green or -brown, sometimes much-branched or becoming woody, velutinous to naked eye and usually viscid, villous/pilose usually predominantly glandular-headed, denser on young stems. Leaves alternate, ovate-lanceolate to rhomboidal, (4–)8–15 × (3.2–)6–11 cm, bases usually obliquely cuneate, margins sinuate, with up to 3 shallow obtuse lobes, apices acute or acute/obtuse, young leaves densely villous/pilose becoming moderately pilose, especially dense on lower surfaces, midribs and veins; petioles (3–)6–10.5 cm long. Flowers solitary, axillary, erect; pedicels (3–)6–8(–10) mm and erect in flower, elongating to 10–35 mm in fruit when pendulous, densely villous. Calyx cylindrical becoming slightly bulbous centrally, moderately pilose externally, hairs glandular and eglandular, 7.8–11 × 1.1–2.2 cm, lobes narrowly triangular, 0.9–2.2 × 0.5–0.8 cm, apices acute, with dense hairs on the margins. Corolla white, sometimes cream or pale yellow with prominent venation, tubular below becoming funnel- or trumpet-shaped above, sometimes double or triple, 14.5–19 cm long, glabrescent externally; tube 7–10 mm broad at base, usually with 5 terminal acuminate to caudate teeth ± 6 mm long. Stamens with filaments free for 3.6–6 cm; anthers oblong, yellow, (8–)10–11 mm long. Ovary ovoid, dark brown, 6–7 mm long, softly spinose; style 10–14.8 cm long; stigma ± 3 mm diameter. Fruit pendulous, broadly ovoid or globose brownish capsules, 2.5–4.5 × 2.5–5.5(–6) cm (including the spines), dehiscencing irregularly or by 2 valves; valves covered with sharply pointed equal and often greenish spines, 4–12 × ± 0.7 mm basally, densely villous/pilose throughout with spreading glandular and eglandular hairs; subtended by the persistent cupular remains of the calyx 1.4–2 cm deep and 2.8–4.5(–6) mm diameter, with dense short glandular and eglandular hairs internally; fruiting pedicels elongated, stout, pilose/villous. Seeds pale brown to brown, reniform, 4–5 × 3.5–4.5 mm with lateral ridge, foveolate, with creamish elaisome. Fig. 8/6–10, p. 41.

KENYA. Tana River District: Kora National Reserve on the Tana River, 16 km W of Project Camp, 22 July 1983, *Cunningham van Someren* 995! & Marenge, bank of Tana River, 25 Apr. 1983, *Hemming* 83/17!; Garissa District: Mulanjo, 15 Aug. 1976, *Kibuwa* 2479!
TANZANIA. Lushoto District: Silviculture Experimental Nursery, 18 Nov. 1981, *Kisena* 1!
DISTR. **K** 7; **T** 3; native to tropical America, cultivated as an ornamental and now a widespread weed throughout the tropics
HAB. Usually along rivers, occasionally on disturbed soil in ruderal sites; 300–700 m
CONSERVATION NOTES. Widespread; least concern (LC)

SYN. ?*D. guayaquilensis* Kunth, Nov. Gen. Sp. Pl. 3: 8 (1818). Type: Ecauador, Guayaquil, *Humboldt, Bonpland & Kunth* 3855 (P-Bonpl., ?holo., fiche HBK 59/9!)
   D. metel Dunal in DC., Prodr. 13(1): 543 (1852), *non* L. Type: Mexico, Victoria, *Berlandier* 2156 (G-DC, holo.)

NOTE. *Datura innoxia* is sometimes known as the Downy Thorn Apple, especially in Europe. It is the rarer of the three *Datura* species found in East Africa and is another important source of tropane alkaloids, and known for both its toxic and medicinal properties.
   Heine (F.T.W.A., 2nd. Ed: 326 (1963)) considered the spelling of *inoxia* as used by Miller *l.c.* and subsequent authors to be an orthographic error. Miller apparently based this epithet on the pre-Linnean polynomial of Boerhaave (Ind. alt. hort. Acad. Lugd.-Bat.,1: 262 (1720)) in which the correct spelling of *innoxia* was used (Heine *l.c.*). Heine's decision has therefore been adopted here. The typification of *D. innoxia* is dealt with by Barclay in Bot. Mus. Leafl. Harvard Univ. 18: 254 (1959). The species *D. metaloides* Dunal (DC., Prodr. 13(1): 544 (1852)), [which is cited as *D. meteloides* (DC. MSS) Novae Hispaniae region. *D. metel* Moc. & See., pl. Mex. ined. ic. et mss. t. 919 collect transl. Candoll.] has been variously treated as a synonym of *D. innoxia* (e.g. Barclay, 1959) or as a distinct species (e.g. Safford (1921) and Satina & Avery (1959)). In line with Haegi's (in Austr. Journ. Bot., 24: 422 (1976)) opinion that more work is required on the typification of this taxon, it has been omitted from list of synonyms.

   3. **Datura metel** *L.*, Sp. Pl.: 179 (1753) & Syst. Nat. 2, ed. 10: 932 (1759); Richard, Tent. Fl. Abyss. 2: 94 (1850); C.H. Wright in F.T.A. 4, 2: 256 (1906); U.O.P.Z.: 225 (1949); Satina & Avery in Chronica Botanica 20: 32 (1959); Verdc. & Trump, Common Poisonous Pl. E. Afr.: 164 (1969); E.P.A. 2: 882 (1963); Heine in F.W.T.A. 2nd ed., 2: 326 (1963); U.K.W.F. 2nd ed.: 244 (1994); Hepper in Fl. Egypt 6: 112 (1998) & in Fl. Egypt 3: 46 (2002); Gonçalves in F.Z. 8(4): 34 (2005); Thulin in Fl. Somalia 3: 220 (2006); Friis in Fl. Eth. 5: 158 (2006). Type: "in Asia, Africa", Herb. Hort. Clifford 55: *Datura* 2α (BM!, lecto. designated by Timmerman in Pharm. Journ. 118: 572 (1927)) [See also Hadkins *et al.,* J.L.S. 125: 298 (1997) & Jarvis, Order out of Chaos: 476 (2007)]

   Annual herbs up to 1.2 m high, occasionally bushy or shrubby, erect, often malodorous; branches greenish-brown, brown or purple, often becoming woody, smooth, glabrescent to pilose with eglandular hairs. Leaves alternate to opposite, greenish yellow to dark green, ovate, ovate-lanceolate, lanceolate or rhomboidal, 7.5–13(–19) × 3.2–10.5(–17) cm, bases obliquely cuneate, margins sinuate to sinuate-dentate with up to 3 shallow obtuse lobes, apices acute, glabrescent to moderately pilose, denser on lower surfaces, midribs and veins; petioles 2.2–10 cm long. Flowers solitary, axillary, erect, sometimes scented; pedicels (4–)6.5–12 mm and erect, pilose, elongating to 10–29 mm in fruit. Calyx cylindrical, (3.2–)6–12 × 6–14 cm, pilose externally, lobes usually broadly occasionally narrowly triangular, (6–)12–16 × 3.5–8 mm, apices acute, with pilose margins; basal ridged collar 4.5–6 × 7–11 mm. Corolla white, cream, yellow, mottled- to deep purple with prominent veins, tubular below flaring for upper third and becoming funnel- or trumpet-shaped, occasionally double or even triple, (11.8–)13–18 cm long flaring to 3–10 cm diameter, glabrescent to sparsely pilose externally, flared part with 5(–10) acute and often curved tips (3–)5–14 × 1–3 mm at base. Stamens enclosed; filaments free for 2.5–4.8 cm; anthers oblong, cream or yellow, to yellowish orange, (10–)12–16 × 1.5–3 mm. Ovary dark brown, 4–8 × 3.5–8 mm, tuberculate; style (9–)10–13.8 cm long; stigma 2–2.5 × 1.7–2.5 mm. Fruit erect or nodding, globose to ovoid tuberculate capsules, green becoming brownish, 2–3 × 2–3 cm (including spines), dehiscing irregularly or by 2 or 4 valves; valves covered with conical, ridged and blunt (obtuse) unequal tubercles, 2–6 × 1–3 mm basally, pilose with short appressed hairs; subtended by the persistent discoid cupulate remains of the calyx 1.8–2.8 cm diameter and 3–9 mm broad; fruiting pedicels elongated, erect or nodding, stout. Seeds greenish/yellow to light brown, reniform or D-shaped with distinct lateral ridge, 2.8–5 × 2–4 mm, minutely and densely foveolate, with dark brown elaiosome. Fig. 8/11–13, p. 41.

UGANDA. Ruwenzori, 1893–1994, *Scott Elliot* 7288!

KENYA. Mombasa District: Mombasa Island, 27 May 1934, *Napier* 6299!; Lamu District: Witu, *F. Thomas* 154!; Kilifi District: near Mazeras, 14 Mar. 1902, *Kassner* 280!

TANZANIA. Tanga District: Nidume Beach, 18 Oct. 1958, *Faulkner* 2201!; Mpwapwa District: Mpwapwa township, 17 May 1976, *Magogo* 825!; Rufiji District: Mafia Island, Baleni, 1 Sept. 1937, *Greenway* 5205!

DISTR. U 2; K 4–7; T 3, ?4, 5, 6, ?8; P, Z; native to South America, possibly in the Antilles, now a widespread weed

HAB. Weed of roadsides and other ruderal sites, particularly in sandy soil; 0–1100 m(–2400 m fide U.K.W.F., and Ruwenzori?)

CONSERVATION NOTES. Widespread; least concern (LC)

SYN. *D. fastuosa* L., Syst. Nat. ed. 10: 2 (1759) & Sp. Pl. ed 2: 256 (1762); Nees von Esenbeck in Trans. Linn. Soc. 17: 74 (1834); Dunal in DC., Prodr. 13(1): 542 (1852); P.O.A. C: 356 (1895); Hiern in Cat. Afr. Pl. Welw. 3: 753 (1898); Durand & Durand in Syll. Fl. Cong.: 396 (1909); C.H. Wright in F.T.A. 4, 2: 256 (1906); U.O.P.Z.: 225 (1949); Schönbeck-Temesy in Fl. Iran 100: 46 (1972). Type: "Habitat in Aegypto", Herb. Linn. 243.3 (left-hand specimen) [LINN!, lecto. designated by Schönbeck-Temesy in Fl. Iran 100: 46 (1972) and verified by Hadkins *et al.* (in J.L.S. 125: 304 (1997); see also Jarvis, Order out of Chaos: 475 (2007)]

    D. alba Nees in Trans. Linn. Soc. 17: 73 (1834); Dunal in DC., Prodr. 13(1): 541 (1852); P.O.A. C: 356 (1895) based on *Datura metel* L., *nom. illegit.*

    D. bojeri Delile in Ind. Sem. H. Monsp.: 23 (1836) & Ann. Sci. Nat. Bot, ser. 2, 7: 286 (1837): Dunal in DC., Prodr. 13(10): 540 (1852). Type: Hort. Montpellier, from seed collected in Mauritius, ? *Bojer* s.n. (G-DC, syn.); Philippines, *Cuming* 2404 (G-DC, syn.) ? [NB. "Type designated by original description", fide Gonçalves, 2005]

    D. fastuosa L. var. *alba* (Nees) C.B.Clarke in Hook., Fl. Brit. Ind. 4: 243 (1883); Hiern in Cat. Afr. Pl. Welw. 3: 753 (1898); C.H. Wright in F.T.A. 4, 2: 257 (1906); ? E.P.A. 2: 882 (1963)

    D. alba Nees var. *africanum* Mattei in Boll. Ort. Bot. Palermo 7: 108 (1908). Type: Somalia, Goscia, Giumbo, *Macaluso* 188 (PAL, holo.; PAL, iso.; photo, K!)

NOTE. Nees' protologue of *D. alba* includes extensive notes, observations and references; he was clearly convinced that *D. alba* differed from *D. metel* and deserved specific recognition largely through their differences in stem pubescence. However, he appeared to base his new species on Roxburgh's *D. metel* (Fl. Ind., 2: 238 (1824), which in turn referred to Willdenow's *D. metel* (Sp. Pl., 1, Pt 2: 1009 (1798)). This in turn referred to Linnaeus' original description (Sp. Pl.: 179 (1753)) whose holotype is the Hort. Clifford specimen cited above (also cited by Willdenow), thereby making Nee's species a *nomen illegitimum*. Willdenow also gave *D. alba* as a synonym of *D. metel*. Although Nees cited Wall. Cat. Suppl. n. 260 after the specific name, no specimens bearing this number could be found in the Supplements to Wallich's catalogue.

    D. metel is thought to be the first *Datura* species to have reached the Old World. It has long been cultivated, possibly as early as the 10th century, for both its drug potential and its ornamental value, but is no longer known in the wild in its native habitat. Symon & Haegi (Solanaceae III: 205, 1991) suggested that it was a well-established cultivated species with a range of forms in its place of origin, and that these forms or cultigens arrived 'ready-made' in Europe. The range of colour forms in both the floral and vegetative parts, and the double and triple corolla variants of the single white-flowered form have led to the description of many cultigens and varieties, and to much of the taxonomic confusion surrounding this species. Child & Shaw (in BSBI News, 82: 55 (1999)) for example, distinguished five varieties and five formae within *D. metel*, largely on the basis of differing corolla colours.

    All plant parts are narcotic and there are reports of the seeds being used for smoking (T ?4 & 6). Plants of this species are grown to keep snakes away in T 5. The fruiting specimen *Gachathi & Opon* 130/81 collected from the lake shore at Mbita Point Field Station, South Nyanza District (K 5) is thought to belong to *D. metel* despite the only mature capsule being abnormally large (± 5 × 5 cm) and very sparsely spinose. The spines are however, very short, pilose throughout, pyramidal and blunt, though a flowering sample is necessary to confirm this identification.

10. BRUGMANSIA

Pers., Syn. Pl. 1: 216 (1805); Lockwood in Bot. Mus. Leaflet 23(6): 273–281 (1973); Persson, Knapp & Blackmore in Solanaceae IV: 171–187 (1999); Hunziker in Gen. Solanaceae: 153–156 (2001); Mansfeld, Encycl. Agric. & Hort. Crops: 1847 (2001)

Datura L., Sp. Pl. 179 (1753) pro parte
Methysticodendron R.E. Schult. in Bot. Mus. Leafl. 17(1): 2 (1955)

Small trees or shrubs. Leaves alternate, glabrous to pubescent, hairs eglandular. Flowers usually solitary, occasionally in short monochasial cymes, pendulous or nodding, usually fragrant, opening diurnally and remaining open during anthesis; pedicels elongating during fruiting. Calyx elongate and tubular, usually enclosing the lower half of the corolla, often zygomorphic, five-dentate or spatulate through splitting irregularly on one side, not circumscissile, often persistent. Corolla tubular below becoming funnel- or trumpet-shaped above; tube long and slender, the lobes fused almost to the apex where the margins 5–10-toothed with the lobes often recurved, each with three prominent veins and the teeth cuspidate or caudate. Stamens usually inserted mid-way on corolla tube and alternating with lobes, included; filaments glabrous above, villous from point of adnation where broadest, tapering towards base and anthers, becoming pilose below; anthers linear, basifixed, villous, free or connivent. Ovary superior, glabrous, bilocular; style long, filiform, usually exserted beyond anthers but included; stigma ovoid to ellipsoid, clasping the apical part of style and appearing bilobed in profile. Fruit a large, smooth berry, indehiscent, usually enclosed by calyx remnants; fruiting pedicels elongated. Seeds numerous, large often triangular, irregular or subreniform, with a thick (suberose) corky testa; caruncle absent.

A genus in which 5 to 14 species have been variously described, though recent work suggests that the true number lies between six (*cf.* Hunziker, 2001) and eight (*cf.* Persson *et al.*, 1999). They are all native to disturbed habitats in Andean South America but have been widely planted as ornamentals and hedging plants in many tropical and subtropical parts of the world. Some of the species are considered to be natural hybrids and there are many races and cultivars. Indeed, although the group originated in north-west South America, some authors now consider that the species no longer occur in the wild and should all be considered as cultigens (eg: Bristol, Bot. Mus., Leafl.: 229–248 (1966)).

As they contain tropane alkaloids, Brugmansias are widely used for their potent psychotic and medicinal properties in South America. They have often been included as a section of the genus *Datura* L., but most authors now consider *Brugmansia* to be generically distinct (*cf.* Lockwood, 1973; Persson *et al.*, 1999). The plants are commonly known as Floripondios or Tree Daturas in the Americas, as Angel's Trumpets in Europe and as Moonflowers in Africa. Fruits are rare in Africa; the plants are self-incompatible and pollinated by hummingbirds or moths in their native habitat.

Flowering pedicels pendulous, pilose to velutinous; calyx
    spatulate, narrowed to an acute point; corolla tubes not or
    barely extended beyond calyces; corolla lobes often caudate
    with narrowly triangular tails up to 5 cm long .... 1. *B.* × *candida*
Flowering pedicels horizontal to nodding, glabrous; calyx dentate
    with five triangular calyx lobes; corolla tubes extended
    (1–)2–6 cm beyond calyces; corolla lobes broadly triangular
    with short (2–3 mm) cuspidate teeth .... 2. *B. suaveolens*

1. **Brugmansia × candida** *Pers., Syn. Pl.* 1: 216 (1805); Troupin, Fl. Rwanda 3: 359 (1985); Hepper in Fl. Egypt 6: 117 (1998); Gonçalves in F.Z. 8(4): 5 (2005). Type: Peru, *Pavon* s.n. (BM, holo., fide Hepper, 1998)

Small trees, shrubs or undershrubs, 3–6 m high, with spreading branches. Leaves dark green, succulent, ovate-lanceolate to lanceolate, 19–26(–50) × 9.5–15(–25) cm, bases obliquely cuneate, margins entire, sinuate or sinuate-dentate with up to 6 acute antrorse teeth, apices acute, glabrescent to pilose above, pilose/velutinous below especially dense on lower midribs and veins, hairs eglandular; petioles 5–9(–25) cm long, pilose. Flowers solitary, pendulous, white with green veins, sometimes becoming yellowish or greenish as mature, usually fragrant; pedicels 3.5–5 cm long, pilose to velutinous, pendulous. Calyx tubular, zygomorphic, spatulate, green, 8–11.5(–14) cm long, pilose or velutinous externally, glabrescent, splitting into spathe 3–6 cm from base, tapering apically into narrowly triangular densely pilose projections up to 6 mm long. Corolla tubular at base becoming trumpet-shaped above, (14–)20–32 cm long; basal tube 12–13.5 cm long, often inflated to fill the calyx cavity, not usually extended beyond the calyx; petals fused almost to the apex, where flared with the margin 5-toothed, the narrowly triangular lobes becoming caudate, extending up to 5 cm, and often recurving. Stamen filaments free for 4–6 cm; anthers grey/brown, linear, 2.1–3 cm long, 2 mm broad, with dense villous eglandular hairs, not connivent. Ovary pale green, elliptic to conical, ± 16 mm long and 4 mm wide, glabrous; style 15–19 cm long, exserted ± 10 mm beyond anthers but included; stigma 3–5 mm long and 2–2.8 mm wide. Fruit green to yellowish-green, ovoid to fusiform, 15–20 × 3–4 cm; pedicels elongated. Seeds angular, ± 6 mm long and 8 mm wide. Fig. 9/7, p. 48.

TANZANIA. Lushoto District: Amani Nursery, 5 July 1940, *Greenway* 5960!
DISTR. **T** 3 (cultivated); probably native to Peru, cultivated in many African countries and occasionally occurring as an escape
CONSERVATION NOTES. Widespread; least concern (LC)

SYN. *Datura arborea* sensu Ruiz & Pav., Fl. Peruv. 2: 15, t. 128 (1799), *non* L. Type: Peru, Passim near Cerado, Chancay and Huanuci Provinces, ? *Pavon* s.n. (BM or MA, holo.) or Ruiz & Pav., Fl. Peruv. 2: 15, t. 128 (1799), lecto.
    D. candida Pasq., Cat. Ort. Bot. Nap.: 36 (1867), *nom nud.*, based on *Brugmansia × candida* Pers.
    D. candida (Pers.) Safford in Journ. Wash. Acad. Sci., 11: 182 (1921); Heine in F.W.T.A. 2, 2nd Ed: 326 (1963)
    ?*Methysticodendron amesianum* R.E. Schult. in Bot. Mus. Leafl. Harvard Univ. 17(1): 2 (1955). Type: Colombia, Sibundoy, *Schultes & Cabrera* 20079 (GH, holo.)
    ?*Brugmansia amesianum* (R.E. Schult.) D'Arcy in Solanac. Newsl. 2(4): 17 (1986)

NOTE. This is a natural hybrid between the Andean South American *B. aurea* Lagerh. and *B. versicolor* Lagerh. which is thought to have originated on the west and central slopes of the Central Ecuadorian Andes. It is now widely cultivated as an ornamental shrub and as a hedging plant in many countries with tropical or mild climates. Commonly known as the large white-flowered Floripondio the species has also become a successful weed in many African countries as an escape from cultivation. It was first illustrated in Ruiz & Pavon's Flora Peruviana, 2, pl.128 (1799) where it was mistakenly identified as Linnaeus' *Datura arborea* (Lockwood, 1973, ined.). There are many forms and cultivars of this species, with double and even triple-formed corollas being common; the cultivar Culebra has distinct linear-ligulate leaves.

2. **Brugmansia suaveolens** (*Willd.*) *Bercht. & Presl,* Rostl. 1, Solanaceae: 45 (1823); Troupin, Fl. Rwanda 3: 359 (1985); Hepper in Fl. Egypt 6: 117 (1998); Friis in Fl. Eth. 5: 160 (2006). Type: Mexico, *Humboldt & Bonpland* s.n., *Willdenow* 4257 sheet 2 (B-W, lecto. designated here, see Note; photo!)

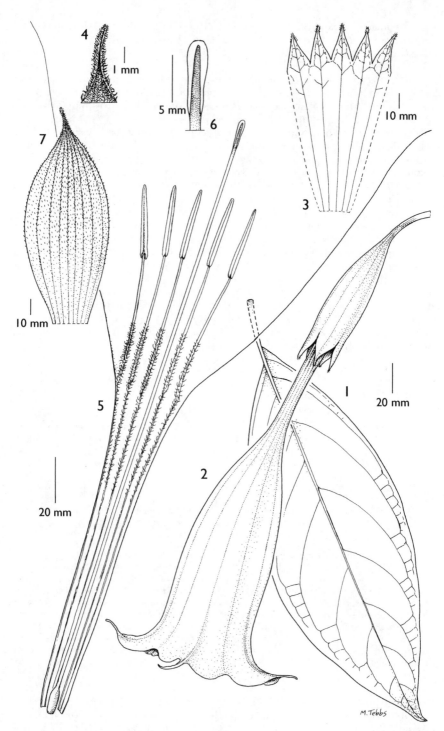

FIG. 9. *BRUGMANSIA SUAVEOLENS* — **1**, leaf; **2**, complete flower; **3**, opened calyx : **4**, sepal lobe apex; **5**, opened flower; **6**, stigma. *B.* × *CANDIDA* — **7**, opened calyx. 1–6 from *Edmonds* s.n.; 7 from *Batty* 875. Drawn by Margaret Tebbs.

Woody herbs, bushes, shrubs or small trees 1.5–6 m high, with spreading or erect branches. Leaves ovate-lanceolate to lanceolate, 11–24(–28) × (4–)6–14.5 cm, bases usually obliquely cuneate, margins entire to sinuate, apices acute to acuminate, glabrescent to pilose on upper and lower surfaces, denser on lower midribs and veins; petioles 2.4–9 cm, glabrescent to pilose. Flowers solitary, nodding at maturity (not pendulous), white to pale green, occasionally yellowish, with green veins, some cultivars bright yellow, red or orange- or pink-tinged, often sweetly-scented; pedicels 2–6 cm long, horizontal in bud, glabrous, nodding at maturity. Calyx light green, inflated, longitudinally ovoid to urceolate, (7–)9–12 × (2.3–)3.3–3.6(–4) cm, apically dentate with (3–)5 broadly triangular acute lobes (0.6–)1–2.6 × (0.6–)0.9–2 cm, usually glabrescent externally, occasionally pilose/villous, often with pilose fringe. Corolla narrowly tubular below, flaring and becoming bell- to trumpet-shaped above, 16–30(–35) cm long; tube 11–15 cm long, not inflated within the calyx and extending 2–10 cm beyond it, upper part fused with margin 5-toothed, the teeth cuspidate and flaring, 3–20 × 5–12 mm. Stamen filaments free for 4–5.5 cm; anthers yellowish-brown to pale brown, 3–3.9 × 1.7–2 mm, usually connivent. Ovary green, conical to obovate, 6–12 × 3–4 mm, glabrous; style 11.8–25.5 cm long, exserted 10–35 mm beyond anthers but included; stigma 6–8 × 2 mm. Fruit green, obovoid (spindle-shaped), fusiform, 10–22 × ± 3.2 cm, often drying in situ with seeds released through outer membranes, pedicels elongating to 7.5 cm and ± 1 cm broad beneath capsule, becoming recurved, woody and striated. Seeds light brown, triangular or irregular often with narrow protrusion, 8–12 × 5–8 mm, roughly verrucate but usually with smooth lateral ridge. Fig. 9/1–6, p. 48.

Uganda. Masaka District: Masaka City (cult.), 27 July 1971, *Lye & Katende* 6504!
Kenya. Kiambu District: Karura forest near Nairobi, 25 Nov. 1966, *Perdue & Kibuwa* 8100!
Tanzania. Lushoto District: W Usambara Mts, Mazumbai Forest Reserve, 25 Apr. 1975, *Hepper & Field* 5134! & Amani, Nderema, 15 July 1956, *Tanner* 2994! & E Usambara Mts, along Derema road NE of Amani, 8 Nov. 1986, *Borhidi et al.* 86553!
Distr. U 4; K 4; T 3; native to SE Brazil, this species is now widely cultivated, and occurs as an escape in Sierra Leone, Ghana, Cameroon, Ethiopia, South Africa and Reunion
Hab. Fairly common as an escape from cultivation in semi-shade and naturalised in forests, especially near rivers and streams and on roadsides; (500–)800–1800 m
Conservation notes. Widespread; least concern (LC)

Syn. *Datura suaveolens* Willd., Enum Hort. Berol.: 227 (1809); Heine in F.W.T.A. 2nd ed., 2: 326 (1963)
   D. gardneri Hook. in Bot. Mag. 72: 4252 (1846). Type: Brazil, Organ Mountains, *Gardner* 560 (K!, ?holo.; B, BM,GH, US, iso.)
   D. suaveolens Willd. var. *macrocalyx* Sendtn., Martius Fl. Bras., 10: 161 (1846). Type: indigenous to Mexico, cult. in hortis Braziliensis, near Rio Belmonte, prov. Spiritus, Princ. Vidensis, *Gardner* 560; Brazil, Prov. Minarum, Congonhas do Campo, *Stephan* s.n.; Antilles, St. Vincent, *Lambert* s.n. (syntypes, localities not given)
   Pseudodatura suaveolens (Willd.) van Zijp in Natuurk. Tijdschr. Ned.-Ind. 80: 28 (1920)

Note. There are three sheets of *D. suaveolens* in B. One (with 4257 written in the left hand corner) simply bears the label with the specific name (under which is written "Slechtendal"), together with a brief description and the collection locality as Mexico followed by the symbol for woody. These details are identical to Willdenow's protologue, with the exception of the leaves being described as "pubescentibus" which is lacking from the latter. The other two sheets have the species name written in the top right hand corner and "Hort. bot. Berol W" written on bottom left hand corners. Both are extremely good specimens of *D. suaveolens* (=*Brugmansia suaveolens*) and are good candidates for selection as the lectotype of this species. From the Paris fiche of Humboldt & Bonpland herbarium there does not seem to be a specimen of this species there. Since these B specimens were presumably seen by Willdenow, it seems desirable to select one of the B specimens as the lectotype, in which case sheet 2 (with *Schlechtendahl* written beneath species name, image ID 157478) is selected here.
   After its introduction to Europe, this species was widely known as *Datura arborea* L. Hooker, however, recognising that this taxon differed from the Linnaean species, and overlooking Willdenow's description of *D. suaveolens*, described it as *D. gardneri* Hook. (Lockwood, 1973 (ined.)). It is often known as the Moonflower in eastern and southern Africa.

11. LYCIUM

L., Sp. Pl. 1: 191 (1753) & Gen. Pl. ed. 5: 88 (1754); Dunal in DC., Prodr. 13(1): 508 (1852); Feinbrun in Collect. Bot. 7: 359–379 (1968); Venter, Taxonomy *Lycium* in Africa, Ph.D. Thesis, Univ. Bloemfontein (2000); Hunziker in Gen. Solanaceae: 158–164 (2001)

Shrubs, sub-shrubs or small trees; stems with long shoots bearing alternate leaves and short shoots terminating in a spine with fasciculate leaves; nodes usually swollen brachyblasts. Leaves fasciculate on brachyblasts or alternate, membranaceous or succulent. Inflorescences axillary, solitary or occasionally paired, rarely in many-flowered terminal fascicles, epedunculate; flowers actinomorphic or zygomorphic, usually hermaphrodite, but a few species functionally dioecious; pedicels filiform, usually erect, rarely deflexed in fruit. Calyx tubular, campanulate or cupulate, with 4–5 equal or unequal calyx lobes which enlarge in fruit. Corolla urceolate, tubular, salverform or infundibular; tube narrow and straight but often constricted above the ovary, with spreading or reflexed imbricate lobes; corolla often detaching after anthesis leaving an annular ring around ovary base. Stamens usually unequal, often comprising two long, two medium and one short, rarely only 4, exserted or included; filaments joined to corolla tube at same or different levels, straight or flexuose, cylindrical or enlarged at point of fusion, often pubescent towards base, usually glabrous where free; anthers usually all fertile though occasionally sterile especially in functionally female flowers, ovoid, obovoid or sagittate, dorsifixed. Ovary 2-locular, ovules numerous; disc annular, inconspicuous and greenish, or prominent when red or orange; style filiform, glabrous, usually exserted; stigma capitate or discoidal, bilobed (style stunted and stigma absent from functionally male flowers). Fruit a many-seeded berry, rarely drupaceous with two pyrenes. Seeds usually numerous, occasionally only one in each locule, small; testa crustaceous, reticulate-foveolate.

A large genus of 75 to 100 species, which are found throughout the world. Around 30 to 40 species are probably native to the Old World especially South Africa, but only *L. europaeum* L. and *L. shawii* Roem. & Schult. have been recorded from East Africa. However, Venter (*pers. comm.*) considers many specimens to have been misidentified and believes that, as discussed below, only *L. shawii* occurs throughout this region.

Lycium shawii *Roem. & Schult.*, Syst. Veg. 4: 693 (1819); Dean, *Lycium* in Africa, Kew ined. (1974); Blundell, Wild Fl. E. Afr.: 188 (1987); Hepper in Fl. Egypt 6: 92 (1998) & in Fl. Egypt, 3: 44 (2002); Venter, Taxonomy *Lycium* in Africa: 197 (2000); Gonçalves in F.Z. 8(4): 40 (2005); Thulin in Fl. Somalia 3: 200 (2006); Friis in Fl. Eth. 5: 110 (2006). Type: North Africa, Shaw, Travels in Barbary: 49, fig. 349! (1738), lecto. designated by Venter (2000)

Shrubs or subshrubs to 3(–4) m high, occasionally small trees to 4.5 m; stems erect, spreading or scrambling, often arching, densely branched, grey, brown or reddish, with glabrous (rarely pubescent) spines 3–14 mm long on short branches or brachyblasts, usually glabrescent, young stems occasionally with short simple eglandular- or occasionally glandular-headed hairs interspersed with stalked glands. Leaves stipulate, usually fasciculate in clusters of up to 14 around spine on brachyblasts, sometimes solitary and alternate, membranaceous, coriaceous or succulent, green, grey or greyish-blue, elliptic, ovate, obovate, oblanceolate or spatulate, 0.5–1.8(–8) × 0.3–0.8(–3) cm, bases cuneate and often decurrent, margins entire, apices obtuse, surfaces glabrescent, sometimes short hairs on veins and margins, with stalked glands on laminas; petioles absent to 1 cm long. Inflorescences of solitary or paired flowers among fascicled leaves, epedunculate; pedicels densely pilose, occasionally glabrescent, erect in flower when 2–5(–12) mm long, erect or splayed in fruit when 4–12 mm long. Flowers sometimes sweetly scented. Calyx

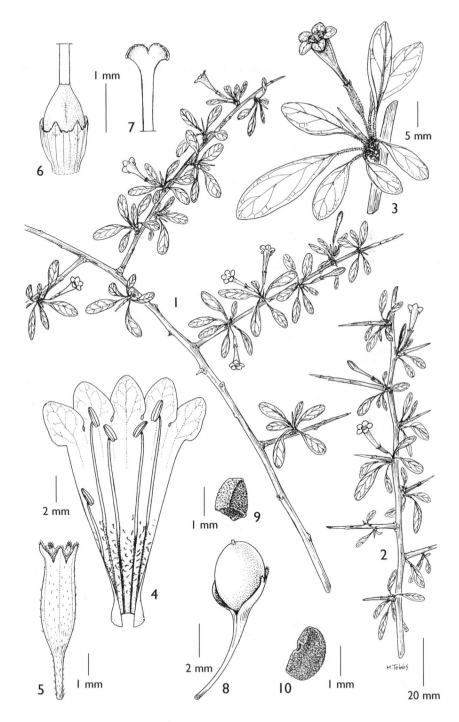

FIG. 10. *LYCIUM SHAWII* — **1**, flowering habit; **2**, short shoot; **3**, solitary flower in axil of fasciculate leaves and spine; **4**, opened flower; **5**, calyx; **6**, ovary; **7**, stigma; **8**, berry enclosed by partially adherent calyx; **9** & **10**, seeds. 1 from *Leippert* 5387; 2 from *Verdcourt* 3824; 3–7 from *Richards* 25154; 8–10 from *Buxton* 1017. Drawn by Margaret Tebbs.

tubular, occasionally cupulate, 2–4(–6) × 1–2.5 mm centrally but always at least twice as long as broad, glabrescent, rarely moderately pilose externally, with 5 unequal lobes 0.3–1.5(–2) × 0.5–2 mm, acute, with ciliate fringes, slightly enlarged in fruit. Corolla white, cream, yellow, pale pink, purple, blue, purplish-white or bluish often fading to white or cream, sometimes tube darker, salverform to infundibular, (7–)9–13 mm, widening to 2–3(–4) mm broad beneath lobes, glabrescent externally, pilose below internally; lobes spreading or recurved, often with purplish or dark central veins extending down the tube, obovate to spatulate, 1.5–3.5(–5) × 1–3 mm, obtuse, glabrescent but sometimes with pilose fringe, giving overall flower diameter of 4–9 mm. Stamens usually 5, unequal, the longer anther or anther pair exserted and the shorter two or three included; filaments free for 1–7(–10) mm; anthers yellow becoming brown, (0.7–)1–1.5 × 0.5–1.3 mm, occasionally one anther abortive when ± 0.3 × 0.2 mm. Ovary smooth, ovoid to conical, 1–2(–3) × 0.6–1.5(–2) mm, glabrous; disc inconspicuous, 0.8–1.3 mm diameter; style (6.5–)8–11 mm long; stigma 0.5–1.3 mm diameter. Fruit a smooth red or orange, occasionally yellow, ovoid to globose, glossy berry, (3–)4–6(–8) × 3–5(–6) mm, glabrous, enclosed basally by adherent and slightly enlarged calyx which splits irregularly often appearing bilabiate as berry matures. Seeds 9–24 per berry, yellowish to brown, discoidal, reniform or angular and irregular, (1.2–)1.8–2.5 × 1–2 mm. Fig. 10, p. 51.

UGANDA. Karamoja District: Kangole, June 1950, *Eggeling* 5990! & idem, July, 1957, ?*Wilson* 380!, & Lokapeliethe, Mathiniko, Sept. 1943, *Dale* 353!
KENYA. Meru District: near Isiolo Township, 22 Apr. 1971, *Kimani* 289!; Masai District: Narok–Ewaso Ngiro, km 11, 11 Dec. 1963, *Verdcourt* 3824!; Tana River District: Tana River Primate Reserve, Woodland road 1.3 km, 15 Mar. 1990, *Luke et al.* TPR 405!
TANZANIA. Mbulu/Singida District: Yaida Valley near Lake Eyassi, Endasiku, 17 Jan. 1970, *Richards* 25154!; Masai District: Kakessio, 7 Apr. 1961, *Newbould* 5830!; Pare District: Lake Kalimawe, 10 Jan. 1967, *Richards* 21938!
DISTR. U 1; K 1–4, 6, 7; T 2, 3, 5, 7, 8; from Egypt, Sudan, Somalia, Eritrea and Ethiopia southwards to Botswana, Namibia and NE South Africa; also widespread from Mediterranean Europe and Arabia to W India
HAB. Rocky outcrops, montane scrub, riverine thickets, *Acacia-Balanites*-grassland, *Acacia-Commiphora* bushland, *Brachystegia-Isoberlinia* woodland, termite mounds, lava plains, stony, alluvial, sandy or volcanic soils (when often associated with *Acacia*, *Grewia* and *Cordia*), and in fallow or degraded land; 15–2000 m
CONSERVATION NOTES. Widespread; least concern (LC)

SYN. *L. barbarum* L., Sp. Pl. 1: 192 (1753) pro parte min. quoad pl. *shawii* fide Feinbrun & Stearn in Israel. J. Bot. 12: 121 (1963)
    L. persicum Miers in Ann. Mag. Nat. Hist., Ser. 2, 14: 12 (1854) & Ill. S. Am. Pl. 2: t. 65 fig. B (1857); C.H. Wright in F.T.A. 4, 2: 254 (1906). Type: Yemen [Aden], *Hooker* s.n. (K!, syn.), *Thomson* s.n. (K!, syn.)
    L. arabicum Boiss., Fl. Orient. 2: 289 (1879); Dammer in E.J. 48: 228 (1912); Feinbrun in Collect. Bot. 7: 366 (1968), *nom. illegit.* Type: Egypt, between Keneh and Kosseir, *Schweinfurth* s.n. (?BM, syn.,); Arabia, at Ouadi Mokatteb and Ramla plains, *Boissier* s.n. (?P, syn.)
    L. merkii Dammer in E.J. 48: 224 (1912); T.T.C.L., as *L. merkeri*: 574 (1949). Type: Tanzania, Mbulu District: Mbugwe [Umbugwe] and Iraku, *Merker* 294 (?B†, syn.) and eastern base of Ol Doinyo Lengai, *Merker* 758 (?B†, syn.)
    L. somalense Dammer in E.J. 48: 225 (1912). Type: Somalia, Analayra, *Cole* s.n. (B†, holo.)
    L. tenuiramosum Dammer in E.J. 48: 225 (1912); T.T.C.L.: 574 (1949). Type: Tanzania, near Kiutiro (not found), *Zimmerman* in herb. Amani 1700 (B†, holo.; ?EA, iso.)
    L. albiflorum Dammer in E.J. 48: 226 (1912). Type: Botswana, near Mamatau, *Seiner* II. 223 (B†, holo.)
    L. withaniifolium Dammer in E.J. 48: 230 (1912). Type: Ethiopia, *Ellenbeck* A/1 & 1183 (B†, syn.)
    L. ellenbeckii Dammer in E.J. 48: 232 (1912); E.P.A.: 855 (1963). Type: Ethiopia, Daroli, *Ellenbeck* 1823 (B†, holo.)
    L. jaegeri Dammer in E.J. 48: 232 (1912). Type: Tanzania, Mbulu District: Mangati, *Jaeger* 253 (B†, holo.)

L. ovinum Dammer in E.J. 53: 352 (1915); Dean, The genus *Lycium*, Kew ined. (1974) sub
  L. boscifolium Schinz. Type: Namibia, *Seiner* III. 202, 311 (B†, syn.)
L. cufodontii Lanza in Miss. Biol. Borana, Racc. Bot.: 202 (1939). Type: Ethiopia, *Cufodontis*
  73 & 99 (FI, syn.)
L. javallense Lanza in Miss. Biol. Borana, Racc. Bot.: 204 (1939). Type: Ethiopia, Javello,
  Cufodontis 561 (FI, holo.)
L. europaeum sensu auctt. e.g.: Greenway in K.T.S.: 537 (1961); U.K.W.F. : 529 (1974);
  K.T.S.L.: 578 (1994), *non* L.

NOTE. The species is notoriously morphologically variable, in part reflecting its extremely
widespread distribution. In eastern Africa heavily grazed plants of this species become stunted
bushes with smaller leaves and flowers, while in Kenya and Tanzania its leaves can be up 8 × 3
cm (*cf. Kibue*, 89 (**K** 6), *Polhill* 118 (**K** 3) & *Milne-Redhead & Taylor* 11061 (**T** 7)), though the
characteristic small leaves are often found on the same plant. A number of varieties of this
species have been described, largely from Arabian and Middle Eastern localities; these are
described or dealt with by Venter (2000) who also gives a number of additional synonyms –
these names have not been encountered on FTEA material. The protologues of the eight
species described from east Africa by Dammer (1912–15) and Lanza (1939) leave little doubt
that they are all synonyms of *L. shawii*, though the Berlin holotypes have been destroyed.

Many East African specimens have been identified as *L. europaeum* L., but according to
Venter this species is confined to North Africa and the Mediterranean. One of the major
features distinguishing these two species is the shape of the flowering calyx, with that of *L.
shawii* being narrowly tubular and at least two to three times longer than broad, while that of
L. europaeum is campanulate to broadly tubular with the length equalling the breadth.
However, although many intermediates have been found on E African herbarium material,
since Venter studied this genus in great detail throughout Africa, her conclusions have been
followed here. It is probable that *Lycium austrinum* Miers included in T.T.C.L. also belongs to
L. shawii. However, the account described its stems as being unarmed or rarely armed with
short-spined stems; its leaves as lanceolate and the flowers borne in clusters of 2–5; it is
possible that the plants identified as this species which was also described as being
widespread but not common in Kolo and Kondoa might be distinct and correctly identified.

Lycium species are often referred to as Box-thorns or Matrimony vines. *Lycium shawii* is
extensively utilised throughout Africa; the leaves are used as a tobacco (**K** 1) and as a green
vegetable (**K** 3), leaf extracts are used medicinally for stomache ache (**K** 1), root extracts as
a cure for coughs, skin rashes or urinary infections (**K** 2), and ground dried berries to
enhance female fertility (**K** 1). Ruffo, Birnie & Tengnäs (in Edible Wild Plants of Tanzania
(2002)) described the use of leaves from plants identified as "*L. europaeum*" as a vegetable
and to treat stomach-ache and constipation; the boiled roots to treat coughs and mouth
sores and the plants for hedging, fodder and ornamental purposes throughout Tanzania.
These plants undoubtedly belong to *L. shawii*, and are also heavily browsed by goats, sheep,
camels and giraffe.

12. **NICANDRA**

Adans., Fam. Pl., 2: 219 (1763), *nom. conserv.*; Dunal in DC., Prodr. 13(1): 433
(1852); Horton, J. Adelaide Bot. Gard. 1(6): 351–356 (1979); Hunziker in Gen.
Solanaceae: 140–143 (2001)

Physalodes Boehm. in Ludwig, Def. Gen. Pl. ed. 3: 41 (1760)
Pentagonia Fabr., Enum. ed.2: 336 (1763)
Calydermos Ruiz & Pav., Fl. Peruv. 2: 43 (1799)

Annual herbs. Leaves membranaceous, alternate. Flowers usually solitary and
showy, axillary to sub-axillary, erect eventually nodding, usually fragrant; pedicels
erect becoming nodding below flowers. Calyx with 5 lobes, joined marginally usually
towards their bases. Corolla with basal narrow part enclosed by calyx lobes and five
quincuncial very shallow bilobed lobes above, upper half of joined corolla recurving.
Stamens included; filaments fused to lower part of corolla tube where enlarged,
geniculate and villous, slender and glabrous to sparsely pilose above where free;
anthers bilobed, often sparsely pilose, basifixed by filaments inserted between

thecae, convergent around stigma. Ovary superior, 3–5(–6)-locular (irregularly divided by 3–6 septa), ovules numerous; disc annular, smooth, often crenulate; style glabrous to sparsely pilose; stigma capitate, lobed. Fruit a berry, mature pericarp thin and translucent, enclosed by enlarged accrescent chartaceous cordate calyx lobes with prominent reticulate venation. Seeds numerous, discoid to orbicular, foveolate.

A monotypic genus usually regarded as being native to Peru and possibly northern Argentina. It has been widely cultivated as an ornamental and is now rarely found in the wild but occurs as a ruderal and has become naturalised in tropical and subtropical regions throughout the world.

Nicandra physalodes (*L.*) *Gaertn.*, Fruct. Sem. Pl. 2: 237 (1791), as "*physaloides*"; Dunal in DC., Prodr. 13(1): 434 (1852); C.H. Wright in Fl. Cap. 4(1): 109 (1904); W.F.K.: 90 (1948); E.P.A. 2: 855 (1963); Fernandes in Garcia de Orta 17(3): 277 (1969); Troupin, Fl. Rwanda 3: 368 (1985); U.K.W.F., 2nd ed.: 244 (1994); Hepper in Fl. Egypt 6: 86 (1998) & in Fl. Egypt 3: 46 (2002); Gonçalves in F.Z. 8(4): 26 (2005); Thulin in Fl. Somalia 3: 198 (2006); Friis in Fl. Eth. 5: 156 (2006). Type: *Atropa physalodes* L., "Habitat in Peru", *Jussieu* s.n., Herb. LINN 246.3 (LINN!, lecto. designated by Schönbeck-Temesy in Rechinger (ed.), Fl. Iranica: 2 (1972)). [See also Jarvis, Order out of Chaos: 340 (2007)]

Annual herbs 0.5–1.5(–2) m high; main stems erect or spreading, smooth, green to brown or purple, slender, sometimes succulent, occasionally becoming woody at the base, sparsely pilose becoming glabrescent. Leaves mid- to dark green or purplish, broadly ovate, ovate-lanceolate to rhomboidal, with 2–8(–10) main acute lobes, 6–12(–38) × 3–8(–36) cm, bases cuneate, often decurrent, margin coarsely sinuate, sinuate-dentate or lacerate, apices acute to acuminate, upper surfaces glabrescent to sparsely pilose and puberulous on veins and midribs, lower surfaces usually glabrescent, margins often pilose; petioles (1–)2–4(–6) cm long. Flowers usually solitary; pedicels dark green to purple, 1–2.4(–5) cm long, erect becoming nodding below flowers, puberulous. Calyx broadly conical, dark green to purple, lobes broadly ovate, 1.1–2(–3) × 0.7–1.4(–1.8) cm, bases sagittate to cordate, apices acute to acuminate, prominently veined, puberulous internally and externally. Corolla blue or mauve becoming white in lower half with five blue or purple spots at base of corolla tube, often connected by purple veins from each lobe lobe through corolla tube, broadly campanulate to cyathiform with narrower basal half enclosed by calyx lobes; lobes fused almost to apex, very shallow and bilobed, 1–5 × 10–20 mm, upper half of corolla recurving to expose basal purple spots, sparsely pilose internally, puberulous externally. Stamens with filaments 4–6 mm long; anthers greenish to yellow, sometimes with distinct purple margins, obovoid to oblong, 2.4–3.5(–4) mm long. Ovary dark green to purplish, obovoid to conical, 2.5–4 × 2.3–3.5 mm; disc whitish but sometimes mottled with purple on lower half, 2.5–4.5 mm broad; style 3–4.5(–5) mm long, glabrous to pilose below; stigma yellow, 0.5–1.5 mm broad. Berries smooth, green to yellow, globose, 1–2 cm diameter, enclosed by enlarged accrescent chartaceous cordate calyx lobes 2.2–3.6 × 1.3–2.2 cm. Seeds mid- to dark- brown, orbicular, 1.2–1.6(–1.8) × 1.1–1.4(–1.6) mm. Fig. 11, p. 55.

KENYA. Trans-Nzoia District: Endebess, 14 Oct.1962, *Bogdan* 5563!; Naivasha District: Lake Naivasha shores, 21 July 1952, *Bally*, 8234!; Nairobi, National [Coryndon] Museum Grounds, 25 June 1952, *Hiza* 4!
TANZANIA. Mbulu District: Lake Manyara National Park, Endabash River Drift, 5 Mar. 1964, *Greenway & Kanuri* 11316!; Lushoto District: Soni–Bumbuli road, N of Kwehangala, 11 May 1953, *Drummond & Hemsley* 2516!; Iringa District: Mufindi, Lugoda Tea Estate, 10 May 1968, *Renvoize & Abdallah* 2077!
DISTR. **K** 3–5; **T** 1–3, 7, 8; Eritrea, Ethiopia, Somalia, Burundi, Angola, Zambia, Malawi, Mozambique, Zimbabwe, Botswana, South Africa and Madagascar

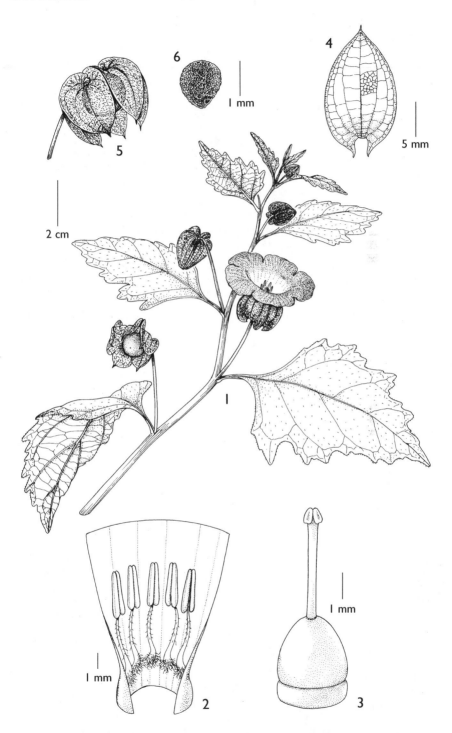

FIG. 11. *NICANDRA PHYSALODES* — **1**, habit in flower and fruit; **2**, opened flower; **3**, gynoecium; **4**, sepal venation; **5**, fruit with berry enclosed by calyx; **6**, seed. 1–3 from *Richards* 23326; 4–6 from *Renvoize & Abdallah* 2077. Drawn by Margaret Tebbs.

HAB. A troublesome weed of cultivation, sometimes becoming a serious pest of gardens, and of wheat, maize or coffee crops; also in forests, forest clearings, secondary bushland, grassland, on wasteland and sometimes on wet or swampy soils. (130–fide U.K.W.F.) 600–2350 m
CONSERVATION NOTES. Widespread; least concern (LC)

SYN. *Atropa physalodes* L., Sp. Pl. 1: 181 (1753) & Syst. ed. 12: 171 (1767) [the second as *physaloides*]
   Nicandra physalodes (L.) Scopoli, Int. Hist. Nat.: 182 (1777)
   Physalis daturifolia Lam., Encycl. Meth., 2: 102 (1786), as *daturaefolia*. Type: cultivated in the Jardin du Roi, Paris, from seeds collected in Peru, *Lamarck* s.n. (P-LAM, fiche LM472/3!, lecto. designated here)
   P. peruviana Mill., Gard. Dict., ed. 8, *Physalis* No. 16 (as "18") (1768); Kuntze, Rev. Gen. Pl. 2: 452 (1891). Type: grown from seeds sent from Peru by *de Jussieu*, ?*Miller* s.n. (not found)
   Calydermos erosus Ruiz & Pav., Fl. Peruv. 2: 44 (1799). Type: Peru, Lima, Chancay & Huanuci, ? *Pavon* s.n. (?MA, holo.)
   Pentagonia physalodes (L.) Hiern., Cat. Afr. Pl. Welw. 1: 752 (1898)
   Physalodes physalodes (L.) Britton in Mem. Torr. Bot. Club 5: 287 (1894), *nom. illegit.*
   Nicandra violacea Lemoine in Rev. Hort. 4: 208 (1906), *nom. nud.*

NOTE. This species is sometimes cited as *N. physalodes* (L.) Scop. Although Scopoli mentioned *Atropa physalodes* beneath his generic protologue in 1777 (Int. Hist. Nat.: 182), he did not make the combination with *Nicandra*, meaning that Gaertner's combination takes precedence. The specific epithet *physalodes* is often misspelt as *physaloides* after Gaertner. The species is commonly known as the Apple of Peru or Shoo-fly; many East African herbarium specimens have been erroneously annotated as being either the Cape Gooseberry or Chinese lantern both of which refer to *Physalis*.
   There are three sheets labelled *Physalis daturaefolia* in Lamarck's herbarium (P-LA), though none are actually labelled as type specimens. The first (LM472/3) is a good flowering specimen, which also has *Atropa physalodes* written beneath the species name; this has been selected here as the lectotype. The second (LM472/4) is mainly vegetative with a flattened fruiting calyx, while the third (LM472/5) is only composed of leaves.
   Leaf size is particularly variable in this species, while seed numbers can also vary greatly, and Horton (1979) found that these ranged from 69 to 638 in just six berries. The flowers usually only open fully for a few hours around mid-day. The stems, leaves and flowers of this species are all reported to be edible, and are sold in local Tanzanian markets (Ruffo, Birnie & Tengnäs, Edible Wild Plants of Tanzania: 476 (2002)).

13. **SOLANDRA**

Sw. in K. Vetensk.-Acad. Handl. Stockholm 8: 300–306 (1787); Dunal in DC., Prodr. 13(1): 533 (1852); D'Arcy in Ann. Missouri Bot. Gard. 60: 675–680 (1973); Hunziker in Biol. & Tax. Solanaceae: 64–65 (1979); Bernadello & Hunziker in Nord. J. Bot. 7: 639–652 (1987); Hunziker in Gen. Solanaceae: 353–356 (2001), *nom. conserv.*

Lianas, climbing shrubs or epiphytes. Stems usually glabrous to pubescent, with fissured bark. Leaves alternate, coriaceous, with simple or branched eglandular hairs, glandular-headed hairs also usually present. Inflorescences 1- to few-flowered clusters on stout woody pedicels at ends of branches. Flowers terminal, large and conspicuous, often fragrant, slightly zygomorphic, borne on stout pedicels. Calyces tubular, usually zygomorphic, 2- or 5- irregularly lobed, sometimes 3-keeled. Corolla infundibuliform or cyathiform with basal cylindrical tube; lobes 5, overlapping in bud, short and round, entire or laciniate, usually recurved. Stamens 5, subequal, exserted or included, inserted on lower half of corolla tube; filaments slender, curved, fused to corolla tube, glabrous throughout or pubescent at the point of corolla fusion; anthers basifixed, longitudinally dehiscent. Style slender, declinate, exserted or barely included, glabrous; stigma globular, bilobed, small. Ovary 2-carpellate, 4-locular, each locule with numerous ovules, partly inferior; with well-developed ovarian nectary wall. Berries conical to ovoid, leathery; calyx persisting and splitting in fruit. Seeds compressed, discoidal or reniform. Embryo curved, not compressed.

Ten species from Mexico and the West Indies to Peru, Bolivia and SE Brazil.

The species are widely cultivated for their spectacular flowers in tropical gardens and in glasshouses throughout the world.

Solandra grandiflora *Sw.* in K. Vetensk.-Acad. Handl. Stockholm 8: 300, t. 11 (1787). Type: Jamaica, *Swartz* s.n. (S, holo.; image S-R–5811!)

Woody climber, glabrous. Leaves coriaceous, elliptic, up to 10.5 × 6 cm, bases cuneate, with apices abruptly shortly acuminate; petioles to 4 cm long. Flowers with calyx 7.5–9.5 cm long, with 5 lobes up to 3.2 × 1.2 cm. Corolla whitish-green with prominent single green veins to each of the lobes, and purplish double veins to the sinuses internally, cyathiform to campanulate above, 15–20 cm long, the narrow basal tube 5–8 cm long, lobes broadly ovate with crenate edges, 2.5–4.5 × 4–7.5 cm. Stamens included; filaments 15 cm long, inserted 6.5–10 cm from base of corolla, completely glabrous; anthers 9–10 × 4–5 mm. Styles ± 18 cm long, glabrous; stigmas capitate, indistinct, 0.8–3 mm diameter. Berries green to whitish-cream, ovoid, ± 5 cm long. Seeds ± 7 × 4 mm. Fig. 14/8 & 9, p. 68.

KENYA. Nairobi City Park (cult.), 14 Sept. & 5 Oct. 1967, *B. Perkins* 13831!

NOTE. This species is indigenous to the Caribbean and is commonly known as the Chalice Vine or the Cup of Gold. It has been hybridized with *S. hartwegii* N.E. Br. (= *S. maxima* (Sessè & Moc.) P.S. Green) in Nairobi, to produce a ± morphologically intermediate sterile hybrid called *S.* × *nairobiensis* ined. (5 Oct. 1967, *Perkins* 13852!). The hybrid has larger leaf and floral parts than *S. grandiflora*; a yellow corolla with only single veins to the lobes; stamens flush with the mouth of the corolla; anthers 9–10 mm long, and filaments lanate at the point of insertion on the corolla tube. Both *S. grandiflora* and *S. maxima* are recorded as sterile in Africa, possibly due to the absence of bats necessary for pollination.

14. CAPSICUM

L., Sp. Pl. 1: 188 (1753) & Gen. Pl. ed. 5: 86 (1754); Fingerhuth, Monogr. gen. capsici: 1–32 (1832); Irish in 9th Rep. Missouri Bot. Gard.: 53–110 (1898); Eshbaugh in Bothalia 14: 845–848 (1983); DeWitt & Bosland, Peppers of the World (1996); Hunziker in Gen. Solanaceae: 232–244 (2001); Bosland & Zewdie in van den Berg *et al.* (eds), Solanaceae V: 179–185 (2001)

Annual or perennial herbs or slender shrubs, sometimes clambering. Leaves alternate or opposite when one often smaller, stipulate. Inflorescences axillary to leaf-opposed, fasciculate with 2–8 flowers or solitary-flowered, pedicellate; flowers actinomorphic; pedicels erect though sometimes apically geniculate in flower, erect or recurved in fruit. Calyx cyathiform to campanulate, often truncate, with 5(–10) lobes often prolonged through the calyx as prominent veins, slightly accrescent and persistent but not enlarged in fruit; sometimes with an annular thickening basally. Corolla campanulate-rotate to stellate with short tube and deeply lobed; lobes valvate in bud, usually spreading after anthesis. Stamens 5, usually equal; filaments joined to corolla tube, sometimes by two conspicuous short, thick lateral basal appendages, glabrous; anthers connivent, basifixed, exserted. Ovary usually sessile, glabrous, bi- or tri-locular, ovules numerous; disc small or absent, annular; style filiform, often exserted; stigma capitate, globose, sometimes bilobed. Fruit dryish or sub-fleshy berries, 2- to 3- locular with large mesocarpellar cells, erect or drooping. Seeds rugose, reniform to suborbicular, compressed, numerous.

Hunziker (2001) considered that the genus comprised a natural assemblage of around 20 species and a few varieties, growing from Mexico to Central Argentina. Five of the species were domesticated, of which two, *C. annuum* and *C. frutescens*, are widely cultivated throughout Africa, while a third, *C. baccatum* has remained largely confined to South America, occasionally occurring as an introduction in Africa.

Capsicum is an economically important genus whose species are major spice and vegetable crops throughout the world. They are universally known as peppers or as cayenne, chilli, sweet, green and red peppers and are probably the most important and widely consumed condiment. The introduction of *Capsicum* to Africa probably occurred through a combination of post-Colombian explorers, missionary work, colonial invasion, and trade (*cf.* Eshbaugh, 1983); by the mid-1800's there were reports of these species in various African countries.

The literature and nomenclature surrounding these species, their varieties and cultivars is vast, complex and controversial with some authors still disagreeing over the treatment of various taxa, and even their common names. The history of their spread from the centres of origin to the rest of the world, and their early taxonomic treatments are given in Irish (1898), while Heiser & Smith (in Econ. Bot. 7: 214–227 (1953)), Heiser & Pickersgill (in Taxon 18: 277–283 (1969) & in Baileya 19: 151–156 (1975)) and D'Arcy & Eshbaugh (in Baileya 19: 93–105 (1974)) for example, all deal with successive treatments of the cultivated peppers and their complicated synonymy. The latest resumé of the generic taxonomy is given in Bosland & Zewdie (2001). These species display considerable variability in the shape and colour of their fruits, with the characteristic pungency being due to capsaicinoids, a mixture of seven phenolic amides or vanillilamides, which are unique to peppers (cf. Hunziker, 2001), and are concentrated in the seeds and inner surfaces of the fruits. Bosland & Zwedie (2001) found that capsaicinoid profiles were not consistent within species and therefore of limited use as chemotaxonomic identifiers. In addition to their pungent culinary value, these substances are also used medicinally - usually internally as a carminative and stimulant, and for flatulence and appetite loss, though they can cause skin irritation and blistering. A summary of the wide range of medicinal uses of these plants is given by Heiser (in Nightshades: 6–27 (1969)). Birds are thought to be immune from the pungent capsaicin in the fruit pods which prevents animals from eating them, but enhances dispersal by fruit-eating birds. The generic name, adapted from the Greek *kapto*, meaning to bite, refers to the hot taste of the fruits. The cultivated species are also widely consumed as a vegetable with the fruits providing a rich source of vitamins A and C, and occasionally grown as ornamentals. Though peppers rarely constitute an important commercial crop anywhere in Africa (Eshbaugh, 1983), in the US the sweet pepper crop in particular is very important commercially, with many varieties being grown (*cf.* Heiser, 1969).

Though some East African specimens have been identified as *C. baccatum* L., this species has not been positively identified from this region (see Excluded species, at the end of this genus).

"Morogoro Chillies" otherwise known as 'Habaneros' or 'Scotch Bonnets', are apparently popular in East African cuisine, but these plants seem to be confined to cultivation. They are cultivars of *C. chinense* Jacq. and though they may occur as escapes in some regions, this species has not been encountered during this revision.

Flowers solitary or in pairs; fruits always solitary, usually
    pendulous, 2.4–5.2 × 1.5–2.2 cm diameter . . . .   1. *C. annuum*
Flowers in (1–)2–4-flowered clusters; fruits usually paired, usually
    erect, 0.8–2 × 0.3–0.9 cm diameter . . . .   2. *C. frutescens*

1. **Capsicum annuum** *L.*, Sp. Pl. 1: 188 (1753); Mill., Gard. Dict. Ed. 8, No.1 (1768); Fingerhuth, Monogr. gen. capsici: 12 (1832); Dunal in DC., Prodr. 13(1): 412 (1852); Hiern, Cat. Afr. Pl. Welw. 3: 751 (1898); Irish in 9th Rep. Missouri Bot. Gard.: 65 (1898); C.H. Wright in F.T.A. 4, 2: 251 (1906); U.O.P.Z.: 170 (1949), as *annum*; F.P.U.: 130 (1962); Heine in F.W.T.A. 2nd ed., 2: 328 (1963); E.P.A. 2: 859 (1963); D'Arcy & Rakotozafy in Fl. Madagascar, Solanaceae: 15 (1994); DeWitt & Bosland, Peppers of the World: 95 (1996); Hepper in Fl. Egypt 6: 73 (1998); Gonçalves in Fl. Cabo Verde 71: 29 (2002) & in F.Z. 8(4): 59 (2005); Thulin in Fl. Somalia 3: 205 (2006); Friis in Fl. Eth. 5: 148 (2006). Type: Herb. Clifford: 59, *Capsicum* 1 (BM!, lecto.) designated by D'Arcy in Ann. Missouri Bot. Gard. 60: 591 (1974). [See Jarvis, Order out of Chaos: 382 (2007) for a discussion of the controversy surrounding the selection of this sheet as the lectotype]

Shrubby herb or small shrub, to 1.5 m, often short-lived; stems ridged, angular, woody, erect, much-branched, pilose when young, glabrescent. Leaves membranaceous, lanceolate to ovate-lanceolate, 2.2–8(–12) × 1–4(–8.5) cm, bases cuneate and decurrent, margins entire, apices acuminate, surfaces sparsely pilose, denser on margins, veins and midribs, with pilose domatia on lower surfaces, hairs as stems; petioles 1–3.5(–5.5) cm, pilose. Inflorescences usually 1-flowered in branch or leaf axil, rarely 2-flowered; flowers white to greenish- or bluish-white; pedicels erect or recurved and slender thickening apically in flower, 7–20 mm long, recurved in fruit when thickened beneath calyx and often woody, 1.2–3.5 cm long, glabrescent. Calyx shallowly cupulate to campanulate, 2–3.2 × (1.6)2–5.5 mm apically, with 5 narrowly triangular lobes 0.5–1.3 × 0.3–0.6 mm, persistent but only slightly enlarged in fruit, sparsely pilose with spreading hairs to glabrous. Corolla stellate, 5–9 × ± 11 mm diameter with short tube ± 1.2 mm long; lobes ovate to triangular, 2.8–5 × 2–4 mm, spreading or reflexed after anthesis, shortly puberulous on margins and lobe apices otherwise glabrescent externally, glabrous internally. Stamens with filaments 1–2 mm long, glabrous; anthers blue to purple, oblong, 1.5–2(–2.5) × 0.6–1(–1.4) mm, basifixed, exserted. Ovary brown, ovoid, 1.5–2 × 1.1–1.6 mm, glabrous; disc 1.5 mm broad and 0.5 mm high; style filiform, 3–5.5 mm long, glabrous, exserted ± 1 mm, straight or geniculate; stigma 0.2–0.5 mm diameter. Fruit usually pendulous, green, yellow, orange, red or blackish, globose, ovoid to narrowly conical or elongated berries, (1.1–)2.4–5.2(–15) × (1.2–)1.5–2.2(–4.5) cm, smooth, often apically acute and sometimes depressed, subtended by broadly cupulate calyx 2–4 × 7–15 mm, which often semi-reflexed away from berry base. Seeds yellow to yellowish-orange, orbicular to discoidal, 3.2–5.5 × 3.2–4.2 mm, with thickened margin, foveolate/reticulate. Fig. 12/7 & 8, p. 60.

var. **annuum** – D'Arcy & Eshbaugh in Baileya 19: 95 (1974); Gonçalves in Fl. Cabo Verde 71: 29 (2002)

UGANDA. Mengo District: Botanic Gardens, Entebbe, 1904, *Dawe* 113!
KENYA. Kilifi District: Mombasa, Rabai Hills, 1885, *W.E. Taylor* s.n.!
TANZANIA. Mbulu District: Manyara National Park near main gate, 9 June 1965, *Greenway & Kanuri* 11833!; Lushoto District: E Usambaras, Sigi, 16 Dec. 1940, *Greenway* 6086!; Tanga District: Kiwande Mission, 19 May 1929, *Samuel* s.n.!
DISTR. U 4; **K** 7; **T** 2, 3; originally from Mexico, cultivated and often naturalised throughout tropical Africa, Europe, SE Asia and Australia
HAB. An escape, uncommon on roadsides and in undergrowth; 350–1050 m
CONSERVATION NOTES. Widespread; least concern (LC)

SYN. *C. grossum* L., Mant. Pl. 1: 47 (1767) & Syst. Nat. ed. 12, 2: 175 (1767). Type: India, specimen not designated
   C. cerasiforme Mill., Gard. Dict. ed. 8: no. 5 (1768); Durand & Durand, Syll. Fl. Cong.: 396 (1909). Type: grown in London from seed collected in West Indies, ? *Miller* s.n. (not found)
   C. abyssinicum Rich. in Tent. Fl. Abyssin., 2: 96 (1851). Type: Ethiopia, Wogerate [Outerate], *Quartin Dillon & Petit* s.n. (P, ?syn.) & s.l., *Dillon & Petit* s.n. (P, ?syn) fide Friis (2006)
   C. annuum L. var. *cerasiforme* (Mill.) Irish in Rep. Miss. Bot. Gard.: 92 (1898)
   C. frutescens sensu Clarke in Hook., Fl. Brit. Ind., 4(1): 239 (1885), *non* L.
   C. longum DC., Cat. Pl. Hort. Monsp.: 86 (1813); Fingerhuth, Monogr. gen. capsici: 23 (1832); Dunal in DC., Prodr. 13(1): 424 (1852); Marloth, Fl. S. Afr. 3(1): 117 (1932). Type: not cited
   ?*C. annuum* L. var. *oblongo-conicum* (Dunal) Cufod., E.P.A.: 860 (1963)

NOTE. This is the most widespread of *Capsicum* species; it was first domesticated in Mexico and parts of Central America and is now only found wild in the southern US, Mexico and the West Indies (Mike Nee, *pers. comm.*). Most of the earlier authors recognised several varieties of *C. annuum* (e.g. Fingerhuth (1832) and Dunal (1852)) mainly on the differences in fruit shape. However, the var. *annuum* is considered to include all of the domesticated forms with larger flowers and fruits than their wild relatives.

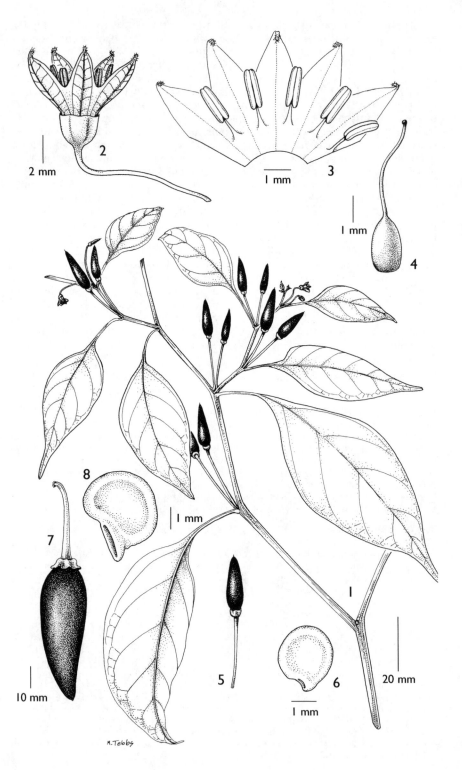

The wild or weedy forms of this species, with much smaller and sometimes globose fruits, have been variously identified as either var. *minimum* (Mill.) Heiser or var. *aviculare* (Dierb.) Hardy & Eshbaugh. More recently, Heiser & Pickersgill (in Baileya 19: 156 (1975)) postulated that the correct name for this spontaneous variety is *glabriusculum* (Dunal) Heiser & Pickersgill, and disputed the validity of many of the synonyms cited by D'Arcy & Eshbaugh (in Baileya 19: 93–105 (1974)). Cufondontis (in E.P.A., 1963) described three varieties of *C. annuum* from Ethiopia, but without any descriptions. In view of the controversy discussed by the above authors it is difficult to speculate on where these taxa belong.

The pungent forms are commonly known as chilli, paprika, red- or cayenne-pepper, while the larger mild-tasting fruits which only contain small amounts of capsaicin are popular vegetables which are known as bell-, sweet- or bull-nose- peppers with most being derived from around five cultivars. The hot varieties, with their high capsaicin content are not only used as spices and for seasoning but also medicinally, while the leaves are eaten as a vegetable. There are remarkably few herbarium specimens of this species.

2. **Capsicum frutescens** *L.*, Sp. Pl. 1: 189 (1753); Mill., Gard. Dict. Ed. 8, No. 9 (1768); Fingerhuth, Monogr. gen. capsici: 17 (1832), as *frutescens* Willd.; Dunal in DC., Prodr. 13(1): 413 (1852); Hiern, Cat. Afr. Pl. Welw. 3: 751 (1898); Irish in 9th Rep. Missouri Bot. Gard.: 65 (1898); C.H. Wright in F.T.A. 4, 2: 251 (1906); Durand & Durand, Syll. Fl. Cong.; 396 (1909); F.P.N.A. 2: 205 (1947); T.T.C.L.: 573 (1949); U.O.P.Z.: 170 (1949); Heine in F.W.T.A. 2nd ed., 2:328 (1963); E.P.A. 2: 860 (1963); Troupin, Fl. Rwanda 3: 360 (1985); DeWitt & Bosland, Peppers of the World: 17 (1996); Hepper in Fl. Egypt 6: 75 (1998); Gonçalves in Fl. Cabo Verde 71: 31 (2002) & in F.Z. 8(4): 60 (2005); Friis in Fl. Eth. 5 : 148 (2006). Type: *Capsicum fruticosum* L. orth. var., Herbarium Amboinense, 21 (1754); Indonesia, "Amboina", Herb. *van Royen*, sheet 908 244150 (L, lecto., image L0053043!), designated by Heiser & Pickersgill in Taxon 18: 280 (1969) [See also Jarvis, Order out of Chaos: 382 (2007)]

Shrub to 2.5 m; stems erect, spreading, usually zig-zagging, much branched, angular, sparse to moderately pilose when young, with simple hairs mixed with stalked glands, glabrescent. Leaves membranaceous, dark green, lanceolate, occasionally ovate to ovate-lanceolate, (1.3–)5–12 × (1.5–)2–5.3 cm, bases cuneate and decurrent, margins entire, apices acuminate, surfaces to sparsely pilose, denser on margins, veins and midribs, with densely pilose domatia on lower surfaces, hairs as stems; petioles (0.5–)1–2.2(–4) cm, pilose. Inflorescences usually fasciculate with 2–4 flowers in each branch or leaf axil, flowers rarely solitary; pedicels erect and slender, thickening apically in flower but prominently drooping beneath flower, 7–18 mm long, completely erect in fruit when 11–28 mm long, usually paired and often woody, glabrescent. Calyx cupulate, truncate with undulating margin, 1.2–3 × 2–3.5 mm apically, lobes virtually absent, persistent but only slightly enlarged in fruit; glabrescent. Corolla white, greenish-white or pale green, rarely yellow, stellate, 5–7.5 mm long and 6–10 mm diameter with tube 1.2–2 mm long; lobes broadly ovate, 2–5 × 1.2–3.2 mm, spreading or reflexed after anthesis, shortly puberulous on margins and lobe apices. Stamens equal, filaments 0.7–1.6 mm long; anthers blue, blue-green or purple, oblong, 1.5–2.5 × 0.7–1.6 mm, exserted. Ovary smooth, brown, ovoid, 1–2.2 × 0.8–1.5 mm, glabrous; disc to 1.3 mm diameter and 0.3 mm broad or absent; style 3.2–4.5 mm long, exserted 0.5–1.5 m; stigma 0.2–0.3 mm diameter. Fruit smooth, red, occasionally orange, ovoid to narrowly conical elongated berries, 8–20 × (3–)5–6(–9) mm, often apically acute, usually borne erect, often deciduous (or eaten by birds) leaving empty cupulate calyces. Seeds yellow to orange-yellow, ovoid, discoidal or reniform to suborbicular, 3–3.5 × 2–3 mm, with thickened margin, foveolate/reticulate. Fig. 12/1–6, p. 60.

Fig. 12. *CAPSICUM FRUTESCENS* — **1**, habit flowering and fruiting; **2**, complete flower; **3**, opened flower; **4**, gynoecium; **5**, erect fruit; **6**, seed. *C. ANNUUM* — **7**, pendulous fruit; **8**, seed. 1 & 6 from *Chancellor* 279; 2–4 from *Kassner* 332; 5 from *Borhidi* 86457; 7–8 from *Greenway* 6086. Drawn by Margaret Tebbs.

Uganda. West Nile District: 1.6 km N of Metu, 17 Sep. 1953, *Chancellor* 279!; Acholi/Bunyoro District: Victoria Nile, 1.6 km upstream from Etobi Lodge, Murchison Falls National Park, 5 Sep. 1967, *Angus* 5856!; Mengo District: Mabira Forest, 12 Nov. 1938, *Loveridge* 57!

Kenya. Teita District: Voi, 7 May 1931, *Napier* 965!; Kwale District: Shimba Hills, Mkurumuji Point area, 28 Mar. 1968, *Magogo & Glover* 564!; Kilifi District: Mtwapa, 13 km N Mombasa, Dec. 1931, *MacNaughton* 127!

Tanzania. Lushoto District: E Usambara Mountains, Kikuhwi Forest Reserve, path between Kikuhwi and Kwamkoro, 3 Nov. 1986, *Borhidi et al.* 86457! & Longuza, Muheza, 24 Oct. 1969, *Ngoundai* 367!; Uzaramo District: Dar es Salaam, University College near Primary School, 14 Aug. 1968, *Mwasumbi* 10376!; Zanzibar, Mkokotoni [Kokotoni], Aug. 1889, *Stuhlmann* 128!

Distr. U 1, 2, 4; K 7; T 3, 6; Z; cultivated and often naturalised in much of tropical Africa; also northern Africa and SW Europe, Indian Ocean islands, India, Papua New Guinea, New Caledonia, Australia

Hab. In secondary vegetation such as abandoned cultivation, forest edges, riverine thickets, road-sides, often in deep shade; 150–1250 m

Conservation notes. Widespread; least concern (LC)

Syn. *C. indicum* Rumph., Herb. Amboin., 5: 247 (1754). Basionym: *C. fruticosum* L. [orth. error for *C. frutescens*], type as above

  ?*C. conoides* Mill. sensu Fingerhuth in Monogr. gen. capsici: 14 (1832). Basionym: *C. conoides* Mill.

  ?*C. conoides* Fingerhuth sensu Dunal in DC Prodr. 13(1): 414 (1852).[NB numerous 'type' specimens cited]

  ?*C. conoides* Mill. var. *oblongo-conicum* Dunal in DC., Prodr. 13(1): 415 (1852). Type: cultivated in Hort. from seed ex Antigua and Lesser Antilles, specimens not designated (?P, holo)

  C. minimum Roxb., Hort. Beng: 17 (1814) & Fl. Ind. 2: 261 (1824); Clarke in Hook. Fl. Brit. Ind., 4(10); 239 (1885). Type: India, specimen not designated

  C. fastigiatum Blume, Bijdr. Fl. Ned. Ind.: 705 (1825). Type: Hort. and locally cult. in Indonesia, no other details given

Note. This is by far the most commonly collected *Capsicum* species in Africa. It is now found wild or as a weed from southern US through Mexico to N & E South America, the Antilles and Argentina. DeWitt & Bosland (1996) suggested that it was first domesticated in Panama from where it spread to Mexico and the Caribbean. The domesticated variety of this species is usually known as var. *pendulum*, while the wild forms are known as var. *baccatum* and var. *tomentosum* and are commonly referred to as bird peppers. Many authors report the difficulty of differentiating herbarium specimens of this species from *C. annuum*. Bailey (in Man. Cult. Plants: 873 (1966)) considered both *C. annuum* and *C. baccatum* to be synonymous with *C. frutescens*, within which he differentiated several varieties often on their fruit size and shape. However, though closely related these are now considered to be distinct species which are distinguishable morphologically, and which also display low inter-fertility.

   Roxburgh's *C. minimum* is often cited as a synonym of *C. frutescens* (Roxb., Hort. Beng: 17 (1814) & Fl. Ind. 2: 261 (1824)). However, the 1824 description mentions paired pedicels and subulate calyx teeth – characters typical of *C. baccatum* and *C. minimum* is more likely to be synonymous with that species.

   Although D'Arcy & Eshbaugh (Baileya: 99 (1974)) synonymised *C. conoides* Mill. (Gard. Dict., No. 8 (1768)) with *C. annuum*, Heiser & Pickersgill (Baileya: 155 (1975)) considered that Miller's species was either a large-fruited form of *C. frutescens* or a spontaneous variety of *C. annuum*; however they concluded that the absence of a type specimen of *C. conoides* precluded its taxonomic placement. Cufodontis (1963) published his variety *C. annuum* L. var. *oblongo-conicum* without any description, basing it on Dunal's variety *C. conoides* var. *oblongo-conicum*. Dunal's variety was based on Fingerhuth's (1832) *C. conoides* rather than Miller's species but Heiser & Pickersgill (1975) suggested that it is probably a synonym of *C. frutescens*. Most of these treatments describe the fruits as being oblongo-conical and borne erect, which are suggestive of conspecificity with *C. frutescens*.

   As well as the fruits being widely used as a condiment, the leaves are also eaten as a vegetable. In K 7 these plants are eaten by elephants and the fruits by birds.

EXCLUDED SPECIES

Capsicum baccatum *L.* **var. baccatum**, characterised by flowers with distinct basal markings; yellow anthers; small globose to broadly ovoid red to orange distally rounded fruits which are usually borne in pairs on long erect pedicels and subtended by subulate reflexed sepal lobes, does not seem to occur in our region. Most recent authors consider that this species embraces both a wild ancestral and a cultivated type (cf. D'Arcy & Eshbaugh in Baileya 19: 95 (1974)). Several varieties have been described, with the var. *baccatum* having been sparingly introduced into Africa where it is found in Chad, Congo Kinshasa, Angola, Mozambique and Zambia.

15. **DISCOPODIUM**

Hochst. in Flora 27: 22 (1844); Dunal in DC., Prodr. 13(1): 478 (1852); Hunziker, Gen. Solanaceae, 186 (2001); Edmonds, *Discopodium* in Africa, in Tax. & Ecol. African plants: 679–691 (2006)

Shrubs or small trees, much branched; stems tessellated or lenticellate when mature, young stems often densely flocculose/villous and appearing brown or ochraceous, glabrescent. Leaves usually alternate, prominently penniveined, domatia often present. Inflorescences of solitary flowers or 2–many-flowered axillary fascicles, epedunculate, pedicels slender. Flowers 5-merous, sometimes aromatic. Calyx broadly cupulate; lobes mucronate with external central ridge. Corolla campanulate-urceolate, tube broadly cylindrical, densely pubescent with short appressed hairs above calyx externally, glabrous internally apart from pilose band between filament bases, lobes narrowly triangular, usually recurved when anthers often exposed, occasionally spreading, densely pubescent with short hairs on both surfaces. Stamens included, usually equal; filaments filiform, broader at the base, adnate to middle of corolla tube, alternating with lobe lobes, free parts glabrous; anthers usually all fertile, occasionally one sterile, bithecate. Ovary globose to conical, bilocular, ovules numerous, placentation axile; disc annular, glabrous, fleshy and sulcate below ovary; style included; stigma discoid/capitate/peltoid with central depression, occasionally bilobed. Fruit a berry. Seeds orbicular or reniform with vesicular testa.

The genus is generally considered to be monospecific but two distinct species occur.

This is one of only two Solanaceous genera which are thought to be truly indigenous to Africa. Subfamilial treatments by D'Arcy (in Hawkes *et al.* Solanaceae III: 75–137 (1991)) placed this genus in the tribe Jaboroseae Miers, though Hunziker (2001) later considered the reasons for this to be 'irrelevant' and placed it in a subtribe of the Solaneae. Olmstead *et al.* (in Nee *et al.* (eds) Solanaceae IV: 111–137 (1999)) too considered that their molecular analysis supported the placement of *Discopodium* in the tribe Solaneae.

Mature leaves small, < 8 × 5 cm; flowers solitary, rarely in
  fascicles of 2(–4); corollas (7–)10–13 mm long; styles
  always glabrous; berries usually ovoid; seeds 1–5 per berry,
  5–6 mm long, reniform, dark brown . . . .   1. *D. eremanthum*
Mature leaves large, usually 12–24 × 5–14.5 cm; flowers usually
  in many-flowered fascicles; corollas 5–7(–8) mm long;
  styles usually pilose or villous (sometimes glabrous in
  T 2); berries globose; seeds 10–34 per berry, 2.5–3.5 mm
  long, orbicular, pale brown . . . .   2. *D. penninervium*

1. **Discopodium eremanthum** *Chiov.* in Racc. Bot. Miss. Consol. Kenya: 89 (1935); U.K.W.F.: 528 (1974): K.T.S.L.: 577 (1994); Edmonds, *Discopodium* in Africa in Tax. & Ecol. African plants: 686 (2006); Friis & Edmonds in Fl. Eth. 5: 150 (2006). Type: Kenya, Mt Aberdare E between Kinangop and Toeiene, *Balbo* 121 (TOM deposited at FT, holo., photo!)

Shrub or small tree, 1–3(–5) m high; stems ascending with terminal foliage, young stems densely flocculose/villous, often glabrescent when mature. Leaves usually coriaceous, elliptic, ovate or ovate-lanceolate, 4–7.8(–15) × 1.4–5(–7.6) cm, base cuneate, margins entire or sinuate, apex acute, upper surfaces glabrescent to pilose especially on midribs and primary veins, lower surfaces tomentose or villous/flocculose, glabrescent, often appearing fuscous brown; petioles 0.4–1.5(–3) cm long. Flowers usually solitary, occasionally in 2–3(–4)-flowered fascicles, often aromatic; pedicels often pendulous, 7–14 mm long, sparsely pilose, glabrescent; calyx green to purplish, (2.5–)3–5 mm long, lobes shallow, 1.2–3.5 × 2–5 mm, usually with distinct ciliate fringe often appearing as tufts on protruding apices. Corolla pale green, yellow or pale purple, (7–)10–13 mm long, tube (5–)6–8 mm long, lobes triangular, (2.5–)4–5 × (1.5–)2.5–4 mm. Stamens with filaments free for 1–3 mm; anthers greenish, yellow or brown, (1.2–)1.7–2.3 × 0.7–1 mm. Ovary green, 1.5–2.5 mm diameter; disc orange or yellow, 3–4 mm diameter; style green, glabrous, (1.2–)2–4 mm long; stigma 0.7–1.5 mm diameter. Fruit yellowish-orange, orange or red, usually longitudinally ovoid, rarely globose, 6–11 × (4.5–)6–9 mm; fruiting calyx cupulate with shallow glabrescent lobes 1.3–3.5 × 2–5 mm, adherent to base of berry; fruiting pedicels erect, 1–2.1 cm. Seeds 1–5 per berry, dark brown, usually reniform, (4–)5–6 × (3.5–)4–4.5 mm. Fig. 13/11–16, p. 65.

UGANDA. Mt Elgon, Sasa, North Bugisu, 22 Mar. 1951, *G.H.S. Wood* 254! & Bugisu, 17 June 1970, *Katende & Lye* 466!
KENYA. Mt Kenya, 9 Sept. 1963, *Verdcourt* 3737!; Naivasha District: Aberdare Range, Kinangop, 17 July 1948, *Hedberg* 1657! & Aberdares, 12 May 1923, *Rammell* 1060!
TANZANIA. Masai District: Ngorongoro, 20 Nov. 1956, *Tanner* 3288!; Moshi District: SW Kilimanjaro, Feb. 1928, *Haarer* 1168!
DISTR. U 3; **K** 3, 4 & ?5; **T** 2; Ethiopia
HAB. Moorlands, upper montane *Hagenia* woodland or *Podocarpus* forest, often in damp or boggy conditions, on tracks through degraded *Arundinaria* forest; (2100–)3000–3500 m
CONSERVATION NOTES. Widespread; least concern (LC)

SYN. *Discopodium sp.* of K.T.S.: 537 (1961)
      D. grandiflorum Cufod. in Senck. Biol.: 46 (Illustr. Suppl. 14): 90 (1965). Type: Ethiopia, Gamu-Gofa, Mt Dita, *Kuls* 773 (FR!, holo.)

2. **Discopodium penninervium** *Hochst.* in Flora 27: 22 (1844); Dunal in DC., Prodr. 13(1): 478 (1852); G.P. 2: 893 (1876); Engl., Hochgebirgsfl. Trop. Afr.: 374 (1892); E. & P. Pf.: 15 (1895); C.H. Wright in F.T.A. 4, 2: 253 (1906); Z.A.E.: 282 (1914); T.S.K.: 158 (1936); I.T.U.: 413 (1952); F.P.N.A. 2: 203 (1947); T.T.C.L.: 574 (1949); K.T.S.: 537 (1961); E.P.A. 2: 856 (1963); Heine in F.W.T.A. 2nd ed., 2: 328 (1963); U.K.W.F.: 528 (1974); Troupin, Fl. Rwanda 3: 368 (1985); K.T.S.L.: 578 (1994); Edmonds, *Discopodium* in Africa in Tax. & Ecol. African plants: 681 (2006); Gonçalves in F.Z. 8(4): 63 (2005); Friis & Edmonds in Fl. Eth. 5: 150 (2006). Type: Ethiopia, Tigre, mountains around Bahara in Haramat, *Schimper* 917 (TUB, holo.; BM!, CGE!, K!, MPU photo!, OXF!, iso.)

Shrub or small tree, usually 1.5–5 m, occasionally to 10 m high; stems soft and somewhat succulent to woody, green, yellow or brownish, varying from villous/flocculose to glabrescent as they mature. Leaves usually membranaceous, ovate to ovate-lanceolate, (8.8–)12–24 × 5–14.5 cm, base cuneate, margins usually entire to sinuate, occasionally sinuate-dentate with 1–6 acute antrorse lobes, apex acute to obtuse, upper surfaces often glabrescent apart from midribs and primary

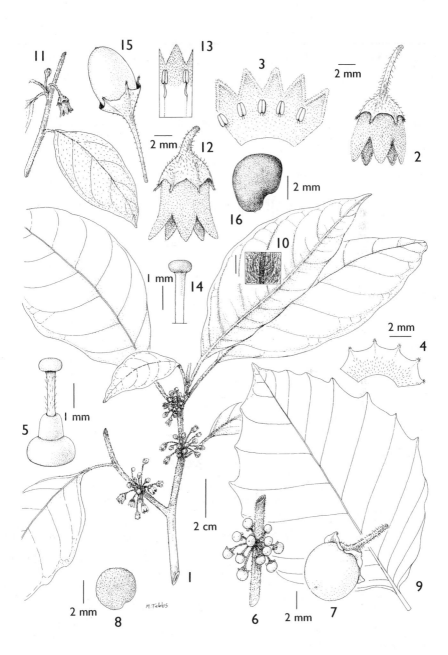

Fig. 13. *DISCOPODIUM PENNINERVIUM* — **1**, habit in flower; **2**, flower; **3**, opened flower; **4**, opened calyx; **5**, disc, ovary, style and stigma; **6**, infructescence; **7**, fruit; **8**, seed; **9**, sinuate-dentate leaf; **10**, leaf indument. *D. EREMANTHUM* — **11**, inflorescence; **12**, flower; **13**, part of open flower; **14**, style and stigma; **15**, fruit; **16**, seed. 1–5 & 10 from *Geesteranus* 5622; 6–8 from *Lucas & Polhill* 83; 9 from *Eggeling* 1625; 11–14 from *Katende* 466; 15–16 from *Hedberg* 1994. Drawn by Margaret Tebbs.

veins which sparsely to moderately pilose, lower surfaces glabrescent, tomentose or villous/flocculose; petioles 1–6(–11) cm long. Inflorescences (3–)6–30-flowered fascicles; pedicels usually erect, 5–12(–15) mm long, densely pilose/villous, glabrescent. Calyx broadly cupulate, 1–2(–4) mm long, densely pilose/villous, glabrescent; lobes broadly triangular or ovate, 0.7–2 × 1.5–2.5 mm, with or without a ciliate fringe which can appear tufted. Corolla white, cream, yellow or greenish, occasionally purplish or yellowish-red, 5–7(–8.5) mm long; tube 2.5–5 mm long, lobes narrowly triangular, 2–3.5(–6) × 0.5–2(–3) mm. Stamens with filaments free for 0.5–2.5 mm; anthers yellow to brown, 1–2 × 0.6–1.3 mm. Ovary 1–2(–3.5) mm diameter; disc 2–3.5 mm diameter; style green, occasionally glabrous (**T** 2 and Ethiopia), usually pilose throughout, rarely villous, usually filiform, occasionally clavate, 1–2.5 × 0.3 mm; stigma green, capitate, 0.8–1.5 mm diameter. Fruit globose, orange, orange-yellow or red, 5–8 mm diameter, fruiting calyx cupulate with calyx lobes 1–2.5 × 2–3 mm wide whose apices are usually semi-reflexed; fruiting pedicels erect or spreading, 8–20 mm long, glabrescent, pilose or densely villous, often becoming woody. Seeds 10–34 per berry, light brown, usually orbicular, 2.5–3.5(–4) × 2.25–3(–3.6) mm. Fig. 13/1–10, p. 65.

UGANDA. Acholi District: Mt Rom, *Eggeling* 2388!; Ruwenzori, Mihunga, 17 Jan. 1939, *Loveridge* 394!; NE Elgon, Feb. 1953, *Tweedie* 1103!
KENYA. Kiambu District: Kirita Forest, 30 Mar. 1961, *Lucas & Polhill* 83! & Limuru, Feb. 1915, *Dummer* 1552!; Kericho District: SW Mau Forest Reserve, Camp 7, 8 Aug. 1949, *Maas Geesteranus* 5622!
TANZANIA. Arusha District: Ngurdoto crater forest, Leopard Point, 27 Nov. 1966, *Richards* 21633!; Moshi District: above Kilimanjaro Timbers, 3 June 1993, *Grimshaw* 93/92!; Lushoto District: Kitivo Forest Reserve, July 1955, *Semsei* 2185!
DISTR. U 1–4; **K** 3–5: **T** 2, 3, 6 & 7; isolated mountains in east, west and central tropical Africa, including Nigeria, Cameroon, Bioko, Congo Kinshasa, Rwanda, Burundi, South Sudan, Ethiopia, and Malawi
HAB. Upland rain-, bamboo- or scrub-forest, including forest edges, also in thickets, grass-swamp, bushed grassland, and as a roadside ruderal; 1400–2500(–3000) m
CONSERVATION NOTES. Widespread; least concern (LC)

SYN. *Solanum cosmeticum* Delile in Rochet, Sec. Voy. Choa: 340 (1846), *nom. nud.* based on Ethiopia, Choa, *Rochet d'Hericourt* 10 (MPU-Delile, holo., photo!; P, iso.) *fide Lester pers. comm.*
   Withania holstii Dammer in P.O.A. C.: 351 (1895); C.H. Wright in F.T.A. 4, 2: 250 (1906); T.T.C.L.: 574 (1949). Type: Tanzania, Lushoto District: Usambara, Magamba, *Holst* 3843 (B†, holo.)
   ?*D. paucinervium* Engl. in V.E. 1(1): 381 (1910); T.T.C.L.: 574 (1949), *nom. nud.*
   D. penninervium Hochst. var. *holstii* (Dammer) Bitter in E.J. 57: 16 (1920); T.T.C.L.: 574 (1949)
   D. penninervium Hochst. var. *nervisequum* Bitter in E.J. 57: 17 (1920). Type: Uganda, Ruwenzori, *Scott Elliot* 7714 (B†, holo.; BM, lecto.!, designated by Edmonds l.c.: 683 (2006))
   D. penninervium Hochst. var. *intermedium* Bitter in E.J. 57: 17 (1920); T.T.C.L.: 574 (1949). Type: Tanzania, Lushoto District: Usambara, Mazumbai [Masumbai], *Braun* 2738 (B†, holo.)
   D. penninervium Hochst. var. *sparsearaneosum* Bitter in E.J. 57: 17 (1920). Type: Cameroon, Manenguba mountains near Bare, *Schäfer* 100 (B†, holo.)
   D. penninervium Hochst. var. *magnifolium* Chiov. in Nuov. Giorn. Bot. Ital. 36: 367 (1929) & E.P.A.: 856 (1963). Type: Ethiopia, Arussi, Mount Galamo, *Basile* 68 (TO, holo. fide Friis, 2006)

NOTE. *D. penninervium* seems to have a fairly restricted distribution in Kenya though it is often common in the habitats in which it is found. It is more widely distributed in Tanzania and is particularly well-distributed within Ethiopia. Bitter (1920) distinguished several varieties of this species on minor variations in leaf pubescence. These variations are all considered to be within the limits acceptable in *D. penninervium*, and these varieties have all therefore been provisionally syonymised with this species. Most of the type specimens cited by Bitter were lost in the bombing of the Berlin herbarium. Only one duplicate has so far been traced, and this specimen (*Scott Elliott* 7714) has been selected as the lectotype of *D. penninervium* Hochst. var. *nervisquum* Bitter.

Some of the Ethiopian isotype specimens of this species (*Schimper* 917) exhibit the glabrous styles typical of *D. eremanthum*, while others are sparsely pilose. A number of specimens collected in **T** 2 (e.g. Arusha District: SW Meru & Ngurdoto Crater; Masai District: Empakaai Crater, Ngorongoro Conservation area; *cf: Frame* 59; *Greenway & Kanuri* 11987; *Haarer* 1168; *Richards* 24652) also have glabrous or glabrescent styles. However, with one exception (*Mooney* 7011 collected from Shoa, Wofasha Forest), all other Ethiopian specimens so far examined have pilose styles.

16. IOCHROMA

Benth. in Edwards, Bot. Reg. 31, t. 20 (1845), *nom. conserv.*; Dunal in DC., Prodr. 13(1): 489 (1852); Shaw in New Plantsman 5(3): 154–192 (1998); Hunziker in Gen. Solanaceae: 220–226 (2001)

Shrubs. Leaves alternate, lamina often decurrent. Inflorescences terminal, subaxillary or axillary, usually in fascicles of 3–many flowers, sometimes in pairs or solitary; pedicels erect or pendent. Calyx urceolate, cupulate or campanulate, enclosing the basal part of the corolla tube, with five short sometimes unequal teeth. Corolla with the long narrow basal part terminating in five to ten short apical lobes. Stamens with filaments fused to lower half to third of corolla tube, filiform above where glabrous, becoming thicker and villous below especially where fused to corolla tube; anthers elongate, basifixed (appearing dorsifixed). Ovary 2-locular, ovules numerous; disc annular, smooth; style filiform but becoming thicker towards the ovary and stigma, glabrous, usually included; stigma capitate or clavate, bilobed. Fruit a berry, either wholly or partially enclosed by accrescent calyx. Seeds numerous, orbicular to reniform, often mixed with sclerotic granules.

15–16 woody species native to humid western South America. Many are planted for their ornamental value in tropical and subtropical gardens.

The *Iochroma* specimens cultivated in Kenya and Tanzania have been variously identified but they predominantly belong to one species namely *I. cyaneum* (Lindl.) M.L. Green. One specimen, possibly belonging to *I. coccinea* Scheid., has been recorded from Nairobi (see below). The plants can, however, also occur as escapes in ruderal habitats.

Iochroma cyaneum (*Lindl.*) *M.L. Green* in Intern. Bot. Congr. Cambr. 1930, Nomencl. Proposals Brit. Botanists 107, No. 7382 (1929); Lawrence & Tucker in Baileya 3: 65 (1955); Gonçalves in F.Z. 8(4): 6 (2005). Type: Herb. Hort. London, introduced from Loja, Ecuador, *Hartweg* s.n.(K! in herb Hooker, syn.); collected from the village of Gonzanamá and the city of Loja, Ecuador, *Seemann* 883 (K!, syn.), *typ. cons.*

Shrubs or woody herbs, 1–4 m tall, often laxly branched; stems pilose or villous with long simple or simple and branched eglandular hairs, denser on young stems. Leaves green, ovate-lanceolate to lanceolate, 5–12(–14) × 2.6–5.4(–6.6) cm, bases cuneate, margins entire to sinuate, apices acute to obtuse, moderately pilose above, floccose beneath with long simple or simple and branched hairs; petioles 1.4–6 cm long. Inflorescences congested terminal, subaxillary or axillary fascicles of 8–24+ flowers; pedicels erect or spreading, pilose or villous, 0.9–1.6(–2.7) cm long in flower. Calyx urceolate or cupulate, enclosing the basal part of corolla tube, 7–12 × 5–9 mm, with five triangular teeth 0.8–2.5 × 0.8–3.5 mm, pilose externally. Corolla usually blue, mauve or purple, sometimes with a whitish throat, tubular or trumpet-shaped, 3.2–4 cm long, with throat 9–12 mm diameter, tube long and often curved, ± 5 mm wide in centre, terminating in five to ten partially reflexed apical broadly triangular lobes 0.8–2 × 1.5–3 mm; basal part of tube tapering towards base, glabrous externally becoming puberulous toward the lobes.

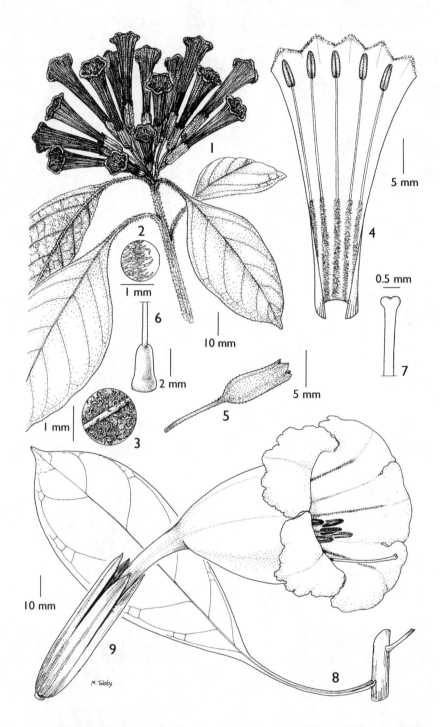

FIG. 14. *IOCHROMA CYANEUM* — **1**, flowering twig; **2**, stem indument; **3**, lower leaf indument; **4**, opened flower; **5**, calyx and pedicel; **6**, ovary and stylar base; **7**, stigma and upper style. *SOLANDRA GRANDIFLORA* — **8**, leaf and petiole; **9**, complete flower. 1–3 from *Kisena* 200; 4–7 from *Greenway* 8772; 8 & 9 from *Hort. Monsp.* s.n. Drawn by Margaret Tebbs.

Stamens usually included, occasionally exserted; filaments 2.5–3.2 cm long; anthers yellow, elongate, 3–4.5 × 1–1.5 mm; often visible in corolla throat. Ovary conical to pyriform, 2.2–5 × 1.1–2.8 mm, glabrous; style 2.7–3.2 cm long, included; stigma 0.3–1 × 0.5–1 mm. Fruit pale greenish-yellow to purplish-green conical berry, 1.8–2.2 × 1.2–1.7 cm, partly enclosed by expanded split calyx, with strong fruity scent. Seeds yellow, discoid, 1 × 1.5 mm, flattened and foveolate; sclerotic granules spherical to ovoid, ± 0.5–1 mm diameter. Fig. 14/1–7, p. 68.

KENYA. Nairobi Arboretum, 21 May 1960, *Odera* 4! & 25 Feb. 1952, *Williams Sangai* 347! & Nairobi, Closeburn, 6 May 1953, *Greenway* 8773!
TANZANIA. Lushoto District: Amani Nursery, 5 July 1940, *Greenway* 5963! & road-side to township, 25 Nov. 1981, *Kisena* 2! & between Boma Office and Silviculture, 13 Dec. 1990, *Sigara* 303!
DISTR. **K** 4; **T** ?2, 3; native to Andean Ecuador, widely grown as an ornamental shrub in Africa
HAB. An escape from gardens and arboreta, occurring on roadsides; 850–1700 m
CONSERVATION NOTES. Widespread; least concern (LC)

SYN. *Habrothamnus cyaneus* Lindl. in Edwards' Bot. Reg. Misc.: 72 (1844)
    Iochroma tubulosum Benth. in Edwards' Bot. Reg. 31: 20 (1845), as *tubulosa*; Hooker in London Journ. Bot., 7: 344 (1848), as *tubulosa*; Dunal in DC., Prodr. 13(1): 490 (1852), *nom. illeg.*

NOTE. Numerous other synonyms are given in Shaw (1998). No fruits were recorded from East African specimens, perhaps again reflecting the lack of necessary pollinators such as hummingbirds; the fruit and seed characters given above have been taken from Shaw (1998). Specimens identified as *I. lanceolata* Miers proved to belong to *I. cyaneum*, rather than to the yellow-flowered species described by Miers.
    Among the names used to identify East African cultivated material are *Cestrum brevifolium* Urb., *Iochroma schlechtendalianum* Dunal and *I. lanceolata* Miers. The latter was described as a yellow-flowered species by Miers, and although Lawrence & Tucker (1955) considered this to be a synonym of *I. cyaneum*, Shaw (1998) reported that the name *lanceolatum* had been misapplied by Hooker and synonymised it with *I. gesnerioides* (Kunth) Miers. He further proposed that many individuals represent hybrids between this latter species and *I. cyaneum*. One specimen (*Greenway* 8769) whose flowers were described as "brownish red" had been identified as the red-flowered species *I. coccinea* Scheid. (a synonym of *I. gesnerioides* (Shaw, 1998)). However, Greenway collected this on the same day and from the same locality – Closeburn, Nairobi - as specimen 8772 whose flowers he recorded as being "almost black purple". Both of these specimens are considered to belong to *I. cyaneum*. I have only seen one good African specimen of *I. gesnerioides* collected as a garden ornamental in Harare, Zimbabwe (*Biegel* 5222!), whose flowers were "chinese lacquer red". Many cultivars of *I. cyaneum* have been produced, and flower colour is particularly variable in this species; it is also known to alter with both maturity and drying. In addition, Shaw (1998) considered that the heterogenicity shown by this species is evidence of a hybrid origin of *I. cyaneum*, and that introgression from other species, including *I. gesnerioides*, may also have occurred, all contributing to the considerable morphological variability displayed by this species.

17. PHYSALIS

L., Sp. Pl. 1: 182 (1753) & Gen. Pl. ed. 5: 85 (1754); Dunal in DC., Prodr. 13(1): 434 (1852); Waterfall in Rhodora 60: 107–114, 128–142, 152–173 (1958) & in Rhodora 69: 82–120, 203–239, 319–329 (1967); Hunziker in Gen. Solanaceae: 202–208 (2001)

Alkekengi Adans., Fam. 2: 218 (1763)
Herschelia Bowdich, Excurs. Mad.: 159 (1825)

Annual or perennial herbs, usually much-branched, sometimes woody basally; stems sometimes viscid, hairs simple, branched or stellate, glandular or eglandular. Leaves alternate, rarely opposite. Inflorescences axillary, usually composed of a single flower, occasionally 2–4(–7) flowered fascicles; flowers erect to pendent, hermaphrodite. Calyx campanulate to cupulate, actinomorphic with 5 lobes or 5-

partite with short lobes; enlarged and persistent in fruit when basally invaginated. Corolla broadly campanulate to rotate or funnelform, rarely urceolate, usually actinomorphic, often with dark basal spots and densely pubescent inner ring alternating with stamens internally, margin entire or tube terminating in 5 short lobes. Stamens 5, usually equal and exserted; filaments fused to lower part of corolla tube where broader, glabrous or sparsely pilose; anthers equal or unequal, oblong, bilobed, basi- or dorsi-fixed, sometimes convergent around stigma. Ovary superior, bilocular, ovules numerous, placentation axile; disc annular, occasionally absent; style filiform, usually glabrous and exserted; stigma discoid-capitate, bilobed. Fruit a berry, mature pericarp thin and translucent, enclosed by enlarged and inflated accrescent chartaceous usually reticulately-veined bladder-like urceolate calyx which 5-angled or prominently 10-ribbed, sometimes with 5 basal auricles, often brightly coloured, the mouth usually almost completely closed by connivent calyx lobes or teeth. Seeds numerous, vesicular to foveolate, sclerotic granules absent.

Between 75 and 100 species, predominantly found in the New World with the greatest concentration of species occurring in Mexico.

A few species are cultivated world-wide for their ornamental showy calyces, or for their edible berries and they are collectively known as Husk-tomatoes or Ground-cherries. Hepper (in Fl. Ceylon, 4: 391 (1987)) described *Physalis* as being an extremely puzzling genus of considerable taxonomic and nomenclatural complexity, while Symon (in Journ. Adelaide Bot. Gard. 3(2): 149 (1981)) thought that many of the nomenclatural problems would remain insoluble until African and Asian names and taxa were studied and compared with those in America. Older literature contains numerous synonyms for each of the *Physalis* species found in the floral region; where the names concerned have not been encountered on any East African material, they have not been included in this treatment. With the exception of *P. peruviana*, it is often difficult to determine which species is actually being described in many of the Floras cited, and the associated nomenclature is extremely complex. It is clearly a genus in urgent need of world-wide revision.

1. Plants densely pubescent, often velutinous; flowers yellow or greenish-yellow with prominent dark blotches, flowering calyces 5–10 mm long with lobes free for 3.5–6 mm; fruiting calyx urceolate, inflated to 2–4.5 × 2–3.5 cm diameter, pubescent externally . . . . 1. *P. peruviana*
Plants moderately pilose to glabrescent; flowers white, cream, yellow to orange-yellow, blotches absent or pale; flowering calyces usually < 5 mm long with lobes free for < 3 mm; fruiting calyx globose to ovoid, inflated to 0.8–2.9 cm diameter, usually glabrescent externally 2
2. Leaf margins entire to sinuate; flowers < 5 mm long, white, cream or yellow without any central blotching; flowering calyx lobes shallow, free for < 1 mm; anthers pale yellow, 0.3–1.3 mm long; style < 2.5 mm long . . . . 2. *P. lagascae*
Leaf margins serrate to sinuate-dentate with prominent teeth; flowers > 6 mm long, yellow to yellow-orange, sometimes with pale purple blotches; flowering calyx lobes free for 1.2–3 mm; anthers purple or purple and yellow, 1.2–4.4 mm long; style 4.1–9 mm long 3
3. Corolla throat with an internal ring of dense long eglandular hairs; mature anthers twisted, 2.8–4.4 mm long; style 5.5–9 mm long; inflated fruiting calyx usually globose; seeds 2.1–2.5 mm long . . . . 3. *P. philadelphica*
Corolla throat glabrescent; mature anthers straight, 1.2–2 mm long; style < 5 mm long; inflated fruiting calyx usually ovoid; seeds 1.2–1.8 mm long . . . . 4. *P. angulata*

1. **Physalis peruviana** *L.*, Sp. Pl. ed. 2: 1670 (1763); Lamarck, Encycl. Méth. Bot. 2: 101 (1786); Dunal in DC., Prodr. 13(1): 440 (1852); Wright in Fl. Cap. 4(1): 106 (1904); C.H. Wright in F.T.A. 4, 2: 248 (1906); Durand & Durand, Syll. Fl. Cong.: 395 (1909); F.P.N.A. 2: 205 (1947), as *P. pubescens*; W.F.K.: 90 (1948); U.O.P.Z.: 411 (1949); Waterfall in Rhodora 69: 141 (1958); Heine in F.W.T.A. 2nd ed., 2: 329 (1963); E.P.A. 2: 859 (1963); U.K.W.F. 2nd ed.: 244 (1994); Hepper in Fl. Egypt 6: 67 (1998); Gonçalves in F.Z. 8(4): 47 (2005); Thulin in Fl. Somalia 3: 202 (2006); Friis in Fl. Eth. 5: 153 (2006). Type: cultivated Hort. Uppsala, "Habitat Limae". Alstoemer, Herb. *Linnaeus* 247.7 (LINN!, lecto.) designated by Fernandes in Garcia de Orta, 17(3): 8 (1969) [and not lectotype selected by Heine in Aubreville & Leroy, Fl. Nouv-Caled., 7: 132 (1976 cited as 1975) fide Jarvis, Order out of Chaos: 742 (2007)]

Perennial herbs to 1.6 m high, sometimes annual, occasionally shrubby and woody basally; main stems prostrate to erect and spreading, usually much-branched, sometimes straggling or trailing, with thick woody tap root; stems light to brownish-green, sometimes angular; all vegetative parts densely villous/pubescent, often appearing light brownish yellow. Leaves alternate, occasionally opposite, usually membranaceous, green to greyish-green, ovate to broadly ovate, 4.5–7(–11) × 3–5.5(–8) cm, bases cordate to sub-cordate, often oblique, margins entire to sinuate, sometimes sinuate-dentate with few shallow obtuse to acute lobes, apices acuminate to acute, pubescent, denser on veins, midribs and lower surfaces; petioles 1.2–5.2 cm. Flowers solitary, axillary; pedicels 6–12 mm and erect, rarely curved apically, 7–16 mm long and usually recurved in fruit, villous. Calyx light green, cupulate to campanulate, 5–10 mm long with five triangular acute lobes 3.5–6 × 1.3–3 mm, villous externally, enlarged and persistent in fruit, with lobes 5–10 × 2–5.5 mm. Corolla yellow or greenish-yellow with black, grey, brown, crimson or purple dark basal spots or blotches and an internal ring of yellowish hairs in throat below anthers, broadly campanulate, 1–1.6 cm long and 0.9–1.9 cm diameter, shortly pubescent externally, margin undulating, ciliate and entire or with five broadly triangular obtuse lobes 0.7–2 × to 3 mm. Stamens often unequal, exserted; filaments free for 2.5–5 mm; anthers usually equal and tinged purple, oblong, bilobed, 3–3.8 × 1–1.7 mm, always exposed in throat of corolla. Ovary smooth, brownish, 1.5–2 × 1.7–3 mm, ovoid, glabrous; disc 0.7–1 × 2.1–3.5 mm; style often exserted, 4.5–7 mm long; stigma 0.3–0.8 mm diameter, often shallowly bilobed. Fruit a smooth greenish berry maturing to yellow, orange or red, globose or ovoid, 8–15 mm diameter, enclosed by enlarged and inflated pubescent, chartaceous, greenish-yellow, reticulately- and often purple-veined bladder-like urceolate 5–10-angled calyx, 3–4.4 × 2–3.5 cm, the mouth closed by connivent calyx lobes. Seeds yellowish to brownish, orbicular to elliptic, 1.6–2 × 1.2–1.7 mm, compressed. Fig. 15/16–23, p. 72.

UGANDA. West Nile District: Paida, hillside beside the Sultan's House, Aug. 1953, *Chancellor* 204!; Kigezi District: Kachwekano Farm, Feb. 1950, *Purseglove* 3329!; Mengo District: Mabira Forest, Nov. 1938, *Loveridge* 41!
KENYA. Mt Elgon, Apr. 1931, *Lugard & Lugard* 641!; Nakuru District: 6 km E of Londiani, along Kericho Road, Nov. 1967, *Perdue & Kibuwa* 9114!; Kiambu District: Muguga Juu, Aug. 1965, *Greenway* 11925!
TANZANIA. Ufipa District: Kito Mountain, Sumbawanga, Apr. 1961, *Richards* 15051!; Mbeya District: 1 km SE of Lomba Local Court, Feb. 1963, *Harwood* 8!; Songea District: Matengo Hills, 11 km N of Miyau, Mar. 1956, *Milne-Redhead & Taylor* 8781!
DISTR. U 1, 2, 4; K 1–6; T 1–8; native to South America but introduced throughout the world to warm and temperate regions, including throughout Africa from Sierre Leone to Ethiopia and S to South Africa; often locally naturalised
HAB. Naturalised and often common weed of shambas, waste and disturbed places, fallow land, plantation areas, river-banks, secondary bushland, forest margins and clearings; 900–2500 m
CONSERVATION NOTES. Widespread; least concern (LC)

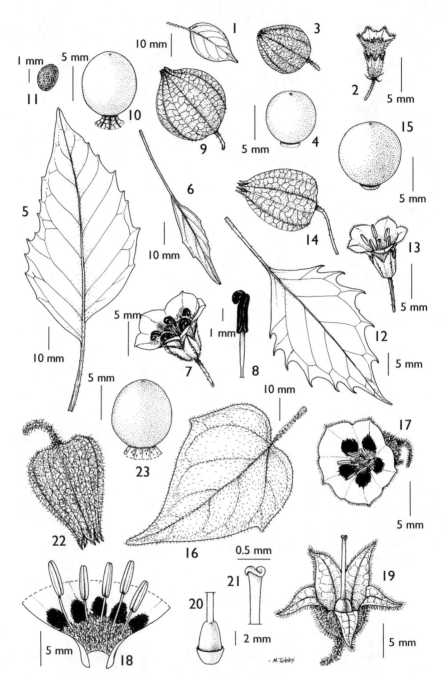

Fig. 15. *PHYSALIS LAGASCAE* — **1**, leaf; **2**, complete flower; **3**, capsule; **4**, berry.
P. PHILADELPHICA — **5 & 6**, leaf variation; **7**, complete flower; **8**, coiled anther; **9**, capsule;
10, berry; **11**, seed. *P. ANGULATA* — **12**, leaf; **13**, complete flower; **14**, capsule; **15**, berry.
P. PERUVIANA — **16**, leaf; **17**, complete flower; **18**, opened flower; **19**, calyx and
gynoecium; **20**, ovary; **21**, stigma; **22**, capsule; **23**, berry. 1–2 from *Polhill & Paulo* 1037; 3–4
from *Purseglove* 568; 5–6 from *Terry* 3517; 7–8 from *van Someren* s.n.; 9–11 from *van Someren*
12086; 12–13 from *Archbold* 2389; 14–15 from *Hepper & Field* 5203; 16, 22–23 from *Greenway*
11925; 17–21 from *Harwood* 8. Drawn by Margaret Tebbs.

SYN. *P. tomentosa* Medic., Act. Akad. Theod. Palat. Phys. 4: 184 (1780). Type: none cited; Medic.
    t. IV ! (page 205) lectotype designated here, drawn/collected by E.Verhelst from plant
    grown/collected in Mannheim, Germany
  P. latifolia Lam., Tabl. Encycl. Méth. Bot. 2: (1793). Type: *Lamarck* s.n. (P-LAM, holo., fiche
    [471/15]!)
  Alkekengi pubescens Moench, Meth.: 473 (1794). Based on and type as *Physalis peruviana* L.
  Physalis esculenta Salisb., Prodr.: 132 (1796). Based on and type as *Physalis peruviana* L.
  P. edulis Sims, Bot. Mag. 27: t. 1068 (1808). Type: cultivated by Gabriel Gillett in Drayton
    Green, source of seed not cited; Bot. Mag. 27: t. 1068! (lecto.)
  ?*P. pubescens* sensu R.Br. in Prodr. Fl. Nov. Holl. 1: 447 (1810) and Robyns in F.P.N.A. 2: 205
    (1947), *non* L.
  Herschelia edulis (Sims) Bowdich, Excurs. Mad.: 159 (1825)
  Physalis peruviana L. var. *latifolia* (Lam.) Dunal in DC., Prodr. 13(1): 440 (1852)

NOTE. Commonly known as the Cape Gooseberry (derived from its introduction from the
  Cape, where it became widely naturalised), ground cherry or jam berry, this species is now
  cultivated for its fruits throughout the world. In East Africa the fruits are widely eaten and
  sold in markets. It is also used as a herbal medicine by midwives in U 2.

2. **Physalis lagascae** Roem. & Schult., Syst. Veg. 4: 679 (1819); Waterfall in
Rhodora 69: 220 (1967); Gonçalves in F.Z. 8(4): 51 (2005); Friis in Fl. Eth. 5: 153
(2006). Type: based on *P. parviflora* sensu Lagasca, Gen. Sp. Nov. Diagn.: 11, n. 147
(1816); type not cited

Annual herb usually < 0.5 m high, occasionally taller, erect, decumbent, prostrate
or creeping; main stems sparsely to densely branched, green or purple-tinged,
pilose with simple hairs (in our area). Leaves light to dark green, 1.7–4.5(–6) ×
1–3(–4.2) cm, smaller above, ovate to ovate-lanceolate, bases cuneate, often oblique,
sometimes decurrent, margins entire to sinuate, rarely sinuate-dentate, apices
acute, sparsely to moderately pilose (in our area), hairs usually short and appressed,
denser on veins, midribs and lower surfaces; petioles 0.5–3.2 cm, pubescence as
stems. Flowers solitary, small, in leaf or branch axils; pedicels 1.5–4 mm long, erect
in flower, pubescent, 3–7 mm long and erect to recurved in fruit. Calyx cupulate,
1.7–3(–4) mm long with five broadly triangular acute lobes 0.5–1 × 0.5–1.5 mm with
ciliate margins, often with long spreading white hairs, enlarged and persistent in
fruit with connivent lobes 0.8–1.8 × 0.8–2 mm. Corolla white, cream, pale yellow or
green, tubular/campanulate, 2.9–5 × 3–4 mm diameter, shortly pubescent
externally, margin ciliate, undulating or with five broadly triangular obtuse lobes
0.2–1 × 0.7–1.1 mm. Stamens often visible in corolla throat; filaments free for
1.5–2.5(–3) mm; anthers pale yellow, ovoid to oblong, bilobed, 0.5–1.3 × 0.3–0.5 mm.
Ovary brownish, ovoid, 0.7–1.1 × 0.5–1.1 mm, smooth, glabrous; disc 0.7–1.5 mm
diameter; style often exserted, 2–2.5 mm long; stigma 0.2–0.3 × 0.2–0.5 mm. Fruit
pale green to yellow, globose, rarely ovoid, 5–10 mm diameter, smooth, subsessile or
on inverted gynobase (–0.75 mm) of enlarged and inflated reticulately-veined pale
green globose calyx, 0.8–1.8 × 0.7–2 cm, the mouth closed by connivent calyx lobes,
veins often clothed with long spreading hairs. Seeds yellow, orbicular, 1.2–1.8 ×
1.1–1.7 mm, compressed. Fig. 15/1–4, p. 72.

UGANDA. Karamoja District: Napak Mountain, June 1957, *Wilson* 359!; Busoga District: locality
    not cited, Dec. 1905, *E. Brown* 365!; Ankole District: Buyaruguru, Feb. 1939, *Purseglove* P568!
KENYA. West Suk District: Wei Wei, 16 June 1978, *Meyerhoff* 12!; North Kavirondo District:
    Kakamega Forest, Rondo sawmill, May 1971, *Tweedie & Mabberley* 3976! & Mumias, 12 May
    1979, *Bridson* 93!
TANZANIA. Tanga District: Ngua Estate, W slope of Usambaras, 19 July 1953, *Drummond &
    Hemsley* 3358!; Kigoma District: Kakombe Valley, Gombe Stream Reserve, 20 Dec. 1963,
    Pirozynski P32!; Iringa District: Kidatu, Mar. 1971, *Mhoro* 679!
DISTR. U 1–4; K 2, 4, 5, ?7; T 1–4, ?5, 6, 7; probably native to Mexico but now a cosmopolitan
    weed throughout tropical and subtropical regions

HAB. Weed of cultivation, plantations and abandoned areas, also on river-banks, disturbed
sites, edges of paddy-fields, grass patches in floodplains; common to local; 40–1850 m
CONSERVATION NOTES. Widespread; least concern (LC)

SYN. *P. parviflora* sensu Lagasca, Gen. Sp. Nov. Diagn.: 11, n. 147 (1816), *non* Zucc. (1806) *nec*
      R.Br. (1810), *nom. illegit.*
    P. minima sensu auctt. plur. e.g.: Nees von Esenbeck in Linnaea 6: 479 (1831) & in Trans.
      Linn. Soc. Bot. 17(1): 69 (1837); P.O.A. C: 351 (1895); Durand & Schinz, Fl. Ind. Cong.
      1: 206 (1896); C.H. Wright in Fl. Cap. 4(1): 106 (1904) & in F.T.A. 4, 2: 247 (1906);
      Durand & Durand, Syll. Fl. Cong.: 395 (1909); W.F.K.: 90 (1948); F.P.U.: 130 (1962);
      E.P.A. 2: 858 (1963); U.K.W.F., 2ⁿᵈ ed.: 244 (1994), *non* L.
    P. divaricata sensu auctt. e.g.: Schönbeck-Temesy in Fl. Iran. 100: 24 (1972); Nasir *et al.* in
      Fl. Pakistan 168: 25 (1985), *non* D. Don
    P. micrantha sensu auctt. e.g. Schulz in Urban, Symb. Antill. 6: 147 (1909); Heine in
      F.W.T.A. 2, ed. 2: 329 (1963); Morton in Journ. W. Afr. Sci. Ass. 9(1): 71, 74 (1964);
      Troupin, Fl. Rwanda 3: 370 (1985); ?fide Hepper in Fl. Ceylon, 4: 393 (1987), *non* Link
    P. lagascae Roem. & Schult. var. *glabrescens* O.E. Schulz in Urban, Symb. Antill. 6: 147
      (1909). Type: Cuba, Havana near Calabazar, *Wright* 3636 (K!, ?isosyn. as *P. minima*,
      without any collection details); Cuba, around Rincon, *Baker & Wilson Herb. Cuba*, 278
      (syn.); *Wilson Herb. Cuba* 1068 (syn.); Trinidad, *Bot. Gard. Herb.* 6747 (syn.). No type
      specimen localities given

NOTE. This taxon is widely known as *P. minima* L. – an appropriate epithet in view of its
characteristic small leaves, flowers, berries and fruiting calyces. However, typification of this
Linnaean species is extremely complex with the lectotype of *P. minima* (Hort. Cliff. n. 62, *Physalis*
5 (BM!)) being conspecific with *P. angulata*, with which it has been synonymised. Moreover, most
of the descriptions and plates on which Linnaeus based this species (e.g. Herm. Lugdb.: 569, t.
571; and Rheede Mal. 10, t. 140, f. 71) also refer to *P. angulata* rather than the small-flowered
prostrate plant described here. Numerous authors have discussed the difficulties of correctly
naming and typifying this taxon (*cf.* Fernandes in Garcia de Orta, 17: 282 (1969)), and some of
the synonyms cited in the literature for *P. minima* belong to *P. angulata*. Though the name *P.
lagascae* has been used here in line with various recent generic treatments, there are no known
type specimens and the protologue does not entirely reflect the morphology characterised by
East African specimens. Roemer & Schultes (1819) described the branches of their taxon as
villous, repeating the phrase used by Opiz (1817) when describing *P. parviflora* on which *P.
lagascae* was based. While the majority of East African specimens are sparsely to moderately
pilose, there are a number of specimens from Kenya and Tanzania which are densely villous even
to the naked eye. In other features these specimens are very similar to the species described
above and they have been included under *P. lagascae*. Indeed, hairy and glabrescent varieties of
this species have been recognised as var. *lagascae* and var. *glabrescens* O.E. Schulz respectively in
Central America (*cf.* Schulz, 1909; Waterfall, 1967). These villous specimens could, however,
belong to the taxon described as *P. pubescens* L. in Flora Zambesiaca (apparently based on a single
specimen from Zambia). It is possible that following more detailed work on its pubescence
variation, a name change might become necessary for *P. lagascae*.
    The fruits of this species are edible, and the boiled leaves are eaten as a vegetable in **U 1**,
K 2 and Ethiopia; it is considered to be a medicinal plant throughout its range.

3. **Physalis philadelphica** *Lam.*, Encycl. Méth. Bot. 2: 101 (1786); Dunal in DC.,
Prodr. 13(1): 450 (1852); F.P.U.: 55 (1962); Waterfall in Rhodora 69: 213 (1967) pro
parte; Hudson in Solanaceae Biol. & Syst.: 416 (1986); Hepper in Fl. Egypt 6: 72
(1998); Gonçalves in F.Z. 8(4): 48 (2005). Type: cultivated "Jardin du Roi ", France,
originally from North America, 1784 (P-LAM, holo.; fiche LM 471/5!)

Annual herb < 0.5 m high; stems brownish-green or purple-tinged, sometimes
woody and/or angular, sparsely to moderately pilose with simple hairs, glabrescent.
Leaves membranaceous, green to greyish-green, lanceolate, ovate-lanceolate to
deltate, 2.8–5.6(–10) × 1.2–3.2(–5.2) cm, bases cuneate, often oblique, sometimes
decurrent, margins sinuate-dentate with 2–9 acute antrorse lobes, rarely sinuate,
apices acute, pilose (sometimes denser on veins, midribs and lower surfaces) to
glabrous, stipulate; petioles (0.6–)1.5–4.5 cm. Flowers solitary, axillary; pedicels
erect in flower and recurved in fruit, 3–8 mm long, sparsely pilose, 8–11 mm long.

Calyx cupulate/campanulate, 3–4(?–7) mm long with five broadly triangular acute lobes 1.5–3 × 1.6–2.4(–4) mm with ciliate margins, sparsely pilose to glabrous externally, enlarged and persistent in fruit with connivent lobes 4–5 × 5–6 mm. Corolla yellow to orange-yellow with pale or sepia basal spots or blotches and an internal ring of dense long eglandular hairs exposed by recurvature of throat, broadly campanulate, 0.8–1.5 cm long and 1.2–1.8 cm diameter, shortly pubescent externally, margin undulating, ciliate and entire or with 5 short broadly triangular obtuse lobes ± 3 × 4 mm. Stamens often unequal, exserted; filaments free for 3–5 mm; anthers purple and yellow, twisted and/or curved at maturity, oblong, bilobed, 2.8–4.4 × 0.9–1 mm, always exserted, maturing at different times and of different lengths within the same flower. Ovary pale green to brownish, ovoid, 1–2 × 1–2 mm, smooth, glabrous; style exserted up to 2.5 mm, 5.5–9 mm long; stigma 0.3–0.8 mm diameter. Fruit maturing to ?purple, globose, 0.9–1.1(–1.4) cm diameter, smooth, sessile on invaginated pedicel base, enclosed by enlarged and inflated reticulately-veined globose calyx which 2.2–2.5 cm diameter, the mouth closed by connivent lobes. Seeds 9–11(–16) per berry, yellow to orange, orbicular, 2.1–2.5 × 1.7–2.2 mm, compressed. Fig. 15/5–11, p. 72.

Kenya. Nakuru District: Ol Joro Orok, 1957, *van Someren* s.n.!; Fort Hall District: Makuyu, 26 June 1960, *van Someren* EA 11981!; Meru District: Meru near Teachers' Training Centre, 1951, *Hancock* 156!
Distr. **K** 3, 4; native to North America, sporadic weed in Sudan, Zambia, Zimbabwe and South Africa
Hab. Weed of gardens, fields and cultivation, especially of crops; occasional to common; 1700–2000 m
Conservation notes. Widespread; least concern (LC)

Syn. *P. angulata* Spreng., Syst. Veg. 1: 697 (1825) pro parte, based on *P. philadelphica* Lam., Type as above
    P. philadelphica Lam. var. *minor* Dunal in DC., Prodr. 13(1): 450 (1852). Type: Mexico, around Tampaco de Tamaulipas, *Berlandier* 20 (P, holo.)
    ?*P. ixocarpa* Hornem. sensu Agnew in U.K.W.F. 2nd ed. : 244 (1994)

Note. This species is commonly known as the tomate, miltomate or husk tomato; it has been in cultivation since pre-Colombian times and is still widely cultivated as a fruit, though it is rarely eaten raw. Wild and domesticated varieties have been recognised (viz. var. *philadelphica* and var. *domestica* – cf. Hudson, 1986) with morphological intermediates being common and the species exhibiting the wide variability characteristic of domesticated plants. The diagnostic features given by Agnew (1994) for *P. ixocarpa* suggest that the plants described belong to *P. philadelphica*.

4. **Physalis angulata** *L.*, Sp. Pl. 1: 183 (1753); Lam., Encycl. Méth. Bot. 2: 101 (1786); Dunal in DC., Prodr. 13(1): 448 (1852); C.H. Wright in F.T.A. 4, 2: 248 (1906); F.P.N.A. 2: 205 (1947); Waterfall in Rhodora 60: 162 (1958) & 69: 216 (1969); Heine in F.W.T.A. 2nd ed., 2: 329 (1963); U.K.W.F., 2nd ed.: 244 (1994); Hepper in Fl. Egypt 6: 41 (1998) & in Fl. Egypt 3: 41 (2002); Gonçalves in F.Z. 8(4): 50 (2005); Thulin in Fl. Somalia 3: 202 (2006); Friis in Fl. Eth. 5: 152 (2006). Type: "Habitat in India utraque", Herb. *Linnaeus* 247.9 (LINN!, lecto.) designated by Fernandes in Bol. Soc. Brot., Ser. 2, 44: 352 (1970) [rather than the different lectotype designated by D'Arcy in Woodson & Schery in Ann. Miss. Bot. Gard. 60: 662 (1974), fide Jarvis Order out of Chaos: 741 (2007)]

Annual herb up to 60 cm (rarely to 1 m) high; main stems erect, branched, greenish-brown or purple-tinged, sparsely pilose on young parts with short simple hairs, becoming glabrescent. Leaves soft, dull green, alternate, ovate-lanceolate to ovate, rarely linear-lanceolate, 4.8–8.4 × 2.2–5.3 cm, bases cuneate, often oblique, sometimes decurrent, margins irregularly serrate with antrorse acute, often deep teeth, rarely sinuate or sinuate-dentate, apices acute, sparsely pilose, denser on veins, midribs and lower surfaces, becoming glabrescent; petioles 1.5–5 cm. Flowers solitary, in leaf or branch axils; pedicels filiform, (4–)9–12 mm long, erect in flower, (7–)10–26 mm long, erect to recurved in fruit. Calyx cupulate, 3–4.5 mm long, shortly pilose, with 5 narrowly triangular acute lobes 1.2–2.5 × 0.6–1.5 mm, margins ciliate, enlarged, persistent and glabrescent in fruit with lobes 3–6.5 × 2–2.2 mm. Corolla yellow, sometimes with pale purple markings, campanulate, 0.6–1 × 0.5–1.2 cm diameter, unpleasantly aromatic, margin ciliate with five broadly triangular obtuse lobes 1–2 × 2.5–6 mm. Stamens sometimes visible in corolla throat; filaments free for 2.6–3.5 mm; anthers purple or yellow and purple, oblong, bilobed, 1.2–2 × 0.4–0.7 mm. Ovary brownish, ovoid, 1.5–2 × 1–2 mm, smooth, glabrous; disc ± 2 mm diameter; style sometimes exserted, 4.1–5 mm long; stigma 0.2–0.4 × 0.4–0.8 mm. Fruit pale green to yellow, usually globose but distinctly 5-angled, 0.7–1.1 cm diameter, smooth, subsessile or on inverted gynobase of enlarged and inflated reticulately-veined, glabrescent, pale green, ovoid calyx, 2.2–3 × 1.8–1.9 cm, the mouth closed by connivent lobes. Seeds yellow-orange, discoid to orbicular, 1.2–1.8 × 1.1–1.4 mm, compressed. Fig. 15/12–15, p. 72.

Tanzania: ?Lushoto District: Magoma in Luengera Valley, 2 Aug. 1978, *Archbold* 2389!; Pangani District: Mtaru, Mseko, Mwera, 7 June 1956, *Tanner* 2909!; Uzaramo District: Dar es Salaam, Buguruni Police Station, 21 July 1969, *Mwasumbi* 10575!; Zanzibar: km 27, Chiracke (?Chuaka), 9 Sept. 1961, *Faulkner* 2947!

Distr. **T** 3, 6, 8; **Z**; native to tropical America, now a common weed of temperate and tropical areas, including most of Africa

Hab. Ruderal of waste places, damp areas, swamp edges, riverine sand banks or weed of cultivated crops (e.g. rice); 90–1600 m

Conservation notes. Widespread; least concern (LC)

Syn. *P. minima* L., Sp. Pl.: 183 (1753); Lam., Encycl. Méth. Bot. 2: 102 (1786); U.K.W.F.: 529 (1974) & 2nd ed.: 244 (1994). Type: "Habitat in Indiae aridis sordidis", Hort. Cliff. n. 62, *Physalis* 6 (BM!, lecto.) designated by Edmonds in Jarvis, Order out of Chaos: 742 (2007) (see notes under *P. lagascae* and below)

  P. micrantha Link, Enum. Pl. Hort. Berol. 1: 181 (1821). Type: Sri Lanka, *Herb. Hermann* 3: 8 No. 97, (BM!, lecto.) designated by Hepper in Fl. Ceylon, 6: 393 (1987)

  ?*P. dubia* Link, Enum. Hort. Berol. 1: 181 (1821), *non* Gmel. No extant type according to Waterfall (p. 217, 1969)

  ?*P. divaricata* D. Don, Prodr. Fl. Nepal: 97 (1825); Dunal in DC., Prodr. 13(1): 444 (1852). Type: Nepal, Bassaria, *Hamilton* s.n. (?BM, holo., not found)

  P. abyssinica Nees in Linnaea 6: 448 (1831), *nom. nud.* based on Herb. Hort. Reg. Berlin, specimen not cited

  ?*P. linkiana* Nees in Linnaea 6: 471 (1831). Type: Brazil, Herb. Hort. Reg. Ber. under the names *Physalidis dubiae* & *Ph. novae* sp. (?B†, holo.)

  P. ciliata Sieb. & Zucc., Fl. Jap. Fam. Nat. 2: 22 (1846). Based on and with type as *P. angulata*

  ?*P. capsicifolia* Dunal in DC., Prodr. 13(1): 449 (1852). Type: Guyana, *Schomburgh* 226 (G-DC in h. Moricand 226, syn.); Martinique, *Sieber* 69 (P-Boiss., syn.); West Indies (P, syn.); Himalayas, *Edgeworth* 113 (P, syn.)

  P. hermannii Dunal in DC., Prodr. 13(1): 444 (1852). Type: India, cultivated Malabar & Sri Lanka, *Hermann* s.n. (?BM, syn.); Java, *Zollinger* 362; South Africa, Cape of Good Hope, *Drège* ?s.n. (G-DC, P, ?MPU, all syn.)

  P. angulata L. var. *capsicifolia* (Dunal) Griseb., in Syst. Veg. Karaiben, 96 (1857)

  ?*P. rydbergii* Rusby in Bull. New York Bot. Gard. 4: 423 (1907), *non* Fern. Type: Bolivia, *Bang* 2520 (?US, ?holo.)

  P. minima sensu Symon in J. Adelaide Bot. Gard. 3(2): 152 (1981) & 8: 18 (1985); Mansfeld, Encycl. Agr. & Hort. Crops: 1800 (2001), *non* L.

Note. This is often known as the "cut-leaved ground cherry"; the berries are eaten or made into preserves and pies and it is used as a vegetable in **T** 3. Specimens collected in Zanzibar (*Faulkner* 2937 & 3129) were considered by Brenan & Meikle in 1965 to be the first African specimens of true *P. minima* – which is now considered to be a synonym of *P. angulata*. Agnew (1994) included *P. minima* as a distinct species, but in his first edition (1974) he mentioned that it might only be a variety of *P. angulata*, distinguishable by being more hairy and having smaller fruits. In his second edition he recorded that "*P. minima*" occurred in **K** 3–5 up to 2170 m; the variation that he noted could be partly associated with higher altitude plants. No herbarium specimens of typical *P. angulata* from Kenya were seen during this revision.

18. **SOLANUM**

Solanum L. sections *Oliganthes, Melongena* and *Monodolichopus* by Maria S. Vorontsova
& Sandra Knapp[3]

L., Sp. Pl. 1: 184 (1753) & Gen. Pl. ed. 5: 85 (1754); Moench, Meth.: 473–476 (1794);
Dunal, Hist. *Solanum*: 1–248 (1813) & Synopsis: 5 (1816) & in DC., Prodr. 13(1):27
(1852); G. Don, Gen. Hist. Dichlam. Pl. 4: 400–442 (1837); Dunal in DC., Prodr.
13(1): 27–387 (1852); E. & P. Pf.: 21–25 (1895); Rev. Gen. Pl., 3: 224–228 (1898);
Dammer in E.J. 28: 473–477 (1901), 38: 176–195 (1906), 48: 236–260; 53: 325–352
(1915); Bitter in F.R. 10: 542–548 (1912) & 18: 301–307 (1922) & in Beheifte 16:
3–320 (1923) & in E.J. 49: 560–569 (1913), 54: 416–506 (1917) & 57: 248–286 (1921);
Heiser in Nightshades: 30–61 (1969); Hawkes & Edmonds in Fl. Europaea 3: 197–199
(1972); D'Arcy in Ann. Missouri Bot. Gard. 60(3): 680–760 (1973); Purdie, Symon &
Haegi, Fl. Austral. 29: 69–175 (1982); Symon in Journ. Adelaide Bot. Gard. 4: 1–367
(1981) & 8: 20–158 (1985); Knapp & Jarvis in J.L.S. 104: 325–367 (1990); D'Arcy &
Rakotozafy in Fl. Madagascar, 176: 37–134 (1994); Hepper in Fl. Egypt 6: 4–54 (1998);
Hunziker in Gen. Solanaceae: 270–315 (2001)

Lycopersicon Mill., Gard. Dict., abr. ed. 4 (1754)
Melongena Mill., Gard. Dict., abr. ed. 4 (1754)
Bassovia Aubl., Hist. Pl. Guiane, 1: 217, t. 85 (1755)
Dulcamara Moench, Meth.: 514 (1794)
Cyphomandra Sendtn. in Flora 28: 162 (1845)

Annual or perennial herbs, woody shrubs, lianas or trees; stems stoloniferous,
rhizomatous, tuberiferous or with gemmiferous roots; branches glabrous or with
simple, branched, stellate, dendritic or echinoid multicellular, eglandular or glandular-
headed hairs, sometimes with prickles. Leaves simple and entire to lobed or
compound, alternate or opposite, stipulate, sometimes with prickles; petiolate or
sessile. Inflorescences terminal, axillary, leaf-opposed or extra-axillary racemose or
paniculate cymes, 2–300+ flowered, rarely 1-flowered; pedunculate to epedunculate,
flowers sessile or pedicellate, pedicels sometimes articulate; flowers white to purple,
rarely yellow, often with basal translucent, green, yellow or purple star, actinomorphic
or zygomorphic, (4–)5(–6)-merous; pedicels erect to reflexed, glabrescent to
pubescent. Calyx cyathiform to campanulate, with 5(–10) triangular acute lobes often
prolonged through the calyx as prominent veins, slightly to fully accrescent and
persistent in fruit; sometimes with an annular thickening basally. Corolla campanulate-
rotate to stellate with short tube and shallowly to deeply lobed; lobes valvate in bud,
often spreading after anthesis. Stamens usually 5, equal or unequal; filaments joined to
corolla tube, glabrous to pubescent; anthers connivent, basifixed, exserted, dehiscing
by apical pores which often develop into short or long lateral slits. Ovary usually sessile,
ovoid or pyriform to globose, glabrous, bi- or tri-locular, ovules numerous; disc small or
absent, annular; style filiform, glabrous to pilose, often exserted; stigma capitate,
globose, sometimes bilobed. Fruit erect or drooping, dryish or sub-fleshy, globose to
obovoid berries, often depressed, 2- to 3- locular, smooth. Seeds few to many, reniform
to suborbicular, compressed laterally, rarely winged, rugose; sclerotic granules present
or absent. Type species: *Solanum nigrum* L.

Between 1000 and 2000 species, though the accepted number is now thought not to
exceed 1400. The genus is considered to be of New World origin, exhibiting its greatest
diversity in the Neotropics particularly in Central and South America; D'Arcy (in Hawkes *et
al.* (eds), Solanaceae III: 98 (1991)) suggested that over 500 *Solanum* species were endemic
to the New World.

---

[3] The Natural History Museum, Cromwell Road, London, SW7 5BD [*now of The Herbarium,
Royal Botanic Gardens, Kew]; species 26–45, 52–62, 65–66.

Solanum is one of the largest flowering plant genera and occurs in tropical and temperate regions throughout the world. The generic name is believed to be derived from the latin 'solamen' meaning comfort, and to allude to the reputed narcotic properties of the generic type *Solanum nigrum* L. *Solanum* species, which are often called Nightshades, exhibit considerable morphological diversity and a range of ecological preferences. Jaeger & Hepper (in D'Arcy (ed.), Solanaceae, Biology & Systematics: 44 (1986)) estimated that around 110 *Solanum* species occur in Africa and its adjacent islands, of which some 20% are probably the result of recent introductions – either deliberate or casual. Approximately 70 *Solanum* species including infra-specific taxa have been recognised in this treatment for Tropical East Africa, of which some are introductions and are only known under cultivation.

As with Solanums elsewhere, the African taxa are often extremely difficult to delimit; many exhibit considerable infra-specific variation and there is widespread confusion over identification and names. Both flowers and fruits are often necessary for accurate identification, together with notes on their respective colours and the fruit texture. In addition, notes on their habit and habitats are also useful. Often a combination of characters is required for definitive assessment. There are a number of species-complexes; they include the *S. anguivi* Lam. group and the *S. incanum* L. group – both belonging to the subgenus *Leptostemonum*, and the Black Nightshades – of the subgenus *Solanum* sect. *Solanum* which centres around the generic type *S. nigrum*.

Of the *Solanum* species found in the floral region, many are important food plants, such as the potato, *S. tuberosum* L.; the African eggplants, *S. aethiopicum* L., *S. macrocarpon* L., and *S. melongena* L.; and the African nightshades, *S. nigrum sensu lato*. Others are cultivated as ornamentals, either for their showy flowers such as the Jasmine Nightshade, *S. laxum* Spreng., or for their colorful berries, e.g., the Jerusalem cherry, *S. pseudocapsicum* L., while some constitute troublesome weeds of disturbed habitats. The genus is also a source of toxin - largely in the form of steroidal alkaloids such as solanidane; of medicinal drugs with *S. campylacanthum* L., for example, being one of the most widely used African medicinal species; and of drug precursors such as solasodine used in the production of corticosteroids and found in around 200 *Solanum* species (*cf.* Hawkes in Nee *et al.* (eds), Solanaceae IV: 5 (1999)). A number of species figure prominently in ethnobotanical practices throughout the FTEA region. Indeed, Bukenya & Carasco (in Nee: 345–360 (1999)) reported that most of the 27 *Solanum* species found in Uganda play important social and economic roles among local populations, being used for magic, spiritual rites, and fertility cults as well as for food, medicine, and ornamentals.

The genus has been the subject of innumerable treatments and revisions since it was first described in 1753. A useful summary of earlier subgeneric classifications of *Solanum* is given in Symon (in Journ. Adelaide Bot. Gard. 4 (1981)). Notable revisions include those of Dunal (Hist. *Solanum* (1813)) & in DC., Prodr. 13(1) (1852); D'Arcy (in Ann. Missouri Bot. Gard. 59: 262–278 (1972) & in Hawkes *et al.* Solanaceae III: 75–137 (1991) and Child & Lester (in van den Bergen *et al.* (eds), Solanaceae V: 39–60 (2001)). A number of new infrageneric taxa later proposed by Child (in F.R. 109: 5–6 & 407–427 (1998)) have since been synonymised by other authors. In 1991, D'Arcy subdivided the genus into seven subgenera containing approximately 62 sections; he then considered that the subgenus *Leptostemonum* (Dunal) Bitter displays its greatest diversity in the Americas but that diversification of this subgenus had occurred in Africa. He also suggested that minor centers of generic diversity had evolved in Africa, Madagascar and Macaronesia where the genus *Solanum* has evolved a number of distinctive sections many of which are endemic to these regions.

Within the last two decades, infra- and inter-generic groupings have been greatly influenced by molecular and cladistic studies. However, Olmstead & Bohs (in Spooner *et al.* (eds), Solanaceae VI: 255–268 (2006)) estimated that only around 30% of *Solanum* species had been analysed to 2006, though they noted that several major efforts to elucidate both the taxonomy and the molecular phylogeny of this genus world-wide are currently in progress. The most significant genera affected by molecular analyses have been *Cyphomandra* and *Lycopersicon* which are now accepted as belonging to the genus *Solanum* itself. Spooner *et al.* (in Amer. Journ. Bot. 80: 676–688 (1993)) used data from *chloroplast* DNA restriction site analysis to demonstrate that chemical, molecular and morphological data provided overwhelming evidence for the cladistic relationship of *Solanum* subg. *Potatoe* and *Lycopersicon*, and subsequently transferred all *Lycopersicon* epithets to *Solanum*. These authors further proposed the adoption of the epithet *Solanum lycopersicum* L. for the cultivated tomato. Subsequently, Bohs & Olmstead (in Nee *et al.* (eds), Solanaceae IV: 97–110 (1999)), using *cp*DNA to derive *ndh*F gene sequences confirmed earlier analyses which suggested that *Cyphomandra* should be included within *Solanum* and Bohs (in Taxon 44: 583–587 (1995)) later transferred the genus and all its species into *Solanum*. Bohs & Olmstead (in Syst. Bot. 22: 5–17 (1997)) then used chloroplast *ndh*F sequences to verify that both *Lycopersicon* and *Cyphomandra* nested within *Solanum*, while Olmstead & Palmer (in Syst. Bot., 22: 19–29 (1997)) showed that *cp*DNA restriction site variation strongly indicated that both *Lycopersicon* and *Cyphomandra* were derived from within *Solanum* and should be relegated to subgeneric or sectional status. Nevertheless several authors either retained these two genera as distinct (e.g. Hunziker in Genera Solanaceae, 2001) or argued that they should be retained for practical purposes (e.g. Nee in Solanaceae IV: 285–333 (1999)). Indeed, D'Arcy (in Hawkes *et al.* (eds), Solanaceae III: Taxonomy, Chemistry, Evolution: 81 (1991)) had already proposed that the conservation of the name *Lycopersicon esculentum* Mill. over other *Lycopersicon* names for the tomato by the 1987 International Botanical Congress could be construed as support for its separate generic status. Hunziker (2001) vehemently rejected the inclusion of both *Lycopersicon* and *Cyphomandra* within *Solanum*, arguing that though closely related, they are clearly distinguishable morphologically. He tabulated a number of characters by which *Solanum* could be delimited from *Cyphomandra*, and also considered that protoplasmic fusion experiments supported the recognition of *Lycopersicon* and *Solanum* as distinct genera. Nee (1999) considered that *Cyphomandra* should be placed in *Solanum* sect. *Pachyphylla* Dunal of the subgenus *Bassovia* (Aubl.) Bitter while *Lycopersicon* belongs to the sect. *Petota* of the subgenus *Solanum*. This FTEA treatment accepts the increasing molecular evidence supporting the inclusion *Cyphomandra* and *Lycopersicon* in *Solanum*, thereby following the F.Z. (but not the Fl. Eth.) *Solanum* account.

Many of *Solanum* species revisions have been regional and especially concerned with the New World components. The latter were tackled by Nee (1999) for example, while Symon (Journ. Adelaide Bot. Gard., 4: 1–367 (1981) & 8: 20–168 (1985)) revised those in Australia and New Guinea. Meanwhile Whalen (in Gentes Herb. 12(4) (1984)) produced an invaluable conspectus of species groups in the subgenus *Leptostemonum*. The most important revisions of African Solanums include those of Dammer (in E.J. 28: 473–477 (1901), 38: 176–195 (1906), 48: 236–260; 53: 325–352 (1915)); Bitter (in F.R. 10: 542–548 (1912) & in E.J. 49: 560–569 (1913), 54: 416–506 (1917) & 57: 248–286 (1921)); Jaeger (Systematic Studies in the genus *Solanum* in Africa (1985, Ph.D. thesis, ined.)) and Jaeger & Hepper (in D'Arcy (ed.), Solanaceae Biology & Systematics, 41–55 (1986)) all of which were largely based on traditional taxonomic characters. Both Dammer and Bitter adopted narrow species concepts and were 'splitters', usually describing any infraspecific variation formally as subspecies or varieties. Though their publications include extremely detailed species descriptions they were invariably based on limited herbarium material and many of the specimens on which their new taxa were based have since been destroyed or lost. For example, Bitter described many new *Solanum* species in his *Solana Africana* from specimens in the Berlin-Dahlem (B) herbarium which was destroyed during World War II. Other species, based on specimens in the Polish herbarium Wroclaw (WRSL), known as Breslau in Bitter's time, were similarly destroyed. Though duplicates of many of the species that he described from other geographical areas, particularly from Central and South America have been traced, this is not so for several of the African species especially those belonging to the section *Solanum*. Some of those that have been traced proved to be valid species, but many others are synonyms. The affinity of the remainder can be surmised to some extent from his extensive protologues, though in such variable and closely related species the synonymy of the taxa concerned is often only tentative. Unfortunately, Bitter died without completing his comprehensive monograph of African *Solanum* species. Jaeger's (1985) subsequent survey of the African species provided the latest comprehensive review of *Solanum* species found throughout Africa, but it remains unpublished. His supervisor Richard Lester was in the process of preparing Jaeger's work for publication before his untimely death in 2006, though there are plans to publish this account posthumously. The African subgenus Leptostemonum is currently being revised by Vorontsova and Knapp.

According to Jaeger & Hepper (1986) native *Solanum* species found in the FTEA region belong to the three sections *Afrosolanum* Bitter, *Bendirianum* Bitter and *Solanum* of the subgenus *Solanum*, and the seven sections *Anisantherum* Bitter, *Ischyracanthum* Bitter, *Melongena* (Mill.) Dunal, *Monodolichopus* Bitter, *Oliganthes* (Dunal) Bitter, *Somalanum* Bitter, and *Torva* Nees of the subgenus *Leptostemonum* (Dunal) Bitter. This latter subgenus is probably the most complex subdivision of *Solanum*, comprising around 33% of the species; recent molecular analyses (e.g. Levin *et al.* in Amer. Journ. Bot., 93: 157–169 (2006)) are beginning to yield interesting information on this group, though they have yet to clarify species relationships within it. Non-native species belonging to other subgenera and many other sections are also widely found throughout this area.

SUBGENUS 1. **BASSOVIA**

(Aubl.) Bitter in F.R. 17: 329 (1921)

Bassovia Aubl., Hist. Pl. Guiane, 1: 217 (1755)

Perennial erect or climbing herbs, rarely shrubs or small trees, never armed; usually glabrescent but occasionally with simple or irregularly branched but rarely stellate hairs; often unpleasantly scented. Leaves large, entire to imparipinnate. Inflorescences few-flowered axillary to extra-axillary cymes, rarely branched and elongated. Filaments short, glabrous; anthers tapering and with small terminal pores or oblong and opening by large terminal pores or slits. Styles sometimes heteromorphic. Fruits globose or conical berries; sclerotic granules sometimes present.

Sect. **PACHYPHYLLA** Dunal, Hist. Solan. 168: (1813) & in Synopsis: 7 (1816) & in DC., Prodr. 13(1): 28, 31 (1852); Don in Gen. Hist. Dichl. Pl., 4: 408 (1937).
Synonyms: *Cyphomandra* Sendtn. in Flora 28: 162 (1845); Dunal in DC., Prodr. 13(1): 387–402 (1852); G.P. 2: 889 (1876); E. & P. Pf.: 25 (1895); E.P.A.: 881 (1963); Bailey, Man. Cult. Plants: 870 (1966); Heiser in Nightshades: 111–115 (1969); D'Arcy in Ann. Missouri Bot. Gard. 60(3): 616 (1973); Gentry & Standley in Fieldiana 24(10): 35–37 (1974); Bohs in Econ. Bot. 43(2): 143–163 (1989); Purdie, Symon & Haegi, Fl. Austral. 29: 68 (1982); Child in F.R. 95 (5–6): 283–298 (1984); Symon in Journ. Adelaide Bot. Gard. 3(2): 141 (1981) & 8: 12 (1985); Troupin, Flora Rwanda 3: 364 (1985); Nee, Fl. Veracruz 49: 63 (1986); Huxley *et al.*, New RHS Dict. Gard. 1: 808 (1992); D'Arcy & Rakotozafy, Fl Madagascar: 19 (1994); Bohs, Fl. Neotrop. Mon. 63: 1–176 (1994); Hepper Fl. Egypt 6: 61 (1998); Hunziker in Gen. Solanaceae: 315–320 (2001); Mansfeld, Encycl. Ag. & Hort. Crops 4: 1834 (2001); Gonçalves in Fl. Cabo Verde 71: 27 (2002) & in F.Z. 8(4): 75 (2005); Friis in Fl. Eth. 5: 110 (2006)
Section *Cyphomandra* Child in F.R. 95 (5–6): 287 (1984)

Shrubs or small trees with a single trunk and large spreading crown, sometimes vines or climbers; glabrous to densely pubescent with simple or branched, glandular or eglandular hairs. Leaves often dimorphic and lobed or pinnate on same plant, coriaceous, often succulent, usually entire, often foetid. Inflorescences scorpioid or racemose cymes, usually pendent, terminal, branched or unbranched; flowers often with unpleasant smell, usually pendent; calyces cyathiform. Corollas campanulate, stellate to urceolate with recurved lobes. Anthers dehiscing by apical pores, with enlarged, curved and differentiated thickened connective bands along backs. Styles glabrous to pubescent. Fruits globose, pyriform or elongate berries with thick pericarp; large sclerosomes sometimes present.

Around 40 species usually found in Neotropical rainforests of Central and S America from Mexico to northern Argentina, with two main centres of diversity on Andean slopes of eastern Peru and Bolivia, and in SE Brazil. Species 1. *S. betaceum.*

SUBGENUS 2. **SOLANUM**

Plants small and herbaceous, occasionally perennial and woody; never armed; glabrous or pubescent with simple to branched or dendritic hairs, rarely with stellate hairs or scales. Leaves simple, entire to sinuate or sometimes pinnate or pinnatilobed. Inflorescences extra-axillary or terminal cymes, pedunculate to sessile, usually lateral, sometimes terminal when remaining so or overtopped by a side-branch; calyces usually with basal tube. Stamens often equal, sometimes anthers or filaments unequal; filaments usually joined to the top of the corolla tube, broader basally, tapering above; anthers oblong, connivent, free or connate, usually basifixed, opening by large terminal slits or pores which often slit laterally with maturity. Style pubescent or glabrous. Fruit fleshy berries, few to many -seeded, with or without sclerotic granules.

Sect. **DULCAMARA** (Moench.) Dumort., Fl. Belg.: 39 (1827)
Synonyms: *Dulcamara* Moench., Meth.: 514 (1794)
    Subsection *Dulcamara* Dunal in DC., Prodr. 13(1): 28 (1852)
    Section *Afrosolanum* Bitter in E.J. 54: 440 (1917)
    Section *Parasolanum* Child in F.R. 95: 142 (1984)

Shrubs, lianas or vines, occasionally sub-shrubs, erect, scandent, creeping, climbing or twining; glabrous or with pubescence of simple and branched (but never stellate) multicellular hairs and stalked glands. Leaves entire to basally trilobed or pinnatilobed, membranous or subcoriaceous, sometimes prominently veined, often with pubescent domatia and coiling petioles. Inflorescences usually terminal, sometimes lateral, many-flowered dichasial cymes but occasionally only simple and few-flowered; pedicels clustered in dense umbels on the main rachides, articulate basally. Corollas stellate with strongly recurved lobes which densely papillate externally. Anthers connate or free; filaments short and glabrous, sometimes articulated. Styles glabrous and exserted. Fruits orange to red, globose or ovoid berries with coriaceous pericarps, on erect to reflexed pedicels; seeds discoid to ovoid; sclerotic granules absent.

Approximately 20 species scattered throughout Eurasia, the Americas, especially in Andean regions, and Africa. Species 2. *S. nakurense*; 3. *S. terminale*; 4. *S. welwitschii*; 5. *S. seaforthianum* and 6. *S. laxum* (cultivated).

Sect. **BENDIRIANUM** Bitter in E.J. 54: 487 (1917); Gilli in F.R., 81: 434 (1970); D'Arcy in Ann. Missouri Bot. Gard. 59: 267, 274 (1972); Jaeger & Hepper in Solanaceae Biol. & Syst.: 45 (1986); Jaeger, Syst. studies *Solanum* in Africa: 283 (1985, ined.); Bukenya & Carasco in Bothalia 25(1): 50 (1995); Friis in Fl. Eth. 5: 115 (2006)

Climbing shrubs; glabrescent to densely pubescent with simple and dendritic (not stellate) hairs, often appearing mealy especially on young stems and the undersurfaces of the leaves. Leaves entire. Infloresences terminal, compoundly branched cymose lax inflorescences; pedicels not multi-clustered, but in groups of 2–3(–5). Corolla rotate-stellate. Filaments unequal with one longer than the other four. Styles heteromorphic.

Bitter considered this to be a monospecific section confined to NE Africa. Species 7. *S. runsoriense*.

Sect. **LYCOPERSICON** (Mill.) Wettstein in E. & P. Pfl., 4, 3b; 24 (1891); Bitter in E.J. 54: 500 (1917); Peralta, Spooner & Knapp in Syst. Bot. Monogr. 84 (2008)
Synonyms: *Lycopersicon* Mill., Gard. Dict., abr. ed. 4: 1754
Solanum subgenus *Lycopersicon* (Mill.) Seithe in E.J. 81: 204 (1962)
Section *Neolycopersicon* Correll in Contrib. Texas Res. Found., Bot. Stud. 4: 39 (1962)
Lycopersicon section *Neolycopersicon* (Correll) A. Child in F.R. 101: 224 (1990)

Annual, binnenial or perennial herbs or vines; stems glabrous to pubescent with simple uniseriate hairs which often glandular-headed. Leaves interrupted imparipinnate to asymmetrically pinnatifid in 2-3 sympodial units; leaflets with crenate, dentate or serrate margins. Inflorescences simple to 1–3+ branched cymes. Corollas stellate, yellow. Anthers strongly coalescent by interlocking hairs forming a tube with a sterile apical appendage, dehiscing by longitudinal slits. Fruits green, whitish, yellow, orange or red berries.

Comparatively few revisions have considered the species related to the tomato; since the relatively recent inclusion of *Lycopersicon* in the genus *Solanum* the majority of treatments have concentrated on molecular aspects and on the wild relatives of the domesticated species. However, the taxonomy of the wild tomatoes and their relatives was comprehensively revised by Peralta *et al.* (2008) using both morphological and molecular data. These authors concluded that the cultivated tomato together with 12 of its wild relatives should be allocated to the separate section *Lycopersicon.*

All species native to western S America with two endemic to the Galapagos Islands. Species 8. *S. lycopersicum* (cultivated)

Sect. **PETOTA** Dumort. Fl. Belg.: 39 (1827)
Synonyms: Subsection *Potatoe* G. Don, Gen. Hist. Dichl., 4: 400 (1837)
   Section *Potatoe* Dunal in DC., Prodr. 13(1): 28 (1852)
   Subsection *Tuberarium* Dunal in DC., Prodr. 13(1): 28, 31 (1852)
   Section *Tuberarium* (Dunal) Bitter in F.R. 10: 531 (1912)
   Subgenus *Potatoe* (G. Don) D'Arcy in Ann. Missouri Bot. Gard. 59: 272 (1973)

Annual to perennial herbs or climbers, often with stolons and tubers; plants often aromatic; pubescence of simple or branched but never stellate hairs, often with glandular heads. Leaves imparipinnate-pinnatifid with interstitial leaflets, rarely simple. Inflorescences terminal or extra-axillary racemose or paniculate cymes, often pendent; pedicels articulated well above base, not subtended by floral bracts. Corollas rotate, rarely stellate. Fruits globose, ovoid or conical green, yellow or orange berries.

Over 230 species, mainly found in the mountains from western North America to Chile with a few species in lowlands of SE Brazil and Argentina. Species belonging to the section *Petota* have been the most extensively studied in the genus. Their classification, revised over many years, was conveniently summarised by Nee (in Nee *et al.* (eds), Solanaceae IV: 300–305 (1999)). Major treatments include those by Correll (The Potato and its Wild Relatives (1961)) and Hawkes (The Potato: Evolution, Biodiversity and Genetic Resources (1990)) who considered all of the species, with more regional treatments including those by Hawkes & Hjerting for south eastern S America (The Potatoes of Argentina, Brazil, Paraguay and Uruguay (1969)) and for Bolivia and Peru by Ochoa (e.g. The Potatoes of South America: Bolivia (1990)). Species 9. *S. tuberosum* (cultivated).

Sect. **SOLANUM** Seithe in E.J. 81: 286 (1962); Edmonds in K.B. 27: 95–114 (1972) & in J.L.S. 75: 141–178 (1977) & 76: 27–51 (1978) & in Hawkes *et al.* (eds.), Biol. & Tax. Solanaceae: 529–548 (1979); Jardine & Edmonds in New Phytol. 73: 1259–1277 (1974); Edmonds & Glidewell in Plant Syst. Evol. 127: 277–291 (1977); Edmonds in Solanaceae Newsletter 2 : 23–28 (1984); Edmonds in J.L.S. 85: 153–167 (1982) & 87: 229–246 (1983) & 88: 237–251 (1984); Edmonds & Chweya, Black Nightshades: 1–113 (1997); Defelice in Weed Technology 17(2): 421–427 (2003); (Edmonds in) Gonçalves in F.Z. 8(4): 67 (2005); Edmonds in Fl. Som. 3: 207 (2006); Edmonds in Fl. Eth. 5: 115 (2006); Olet, Taxonomy *Solanum* sect. *Solanum* in Uganda (Ph.D Thesis, 2004); Manoko, Systematic Study African *Solanum* Sect. *Solanum* (Ph.D. Thesis, 2007)
   Synonyms: Section *Maurella* Dunal in Hist. Solan.: 119, 151 (1813) & in Synopsis: 12 (1816)
   Section *Morella* (Dunal) Dumort., Fl. Belg.: 39 (1827)
   Group *Morella sensu* Don, Gen. Hist. Dichlam. Pl. 4: 411 (1837)
   Subsect. *Morella - Morellae spuriae* Dunal in DC., Prodr. 13(1): 44 (1852) & *Morellae verae* in DC., Prodr. 13(1): 45 (1852)
   Subsection *Morella* Dunal in E. & P. Pf.: 22 (1895)
   Section *Morella* Dunal in Lowe, Man. Fl. Madeira 2: 72 (1872)
   Section *Morella* (Dunal) Bitter in E.J. 54: 493 (1916) & 55: 63–65 (1919)

Annual herbs, perennials or subshrubs, erect or scrambling; sparsely to densely pubescent with patent or appressed simple, uniseriate, multicellular, eglandular- or glandular-headed hairs interspersed with four-celled glands. Leaves entire to lobed.

Inflorescences pedunculate leaf-opposed or extra-axillary simple or forked cymes, condensed and umbellate or lax and appearing racemose, few-flowered; calyces cupulate-stellate to campanulate, shortly tubular with small lobes. Corollas stellate with spreading or recurved lobes. Filaments fused to top of short corolla tube, pilose to villous internally, rarely glabrous; anthers connivent. Style shortly pilose. Fruits small globose to ovoid berries, black, dark purple, red, orange, yellow or green with opaque or translucent cuticles; seeds small, numerous; sclerotic granules present or absent.

Around 40 species, distributed throughout the world from temperate to tropical regions and from sea-level to high altitudes, with centres of diversity in S and C America and some Old World endemics. Species 10. *S. nigrum*; 11. *S. scabrum*; 12. *S. villosum*. 13. *S. americanum*; 14. *S. tarderemotum*; 15. *S. florulentum*; 16. *S. pseudospinosum* and 17. *S. memphiticum*.

NOTE. Dunal (1813) recognised 12 species groups in the genus *Solanum* using a rigid hierarchical system but without mentioning any names for these infraspecific entities. D'Arcy (in Ann. Missouri Bot. Gard., 59: 263 (1972)) considered all of these groups (plus two added in a 1816 publication) to be of equal sectional rank; Knapp (in Taxon 32: 635 (1983)) later similarly concluded that all those preceded by the symbol § in Dunal's work were validly published at sectional rank. Thus, Dunal's use of the name *Maurella* for the species group containing species related to the black nightshade *Solanum nigrum* is the first reference to a sectional subdivision for these species. Dumortier (1827) also recognised this group at the sectional level, though his citation of *Morella* was presumably an orthographic error. Seithe (in E.J. 81(3): 286 (1962)) later corrected this sectional name to *Solanum* in compliance with the International Code of Nomenclature, since *Solanum nigrum* is the generic type species. Many of these species look superficially similar, especially in the herbarium where they are notoriously difficult to separate. The taxonomic difficulties are variously due to the weedy and invasive nature of these plants, their notorious morphological variability and phenotypic plasticity, the existence of a polyploid series with diploid, tetraploid, hexaploid and even octoploid species (Edmonds, 1977). The limited taxonomic value of various micromorphological characters investigated in this species group were discussed in Edmonds (1982, 1983 & 1984). Although some of species belonging to this section have been closely studied in the Europe, Asia, the Americas, Australia and New Zealand, the African species have as yet received little attention. Bitter (in F.R. 10: 542–548 (1912) & in E.J. 49: 560–569 (1913) & 54: 416–425, 493–495 (1917)) accepted four previously described species in Africa whilst also describing a further 12 new species. Some of the species found in the floral area are indigenous to that area or to Africa, while others have been introduced from Eurasia or the Americas. It is interesting that most of the indigenous African taxa are tetraploid (cf. Manoko *et al.*, Syst. Study African *Solanum* sect. *Solanum* chapters 3 and 4 (2007)).

Only selected synonyms of the Eurasian species have been included here. European literature on this species group is extensive, with most of the older regional floras describing numerous forms, varieties and subspecies, often of the same taxa under different names (cf. Filov in Kulturnaja Flora SSSR 20: 370–386 (1958) where 5 subspecies and 16 varieties of *S. nigrum* were recognised). Many of the references cited in this account give extensive lists of synonyms for some of the species included (e.g. Edmonds, 1972; D'Arcy in Ann. Missouri Bot. Gard. 60(3): 735 (1973); Henderson in Contrib. Queensland Herbarium 16: 1–79 (1974)). Apart from Flora accounts, there is a vast literature on these species, reflecting the variability inherent in the species-group and the difficulty of defining species boundaries, especially in the herbarium. Extensive bibliographies of work carried out on various species belonging to the section *Solanum* may be found in some of the papers cited.

Many of the species constitute important leafy vegetables in Africa, with some making vital contributions to local and rural economies; the leaves are usually cooked as a form of spinach, while the ripe berries are often eaten as a fruit. The many medicinal and culinary uses reported for species belonging to section *Solanum* in Africa are considered in Watt & Breyer-Brandwijk (The Medicinal & Poisonous Plants of S & E Africa, 1962) and in Edmonds & Chweya (1997). In addition, these authors also reported the various toxic effects attributed to these species, both to animals and humans. Though Watt & Breyer-Brandwijk (1962) mostly refer to *S. nigrum*, the illustration given is indicative of *S. retroflexum* Dunal which is found in southern rather than eastern Africa.

There has been considerable research on potentially useful chemical components inherent in the section *Solanum* species. These include the steroidal alkaloids diosgenin and tigogenin, together with steroidal sapogenin (cf. Carle, Pl. Syst. Evol. 138: 61–71 (1981);

Hunziker, Genera Solanaceae (2001)), from which steroidal hormones may be partially synthesised. The species also constitute important weeds in many countries of the world, with their berries often contaminating a wide range of crops resulting in reduced yields (cf. Weller & Phipps, Protection Ecology 1: 121–139 (1978/79)). A comprehensive review of the Black Nightshades, including their historic documentation, world distributions, medicinal and potential food usages, modes of distribution and spread, poisonous properties and effects, and damage caused as crop contaminants is given by Defelice (2003).

A key to these species is difficult to both construct and use partly due to overlapping ranges of many of the potentially distinguishing characters. During extensive work on these species, multiple duplicates of each accession were grown over many years, both in the field and the glasshouse. Herbarium material of these plants showed incredible variability in almost all of the characters used to distinguish the species (cf. Edmonds, 1977 & 1979; Jardine & Edmonds 1974). The species can only be identified using a suite of differentiating characters. Though cytological studies are necessary to determine the ploidy level of the species, stomatal and pollen grain diameters can also be of help in pinpointing probable ploidy levels of the taxa (*cf.* Edmonds, 1979; Edmonds & Chweya, 1997). When collecting specimens of species belonging to sect. *Solanum* flowers and ripe fruits should always be included together with a note of their colours.

Sect. **PSEUDOCAPSICUM** Moench, Meth.: 28, 476–477 (1794)
Synonyms: Section *Pseudocapsica* Roem. & Schult. in L., Syst. Veg. 4: 569, 584 (1819)
     Section *Pseudocapsicum* Dunal, Synopsis: 11 (1816) & in DC., Prodr. 13(1): 150 (1852)
     Section *Pseudocapsicum* sensu Bitter in E.J. 54: 497 (1917)

Herbs or shrubs or small trees; glabrous to pubescent with simple or branched hairs. Leaves entire. Inflorescences few-flowered usually axillary, occasionally leaf-opposed, cymes; peduncles short or vestigial; calyces campanulate with long narrow calyx lobes. Corollas stellate. Anthers oblong; filaments glabrous. Styles glabrous. Fruits globose, yellow, orange or scarlet berries; seeds flattened; sclerotic granules absent.

Approximately 15 neotropical species with a disjunct distribution and centres of speciation in Mexico, northern Argentina, Paraguay and Bolivia. Species 18. *S. pseudocapsicum* (cultivated).

Sect. **BREVANTHERUM** Seithe in E.J. 81(3): 297 (1962)
Synonyms: Section *Lepidotum* (Dunal) Seith in E.J. 81(3): 298 (1962)
     Section *Stellatigeminatum* Child in F.R. 109: 412 (1998)
     Subsection *Cliocarpus* (Miers) Child in F.R. 109: 413 (1998)

Shrubs or small trees, rarely climbing or scrambling; all parts densely pubescent/flocculose with hairs predominantly branched, dendritic, stellate or modified to peltate scales. Leaves entire. Inflorescences terminal becoming lateral, large and many-flowered corymbose cymes, multiply forked usually on long simple peduncles, coarctate; calyces campanulate. Corollas stellate to stellate-rotate, small. Anthers short and thick, non-connivent. Style usually pubescent, sometimes glabrous; ripe fruits globose yellow berries to blackish, sometimes sparsely pubescent.

Around 35 species found from southern USA to northern Argentina, with some introduced and becoming naturalised in the Old World. Species 19. *S. mauritianum.*

SUBGENUS 3. **LEPTOSTEMONUM**
(Dunal) Bitter in E.J. 55: 69 (1919)

Spinosa, grad. ambig. L, Sp. Plant.: 186 (1753)
Aculeata, grad. ambig. Dunal, Hist. *Solanum*: 125 (1813), *nomen nudum*
Section *Aculeata* (Dunal) G. Don, Gen. Syst. Dichl. Pl. 4: 423 (1838)
Section *Leptostemonum* Dunal in DC., Prodr. 13(1): 29 & 183 (1852)
Armatae, grad. ambig., C.H. Wright in F.T.A. 4, 2: 209 (1906)
Subgenus *Stellatipilum* Seithe in E.J. 81: 296 (1962)

Usually shrubs or perennial herbs, occasionally vines or small trees; mostly armed with broad-based recurved, acicular or subulate prickles; pubescence of sessile or stalked stellate or sometimes multangulate or echinoid hairs, which often glandular, simple or dendritic. Leaves entire, lobed or pinnatifid; often armed as stems. Inflorescences extra-axillary, simple to branched few- to many-flowered cymes; peduncles often short; flowers actinomorphic or zygomorphic, perfect or heterostylous and the plants andromonoecious or occasionally dioecious, 4–5(–6)-partite; calyces deeply lobed, with stellate pubescence or prickles externally, often accrescent. Stamens equal or unequal; filaments short and glabrous; anthers slender, free, sometimes connivent, tapering apically and dehiscing through small apical pores that occasionally lengthen to slits with age, sometimes heteromorphic. Ovaries 2-locular, sometimes 4-locular through development of secondary longitudinal septa; styles glabrous or with glandular pubescence, straight or curved. Fruits dry, succulent or fleshy berries with thin to thick pericarps, few- to many-seeded; sclerotic granules absent.

Descriptions of stellate trichomes in the MSV/SK treatments follow Roe (Taxon 20: 501–508, 1971). Trichomes with one layer of lateral rays and usually less than 10 rays arranged around the central midpoint are called "porrect", the trichomes with more than one layer of lateral rays (± 12–20 rays) are called "multangulate", and the trichomes with more than ± 20 rays protruding in all directions are called "echinoid".

Sect. **HERPOSOLANUM** Bitter in F.R. 11: 250 (1912)
Synonyms: Section *Aculeigerum* Seithe in E.J. 81(3): 291 (1962) pro parte (cf. Whalen in Gentes Herb. 12 (4): 210 (1984))
[*Solanum wendlandii* group sensu Whalen in Gentes Herb. 12 (4): 208 (1984)]

Lianas or scrambling vines; glabrous or with scattered simple and/or branched (never stellate) hairs; armed with small recurved and sometimes sparse prickles. Leaves entire, lobed or pinnate. Inflorescences terminal at first, pleiochasial multi-branched cymes; flowers dimorphic, andromonoecious. Corolla stellate to pentagonal. Stamens unequal with one filament longer than the other and sometimes with a larger anther. Fruits dull globose often red berries usually borne on a swollen pedicel.

6 species occurring from Mexico through C America to SE Brazil. Species 20. *S. wendlandii* (cultivated).

Sect. **GIGANTEIFORMA** (Bitter) Child in F.R. 109: 415 (1998); Welman in Bothalia 38(1): 39–47 (2008)
Synonyms: Section *Torvaria* (Dunal) Bitter series *Giganteiformia* Bitter in E.J. 57: 255 (1921)
[*Solanum giganteum* group sensu Whalen in Gentes Herb. 12 (4): 212 (1984)]

Moderate shrubs to small trees; leaves and inflorescences often aggregated terminally; white pubescence of sparse to dense small sessile or stalked stellate hairs, often appearing floccose or scurfy; often armed with short recurved prickles, which may be absent or replaced with bristles. Leaves usually entire, with sparse prickles or unarmed; densely floccose below but often glabrous above. Inflorescences many-flowered paniculate or corymbose cymes, usually branched occasionally unbranched when fewer-flowered, terminal becoming lateral; peduncles long, sometimes with small prickles; flowers small, 4- or 5-merous; flowers and pedicel scars closely spaced on floral axes; calyces campanulate. Corollas stellate with narrowly triangular to linear lobes. Anthers equal. Fruits small globose black to red berries, on erect pedicels.

8–10 indigenous species centred in tropical East Africa but with outlying species occurring in W and S Africa; they are found in montane forests and savannah habitats. Species 21. *S. giganteum*; 22. *S. goetzii*; 23. *S. schummanianum*; 24. *S. schliebenii* and 25. *S. tettense.*

Sect. **OLIGANTHES** (Dunal) Bitter in F.R. 16: 3 (1923).
Synonyms: *Oliganthes, grad. ambig.* Dunal in DC., Prodr. 13(1): 30, 282 (1852)
   Section *Oliganthes* (Dunal) Bitter series *Afroindica* Bitter in F.R. Beih. 16: 4 (1923)
   Section *Oliganthes* (Dunal) Bitter series *Aethiopica* Bitter in F.R. Beih. 16: 43 (1923)
   Section *Oliganthes* (Dunal) Bitter series *Austroafricana* Bitter in F.R. Beih. 16: 71
      (1923) pro parte (cf. Whalen in Gentes Herb. 12 (4): 220 (1984)
   Section *Oliganthes* (Dunal) Bitter series *Eoafra* Bitter in F.R. Beih. 16: 102 (1923)
      pro parte (cf. Whalen in Gentes Herb. 12 (4): 220 (1984)
   Section *Oliganthes* (Dunal) Bitter series *Capensiformia* Bitter in F.R. Beih. 16: 62
      (1923)
   [*Solanum anguivi* group sensu Whalen in Gentes Herb. 12 (4): 220 (1984)]

Small to medium perennial scramblers and shrubs, sometimes annual herbs (in cultivation); usually armed with recurved or acicular prickles but these occasionally absent; stellate pubescence of sessile or stalked hairs in which the median rays long or short. Leaves entire or lobed, often with prickles. Inflorescences lateral, usually racemose unbranched cymes, sometimes branched, usually few- but sometimes many-flowered, peduncles often short or vestigial, occasionally flowers solitary; pedicels and calyces often with small prickles; flowers usually hermaphrodite, occasionally heterostylous when the plants weakly andromonoecious. Corollas stellate with narrow lobes. Stamens equal, anthers equal and attenuate. Fruits globose, rarely ellipsoid, soft red to yellow berries with thin pericarp.

Around 40 species centred in Africa from where the section spreads into Arabia, India, S Asia and Malesia. Species 26. *S. aethiopicum*; 27. *S. agnewiorum*; 28. *S. anguivi*; 29. *S. cordatum*; 30. *S. cyaneopurpureum*; 31. *S. forskalii*; 32. *S. hastifolium*; 33. *S. inaequiradians*; 34. *S. lamprocarpum*; 35. *S. lanzae*; 36. *S. malindiense*; 37. *S. mauense*; 38. *S. polhillii*; 39. *S. ruvu*; 40. *S. setaceum*; 41. *S. stipitatostellatum*; 42. *S. taitense*; 43. *S. usambarense*; 44. *S. usaramense*; 45. *S. zanzibarense*[4]

NOTE. Many members of section *Oliganthes* are not easily classified as herbs, shrubs, or trees, and many species appear to be either climbing or erect depending on immediate habitat and plant age. Herbarium specimen data provided in this account is incomplete and further field observations are needed. The presence of immature fruit striping and mature fruit colour could be useful taxonomic characters but records are incomplete.

   Members of this section are usually not andromonoecious and most flowers are functionally hermaphrodite (i.e. all long-styled – unlike those in Section *Melongena*). This distinction is gradual however; the larger-fruited species such as *S. polhillii* and *S. agnewiorum* always exhibit some andromonoecy and distal functionally male flowers (short-styled) occur sporadically throughout the section.

   The species in this section have been circumscribed on the basis of broad morphological continuities and discontinuities following the study of several thousand collections in numerous herbaria. All intermediate specimens and morphological outliers are not listed, but we believe the species concepts presented here reflect the majority of morphological forms encountered and their distribution.

[4] Unpublished manuscripts by Roger Polhill and Richard Lester together with Peter Jaeger's (1985) doctoral thesis "Systematic studies in the genus *Solanum* in Africa" have contributed significantly to the treatment of the sections *Oliganthes*, *Melongena* and *Monodolichopus* and the authors (MSV & SK) would like to acknowledge their work.

Section *Oliganthes* is likely to represent several evolutionary lineages. Members of section *Oliganthes* as delimited here can be distinguished from section *Melongena* by their more slender habit and a greater number of smaller softer fruits which are orange to red.

Sect. **SOMALANUM** Bitter in E.J. 54: 500 (1917).
Synonym: [*Solanum jubae* group sensu Whalen in Gentes Herb. 12 (4): 227 (1984)]

Unarmed shrubs; dense pubescence of stellate hairs on all young parts, rays either equal or with a long median ray, usually eglandular, sometimes glandular. Leaves entire to sinuate, usually closely spaced on short lateral shoots. Inflorescences umbellate cymes, lateral or appearing to terminate the short shoots, 1–8-flowered; peduncle short or vestigial; pedicels slender; calyces campanulate. Corollas stellate. Anthers equal. Styles often curved apically. Berries red, globose, with thin pericarp.

A small section which Bitter (1917) considered to be composed of four dry-country species from NE Africa; Whalen (1984) later increased this to five species, found in dry scrublands, savannas and grasslands. Later authors have suggested that the species might be better placed in the sect. *Oliganthes*. Species 46. *S. jubae* and 47. *S. pampaninii.*

Sect. **ERYTHROTRICHUM** (Whalen) Child in F.R. 109: 419 (1998)
Synonym: [*Solanum erythrotrichum* group sensu Whalen in Gentes Herb. 12 (4): 239 (1984)]

Sprawling or erect shrubs; often strikingly reddish-pubescent with ferrugineous subsessile to stalked stellate hairs with long or short median rays; stems armed with flattened recurved often pubescent prickles . Leaves entire to deeply sinuate-dentate, armed with stright prickles; petioles often broadly winged from decurrent laminas. Inflorescences terminal becoming lateral, unbranched to multiply branched lax helicoid cymes; calyces cupulate/campanulate, often deeply lobed. Corollas deeply stellate with ovate to lanceolate lobes. Stamens equal or unequal; anthers connivent. Styles often heteromorphic. Fruits yellow, orange to blackish globose to depressed-globose pubescent berries, 4-locular; seeds numerous, round, not flattened.

Nee (in Nee *et al.* (eds) Solanaceae IV: 320 (1999)) considered this species group to be the most difficult for species delimitation in the subgenus *Leptostemonum*. Approximately 25 species in C America, the northern S American Andes and Brazil. Species 48. *S. robustum* (introduced).

Sect. **CRINITUM** (Whalen) Child in F.R. 109: 423 (1998).
Synonym: [*Solanum crinitum* group sensu Whalen in Gentes Herb. 12 (4): 246 (1984)]

Robust shrubs to tall trees; lanate/pubescent with multiseriate stalked-stellate and simple hairs when young; armed with deltoid recurved prickles or bristles which may become sparse or absent with maturity. Leaves entire to sinuate-dentate, often coriaceous; lower surfaces often lanate and with scattered prickles. Inflorescences subterminal to lateral, simple to many-forked lax monochasial cymes, few- to many-flowered with elongated rachides which eventually covered with prominent pedicel scars; peduncles, pedicels and rachides often setaceous; buds ovoid, densely clustered; flowers large and showy; calyces campanulate/cupulate, calyx lobes often with bristles. Corollas pentagonal-stellate, lobes shallow, usually connected by interlobe tissue. Stamens usually equal, occasionally unequal; filaments long; anthers long, attenuate, connivent; plants strongly andromonoecious with gynoecia reduced and sterile in many flowers. Fruits large, 4-locular, green to yellowish pubescent berries with leathery pericarp, sometimes becoming glossy and glabrous.

Around 8 species native to S America where the species extend from Venezuela to Peru, into Brazil and the Guianas. Species 49. *S. wrightii* (introduced/cultivated).

Sect. **ACANTOPHORA** Dunal, Hist. *Solanum*: 218 (1813)

Note. Dunal spelt the name of his section *Acantophora* in 1813, but changed it to *Acanthophora* in 1816; all subsequent authors have adopted the latter spelling.

Synonyms: Section *Acanthophora* Dunal in Synopsis: 41 (1816)

Section *Aculeata* (Dunal) G. Don subsect. *Acanthophora* (Dunal) G. Don, Gen. Syst., 4: 434 (1838)

Psilocarpa grad. ambig., Dunal in DC., Prodr. 13(1): 235(1852)

Section *Simplicipilum* Bitter in F.R. Beih. 16: 147 (1923)

[*Solanum mammosum* group sensu Whalen in Gentes Herb. 12 (4): 251 (1984)]

Herbs or small shrubs, sometimes sprawling; pubescence of simple and often glandular-headed hairs, occasionally stellate; armed with many straight or occasionally recurved acicular prickles. Leaves usually lobed or toothed, sometimes pinnate, with many prickles. Inflorescences lateral, simple or branched few- to many-flowered cymes, often with prickly axes; only the lower flowers setting fruit, the distal ones with reduced gynoecia. Corollas deeply stellate. Fruits globose glabrous marbled green berries becoming yellow to red, drying blackish with tough pericarp; seeds sometimes flattened with a surrounding wing.

Around 19 species found throughout tropical and subtropical America, showing their greatest diversity in the northern Andes and SE Brazil and with several species established in the Old World. Species 50. *S. mammosum* (introduced and cultivated); 51. *S. aculeatissimum* (probably introduced); 68. *S. capsicoides* (introduced but of doubtful occurrence in EA); and 69. *S. atropurpureum* (of doubtful occurrence in EA).

Sect. **MELONGENA** (Mill.) Dunal in Hist. *Solanum*: 208 (1813)

Synonyms: *Melongena* Mill., Gard. Dict., ed. 4, abr., 2 (1754)

Section *Aculeata* (Dunal) G. Don subsect. *Melongena* (Mill.) G. Don, Gen. Syst. 4: 43 (1838)

Section *Andromoecum* Bitter in F.R. Beih. 16: 157 (1923)

Section *Andromoecum* Bitter series *Acanthocalyx* Bitter in F.R. Beih. 16: 175 (1923)

Section *Andromoecum* Bitter series *Aculeastrum* Bitter in F.R. Beih. 16: 165 (1923)

Section *Andromoecum* Bitter series *Incaniformia* Bitter in F.R. Beih. 16: 201 (1923)

Section *Andromoecum* Bitter series *Macrocarpa* Bitter in F.R. Beih. 16: 186 (1923)

Section *Stellatipilum* Seithe in E.J. 81: 297 (1962) pro parte

[*Solanum incanum* group sensu Whalen in Gentes Herb. 12 (4): 261 (1984)]

Perennial shrubs or small trees, sometimes annual herbs in cultivation; pubescence of sessile or short-stalked stellate hairs with ± equal rays, often dense and felty; armed with straight or recurved prickles, these sometimes absent. Leaves entire to sinuate, often felty and with prickles. Inflorescences usually unbranched cymes, rarely forked, often minutely prickly, with single to a few hermaphrodite (long-styled) flowers basally and few to many smaller male (short-styled) flowers with reduced gynoecia distally; peduncles very short to vestigial with first flower inserted at or near inflorescence base; calyces and pedicels of hermaphrodite flowers usually with prickles, the pedicels enlarging in fruit when calyces often accrescent; pedicels and calyces of distal staminate flowers unarmed to sparsely prickly; flowers heterostylous, the plants andromonoecious. Corollas rotate-pentagonal to stellate-pentagonal with interpetalar tissue. Stamens equal, anthers attenuate. Fruits usually mottled green berries when young, becoming yellow to orange with fleshy, leathery or hard pericarp; seeds numerous.

Around 12–15 species centred in eastern Africa where found from Sudan to Tanzania, with some species occurring in tropical W Africa, and in S Africa and Madagascar; others are found in the Middle East and S Asia, where *S. melongena* itself was domesticated. Species 52. *S. aculeastrum*; 53. *S. campylacanthum*; 54. *S. dasyphyllum*; 55. *S. incanum*; 56. *S. lichtensteinii*; 57. *S. macrocarpon*; 58. *S. melongena*; 59. *S. nigriviolaceum*; 60. *S. phoxocarpum*; 61. *S. richardii* and 62. *S. thomsonii*.

NOTE. All members of this section exhibit strong andromonoecy (with the single exception of *S. thomsonii* where it is weak), with basal long-styled hermaphrodite flower(s) forming fruit and all other flowers short-styled, functionally male, and usually with smaller corollas. Section *Melongena* is likely to represent several evolutionary lineages. Members of section *Melongena* can be distinguished from section *Oliganthes* by their more robust habit and usually a single yellow large mature fruit.

Sect. **ISCHYRACANTHUM** Bitter in F.R. 16: 142 (1923)
Synonym: [*Solanum arundo* group sensu Whalen in Gentes Herb. 12: 263 (1984)]

Shrubs or trees; dense whitish tomentose indumentum of small sessile to shortly stalked and often closely interlaced stellate hairs especially when young, often becoming glabrous; stems with dense, flattened recurved prickles. Leaves entire; midribs and petioles with straight acicular prickles. Inflorescences simple to branched, few-many-flowered racemose or scorpioid cymes with long axes and short peduncles; calyces campanulate/cupulate with acicular prickles. Corolla flowers stellate, lobes narrowly triangular. Anthers equal, linear to attenuate; flowers often andromonoeious, the distal flowers with short styles, and the hermaphroditic flowers with long exserted styles; gynoecia reduced and sterile in distal flowers. Fruits yellow with tough pericarp; seeds numerous, brown, non-flattened.

A small distinct section of two or three species endemic to NE Africa. Species 62. *S. arundo* and 63. *S. dennekense.*

Sect. **MONODOLICHOPUS** Bitter in F.R. Beih., 16: 265 (1923)
Synonym: [*Solanum thruppii* group sensu Whalen in Gentes Herb. 12: 265 (1984)]

Armed shrubs, subshrubs or perennial herbs; pubescent to glabrescent with sessile or shortly stalked stellate and interlaced hairs, often farinose; prickles acicular and spreading or stout and recurved. Leaves undulate to deeply lobed, often with prickles on petioles, midribs and main veins. Inflorescences lateral, leaf-opposed to extra-axillary, unbranched monochasial cymes, 4–14-flowered; flowers hermaphrodite; calyces campanulate, calyx lobes triangular to linear, often with prickles and accrescent around developing fruits. Corollas pentagonal-stellate, often with extensive interlobear tissue. Stamens unequal, anthers oblong but not tapering, with lowermost filament longer than the other four. Style curved apically. Fruits globose yellow berries with pericarp drying and becoming chartaceous; seeds numerous.

Two species occurring from the Arabian Peninsula though Egypt to Tanzania which according to Whalen (1984) have no clear relatives among other Old World Solanums. The only zygomorphic-flowered African Solanums to exhibit unequal length filaments rather than unequal anthers. Species 65. *S. coagulans* and 66. *S. melastomoides.*

Sect. **ANISANTHERUM** Bitter in E.J. 54: 503–506 (1917)

Unarmed shrubs or woody herbs with dense pubescence of short stellate hairs often mixed with short simple glandular hairs at least on young parts, mature stems becoming glabrous with smooth reddish bark. Leaves entire, often coriaceous. Inflorescences terminal becoming lateral simple to forked lax cymes to 20-flowered; peduncles short becoming woody, rachides always with prominent abscission scars; flowers zygomorphic. Corollas stellate to rotate-stellate. Stamens unequal with one anther elongated and incurved over the stigma. Style sigmoidal, curved apically. Fruits red or orange glossy berries, pericarp sometimes translucent and occasionally with sparse stellate pubescence; seeds numerous.

Two disjunct species, one found in NE Africa and the other (*S. pubescens* Willd.) predominantly in India. Species 67. *S. somalense.*

KEY TO THE SPECIES

The species are keyed out below into the various sections of *Solanum* found in the FTEA region. However, the characters used apply to those found within the FTEA area and hence are not always consistent with those given in the sectional accounts above, which apply to all species occurring throughout their distributional areas. *Solanum* species are extremely difficult to key out due to the incredible morphological plasticity that the majority of species exhibit. The key will be the most effective for the most commonly encountered variants. For more accurate identification it is advisable to consider all available information including morphological and distributional data together with any population variation displayed within a populations, especially when collecting fresh material.

1. Leaves pinnate or imparipinnate, with at least some laminas composed of stalked leaflet pairs on distinct rachides, occasionally accompanied by some simple leaves; pubescence of simple or simple and branched, but never stellate hairs 2

   Leaves simple or lobed without distinct leaflets or rachides; pubescence of simple, branched or stellate hairs 6

2. Annual or perennial herbs, sparsely to densely pubescent, hairs simple or occasionally branched, often glandular; leaves prominently imparipinnate with long rachides and interstitial leaflets; pedicels articulated at or above the middle (Sect. *Petota* & *Lycopersicon*) 3

   Plants twining or climbing shrubs or vines, sparsely pubescent to glabrescent, hairs simple and eglandular; leaves partially or incompletely imparipinnate with upper leaflets sometimes confluent with shorter rachides or with entire mature leaves, interstitial leaflets usually absent; pedicels articulated at or just above the base 4

3. Plants without stolons or tubers, viscid-pubescent, strongly aromatic; flowers yellow; corollas stellate; anthers prominently connivent, fused into a tube by interlocking lateral and internal hairs, each terminating in a sterile conical beak (tomato) .... 8. *S. lycopersicum* (p. 121)

   Plants with stolons or tubers, sparsely to moderately pilose, not viscid or strongly aromatic; flowers white blue or violet; corollas rotate to rotate/pentagonal; anthers not fused or beaked (potato) .... 9. *S. tuberosum* (p. 123)

4. Plants with or without small recurved prickles; calyces 4–6 mm long; corollas 1.4–3.5 cm radius; anthers 7–10.5 mm long; seeds 1–1.2 mm long (Sect. *Herposolanum*) 20. *S. wendlandii* (p. 147)

   Plants always unarmed; calyces 1–2.5 mm long; corollas 0.7–1.4 cm radius; anthers 2.5–4.5 mm long; seeds 2.2–3.6 mm (Sect. *Dulcamara* pro parte) 5

5. Petioles straight; inflorescences often pendent, many(–50)-flowered; flowers purple to deep blue; corollas deeply lobed almost to base; styles glabrous; berries red; seeds pale yellow to cream, with distinct marginal rim and densely covered with silky strands of thickening . . . .   5. *S. seaforthianum* (p. 115)

   Petioles coiled or curved; inflorescences usually < 20-flowered; flowers white or tinged pale blue; corollas lobed for less than half their length; styles pilose basally; berries purple to black; seeds light to dark brown without raised margin or strands of thickening . . . .   6. *S. laxum* (p. 116)

6. Shrubs or trees often with large trunks and crowns; calyces and corollas fleshy; corollas cyathiform; anthers with broad connective forming gibbose thickened bands along backs; fruits usually more than twice as long as broad, on long pendulous pedicels 1.6–5.2 cm long (tree tomato) (Sect. *Pachyphylla*) . . . .   1. *S. betaceum* (p. 104)

   Annual or perennial herbs, climbers or shrubs, rarely trees; calyces and corollas not fleshy, corollas stellate, campanulate or cupulate; anthers without thickened connective; fruits globose to ovoid, usually less than twice as long as broad on erect to reflexed pedicels usually < 2.5 cm long 7

7. Plants with simple or simple and branched but never stellate hairs; always unarmed; anthers oblong to ellipsoid, never tapering above, opening by subapical introrse pores which become longitudinal slits 8

   Plants usually with stellate hairs, rarely with predominantly simple hairs (cf. *S. aculeatissimum*) or subglabrous (cf. *S. macrocarpon*); with or without prickles and/or bristles; anthers linear to bottle-shaped, sometimes oblong but always tapering above, opening by small terminal pores which rarely become longitudinal slits (often in sect. *Oliganthes*) 20

8. Annual or perennial herbs or subshrubs, rarely woody; inflorescences extra-axillary or leaf-opposed umbellate to racemose few-(1–30) flowered cymes; flowers < 10 mm radius; berry pericarps soft, succulent and often fleshy 9

   Lianas, vines, climbing or twining shrubs, occasionally woody herbs; inflorescences terminal to lateral simple to branched spicate to lax cymes or panicles, usually many(–100) but sometimes few-flowered; flowers 9–20 mm radius; berry pericarps coriaceous 17

9. Inflorescences < 3-flowered; peduncles
vestigial or very short (< 8 mm); calyx lobes
2.3–5 mm long; filaments and styles
glabrous; seeds 3–3.8 mm long (Sect.
Pseudocapsicum) . . . .   18. *S. pseudocapsicum* (p. 144)
Inflorescences usually > 3-flowered; peduncles
distinct and up to 5 cm long; calyx lobes
small, 0.4–2 mm long; filaments pilose to
villous internally (except in *S. memphiticum*);
styles shortly pilose; seeds 0.8–2.4 mm long
(Sect. *Solanum*) 10

10. Plants densely viscid with long (up to 1 mm)
glandular-headed hairs; flowering calyx
lobes obovate to spatulate; fruiting calyces
enlarged and conspicuously stellate with
triangular lobes 2.2–5 × 2–3 mm, adherent
to base of berries . . . .   17. *S. memphiticum* (p. 142)
Plants glabrescent to pubescent with short to
long appressed or spreading eglandular-
headed hairs; flowering calyx lobes ovate to
triangular; fruiting calyces only slightly
enlarged with broadly-ovate or -triangular
lobes, usually < 2 × 2 mm, adherent to or
reflexed from base of berries 11

11. Inflorescences condensed umbellate cymes,
4–7-flowered; fruiting rachides absent or if
present < 2 mm; anthers 0.8–1.5 mm long;
berries globose, purple or black with shiny
cuticles, usually borne on erect spreading
pedicels (but often reflexed in Floral
region); fruiting calyx lobes reflexed; seeds
usually < 1.4 mm long . . . .   13. *S. americanum* (p. 134)
Inflorescences umbellate to extended cymes;
fruiting rachides well defined, often long,
lax and extended, rarely short; anthers
usually > 1.5 mm long; berries globose to
ovoid, blackish, purple, yellow, orange, red
or greenish with dull cuticles on reflexed
pedicels; fruiting calyx lobes reflexed or
adherent; seeds usually > 1.4 mm long 12

12. Inflorescences umbellate to racemose; corolla
lobes triangular, as long as broad; calyx
lobes triangular; filaments 1.3–2.5 mm
long; fruiting pedicels often equal to or
longer than peduncles; berries obovoid,
red, orange or yellow . . . .   12. *S. villosum* (p. 131)
Inflorescences lax cymes; corolla lobes usually
ovate, longer than broad; calyx lobes ovate;
filaments always < 1.5 mm long; fruiting
pedicels shorter than peduncles; berries
globose to broadly ovoid, purple, blackish
or green 13

13. Leaves usually large and ovate, 8–12(–25) cm
long, margins usually entire occasionally
slightly sinuate; anthers purple to brownish,
sometimes yellow; berries always purple to
black, broadly ovoid, large, > 14 mm broad　　11. *S. scabrum* (p. 128)

Leaves usually < 8 cm long (except for *S.
tarderemotum* where can be up to 14 cm
long), ovate-lanceolate to lanceolate with
entire, sinuate or sinuate-dentate margins;
anthers yellow to orange, never purple or
brown; berries usually purple to green,
occasionally blackish (*S. nigrum*), globose to
ovoid, < 12 mm broad 14

14. Inflorescences simple cymes 3–6(–10)-flowered,
with fruiting rachides 2–10 mm long; corolla
lobes without a coloured midrib; berries
purple or blackish, falling from calyces when
fully mature leaving dried calyces and
pedicels on plants 15

Inflorescences simple or forked cymes to 32-
flowered with fruiting rachides to 14 mm
long; corolla lobes often with purple
midrib; berries yellowish-green, purple or
black, 4–6 mm diameter, falling still
attached to pedicels when mature 16

15. Plants conspicuously pilose/villous with long
spreading hairs; inflorescences lax cymes
3–5-flowered with angular fruiting rachides
(1–)5–10 mm long; corolla lobes narrowly
ovate or triangular, two to three times longer
than broad; styles geniculate and exserted to
2.5 mm; berries globose usually < 5 mm
diameter; fruiting calyces pentagonal with
acute triangular lobes; with 2–9 sclerotic
granules . . . . 16. *S. pseudospinosum* (p. 140)

Plants pilose to glabrescent with short
appressed hairs; inflorescences condensed
cymes 3–10-flowered with straight fruiting
rachides (0–)2–7 mm long; corolla lobes
broadly ovate, less than twice as long as
broad; styles straight and included; berries
broadly ovoid, 5–11 mm diameter; fruiting
calyx lobes ovate, acute or obtuse; sclerotic
granules absent . . . . 10. *S. nigrum* (p. 124)

16. Inflorescences lax cymes 5–10+-flowered;
filaments 0.5–1 mm long; fruiting calyx
lobes usually ovate; berries with up to 3
sclerotic granules . . . . 14. *S. tarderemotum* (p. 137)

Inflorescences forked (sometimes multiply)
or simple mixed with forked lax cymes,
10–32-flowered; filaments < 0.8 mm long,;
fruiting calyx pentagonal with broadly
triangular calyx lobes; berries with up to 6
sclerotic granules . . . . 15. *S. florulentum* (p. 139)

17. Plants with simple and complexely branched
    hairs; inflorescences terminating in lax
    umbellate fascicles of 2–3(–5) flowers; leaves
    often mealy beneath; corollas lobed for less
    than half length, semi-reflexed to lateral
    after anthesis; filaments often unequal with
    one longer, densely pubescent to glabrous;
    styles heteromorphic, 2.5–4.5 mm long
    when included or 6–9.5 mm long when
    exserted; berries black (Sect. *Bendirianum*)   7. *S. runsoriense* (p. 118)
    Plants without complexely branched dendritic
    hairs; leaves light to dark green never mealy;
    inflorescences usually terminating in
    compact clusters of many flowers; corollas
    often lobed almost to base, strongly reflexed
    to pedicels after anthesis; filaments equal,
    virtually glabrous; styles not heteromorphic,
    5–9 mm long; berries orange to red (Sect.
    Dulcamara pro parte) 18
18. Erect subshrubs; leaves usually < 4.5 × 2 cm;
    petioles not curved or coiled; inflorescences
    simple umbellate cymes, usually 2–5 cm long
    and lateral, 3–7(–11)-flowered; buds broadly
    ovoid; flowering calyces 2.5–5 mm long, with
    lobes 1.2–3.5(–4) mm long . . . .   2. *S. nakurense* (p. 106)
    Scandent, twining or climbing shrubs or
    lianas; leaves > 5 × 3 cm; petioles coiled or
    curved; inflorescences simple or multiply
    branched racemose cymes, 5.5–18 cm long
    and terminal, many (usually > 15)-flowered;
    buds obovoid; flowering calyces 1–2 mm
    long, with lobes 0.5–1 mm long 19
19. Inflorescences forked (often multiply)
    corymbose or paniculate cymes, usually <
    70-flowered; lobes broadly ovate, usually
    two or three times as long as broad; anthers
    free and often splayed (spreading) . . . . . .   3. *S. terminale* (p. 107)
    Inflorescences simple spicate cymes, often >
    100-flowered; lobes ligulate, more than four
    times longer than broad; anthers connate   4. *S.welwitschii* (p. 113)
20. Stamens conspicuously unequal, with *either* one
    long and four short filaments *or* one large
    and four small anthers within the same flower 21
    Stamens equal or subequal; only small, if any,
    differences in sizes between filaments or
    anthers within the same flower 23
21. Shrubs or woody herbs; prickles absent;
    filaments equal, 0.5–1.3 mm long; anthers
    unequal with one elongated, 7–10 mm long
    and curved apically over stigma, remaining
    four anthers 4.8–7.5 mm long; seeds yellow
    to light brown (Sect. *Anisanthrum*) . . . .   67. *S. somalense* (p. 223)
    Herbs (sometimes shrubs); armed with
    prickles; filaments unequal, the four
    shorter 0.5–1.5 mm long with the fifth
    1.8–7.2 mm; anthers equal, 4.5–7 mm long;
    seeds almost black (Sect. *Monodolichopus*) 22

22. Leaves (1.5–)5–14 × (0.8–)1.5–6.5 cm, with 3–6(–11) lobes on each side; long stamen filament 1.5–2 mm longer than others; fruiting calyx heavily armed, accrescent and covering the berry; common weed . . . .   65. *S. coagulans* (p. 220)

    Leaves (1.5–)2–4(–5) × 1.3–2.2(–3.2) cm, entire or with up to 3 lobes on each side; long stamen filament 4–6 mm longer than others; fruiting calyx unarmed, berries mostly exposed; **K** 1 . . . .   66. *S. melastomoides* (p. 222)

23. Large shrub or tree usually with thick trunk to 22 cm diameter; flowers showy, 2.5–4.4 cm radius, blue, purple and white within the same inflorescence; flowering calyx lobes narrowly triangular becoming subulate and strongly recurved; corollas pentagonal-stellate, shallowly lobed, lobes with dense central bands of small sessile stellate hairs separated by glabrescent interstices; anthers 10–16 mm long; calyx tube enlarging, thickening and becoming raised forming a woody rim around berry base (potato tree) (Sect. *Crinitum*) . . . .   49. *S. wrightii* (p. 193)

    Perennial herbs, shrubs or small trees; flowers usually < 2.5 cm radius (except for some sect. *Melongena* species), not showy as above, white, blue or purple but not varying within the same inflorescence or plant; flowering calyx lobes deltate to triangular but not subulate or strongly recurved; corollas deeply lobed, pubescent bands and interlobe tissue as above absent; anthers usually < 10 mm long; calyx tubes not enlarged as above 24

24. Plants with dense ferruginous pubescence of stalked intertwined stellate hairs mixed with brown sessile glands; armed with sharp yellow pyramidal prickles; leaves thick, soft, dark green above, yellow to fuscous below with distinct orange to brown midribs and veins; petioles decurrent and broadly winged, wings to 1.5 cm wide and often extending along internodes; berries rusty tomentose, covered with simple silky appressed hairs which sometimes slough off, fruits becoming black (Sect. *Erythrotrichum*) . . . .   48. *S. robustum* (p. 191)

    Plants glabrescent to densely stellate-pubescent, green, whitish, yellow or ochraceous but not ferruginous, stellate hairs not usually intertwined but if so hairs sessile (cf. *S. goetzei*), unarmed or armed with straight or recurved prickles or straight bristles; leaves membranaceous to coarse, lower leaf pubescence not as above; petioles decurrent or not, but never winged; berries glabrescent to stellate-pubescent usually red, orange or yellow, occasionally becoming blackish 25

25. Plants armed with stout recurved prickles on
    the stems *and* straight acicular prickles on
    the leaves and petioles of the same plant;
    berries hard and leathery with thickened
    (–2 mm) pericarp (Sect. *Ischyracanthum*) 26
    Plants unarmed or with *either* recurved *or*
    straight prickles, rarely mixed on the same
    plant (cf. *S. aculeatissimum*); berries often soft
    and mucilagineous, sometimes hard;
    pericarp thin or thick, not usually thickened 27
26. Leaves dark green, glabrescent and shiny
    above when fully mature, broadly ovate with
    repand or sinuate margins with 0–3 obtuse
    and often deep lobes, bases cuneate; leaf
    prickles to 19 mm long with those on the
    lower lamina often exceeding leaf margins;
    berry pericarp glabrescent ....   63. *S. arundo* (p. 216)
    Leaves light green above and whitish below,
    dull, both surfaces pubescent, often obovate
    with entire to sinuate margins, definite lobes
    lacking, bases usually cordate; leaf prickles to
    9 mm long, not exceeding the leaf margins;
    berry pericarp ± stellate-tomentose ....   64. *S. dennekense* (p. 219)
27. Inflorescences usually multiply forked
    terminal or subterminal pyramidal cymes,
    corymbs or panicles with long peduncles
    (up to 6 cm) and/or basal floral branches,
    many- (usually > 50–200)-flowered 28
    Inflorescences usually unbranched or branched
    only once, few (1–50)-flowered cymes,
    peduncles absent, vestigial or short (0–2 cm),
    rarely to 4 cm when inflorescences usually
    unbranched 32
28. Plants covered with distinct tomentose/floccose
    pubescence which always dense on young
    parts, floral branches and lower leaf surfaces
    which usually appear yellowish/green and
    whitish respectively 29
    Plants covered with mealy or ochraceous
    pubescence which dense on juvenile parts
    with lower leaf surfaces yellow,
    occasionally plants glabrescent (Sect.
    Giganteiforma pro parte) 30
29. Plants unarmed; leaf petioles subtended by two
    small sessile ovate auricular pseudo- stipules;
    peduncles > 10 cm; corolla lobes usually as
    long as broad; filaments 1–1.2 mm long;
    ovaries densely pubescent; berries scabrid;
    seeds 100–150 + per berry, < 2 mm long
    (Sect. *Brevantherum*) ....   19. *S. mauritianum* (p. 146)
    Plants armed with stout conical prickles;
    pseudostipules usually absent, occasionally
    present when elliptic and often single;
    peduncles < 6 cm; corolla lobes at least twice
    as long as broad; filaments 0.3–0.8 mm long;
    ovaries glabrous; berries glossy and
    glabrous; seeds < 40 per berry, 2.3–3.2 mm
    long (Sect. *Giganteiforma* pro parte) ......   21. *S. giganteum* (p. 148)

30. Plants armed with scattered to dense small robust prickles, occasionally unarmed; styles glabrous; ovaries glabrous except for a few stipitate glands towards stylar base ....   25. *S. tettense* (p. 158)

     Plants armed with narrow soft narrow bristles, occasionally unarmed; lower parts of styles stellate-pubescent; ovaries often with scattered hairs basally 31

31. Stems with unbranched glabrous bristles to 11.5 mm long, acute apically or terminating in stellate clusters of short stout rays, occasionally unarmed; leaf midribs and veins, peduncles and inflorescence forks with dense to sparse shorter bristles; leaves usually < 20 × 8 cm, leaf venation not prominent; seeds 3.5–5 mm long ....   23. *S. schummanianum* (p. 154)

     Stems with unbranched or forked pubescent bristles to 7 mm long with acute point; bristles absent from leaves, peduncles and inflorescence forks; leaves 20–40 × 8–14 cm with prominent venation on both surfaces; seeds 3–3.3 mm long ....   24. *S. schliebenii* (p. 157)

32. Shrubs; short shoots often present; prickles always absent; leaves entire, < 5.8 × 4 cm; inflorescences terminal, on lateral shoots or solitary, 1–8-flowered; peduncles absent or vestigial; styles with a dense pubescent collar basally around ovary junction (Sect. *Somalanum*) 33

     Shrubs, small trees or perennial herbs; short shoots absent; prickles usually present, occasionally absent; leaves entire to deeply lobed, usually > 6 × 4 cm; inflorescences with distinct peduncles, and/or rachides and branches, often many(–100)-flowered, occasionally few- or solitary-flowered (cf. sections *Oliganthes* and *Melongena*); styles without basal pubescent collars 34

33. Stems with short shoots and distinct right-angled lateral branches; leaves rough, yellowish; inflorescences up to 8-flowered umbellate cymes or 2–3-flowered or solitary; calyx 2–4(–5) mm long; corollas 8–14 mm radius with narrowly lanceolate acute lobes 5–11 mm long; seeds 2.1–3 mm long; **K** 1 wood/bushland ....   46. *S. jubae* (p. 186)

     Stems usually without short shoots and angular lateral branches; leaves membranaceous, greenish; inflorescences 1–2-flowered; calyx 6–13 mm long, corollas 1.5–2.9 cm radius with broadly ovate apiculate lobes 10–20 mm long; seeds 4–5 mm long; **K** 7 dunes and shores ....   47. *S. pampaninii* (p. 190)

34. Plants always unarmed; stems with dense
    indumentum of interlocking sessile stellate
    hairs; leaves always entire; inflorescences
    simple or simple and forked 5–10 flowered
    lax scorpioid cymes with 2–12 mm peduncles;
    anthers 3–4 mm; berries usually with thin
    translucent pericarp, borne on woody erect
    spreading pedicels; 8–26 seeds per berry,
    deeply reticulate (Sect. *Giganteiforma* pro
    parte) .... 22. *S. goetzei* (p. 151)
    Plants usually armed, occasionally unarmed
    (though always unarmed in *S. aethiopicum, S.
    lanzae, S. macrocarpon*); stem pubescence not
    as above; leaves sinuate to sinuate-dentate or
    entire in sect. *Oliganthes* & *Melongena*;
    inflorescences usually simple lax or
    umbellate cymes, occasionally forked once
    (or more in *S. usambarense*), epedunculate to
    pedunculate; anthers usually 4–11 mm long;
    berries usually opaque, not translucent,
    borne on pendent, recurved or reflexed
    pedicels, occasionally erect but not
    splayed/spreading; (10–30) > 100 seeds per
    berry, punctuate to shallowly reticulate 35
35. Plants conspicuously armed with acicular or
    recurved prickles (–1.5 cm long); young
    stems hispid, hirsute or villous, with simple,
    long (–2.5 mm) eglandular and shorter
    glandular-headed hairs mixed with stellate
    hairs and shortly-stalked club glands; leaves
    broadly ovate to palmatifid, as long as
    broad, deeply sinuate-dentate, the lobes
    often biserrate; flowering calyx lobes
    narrowly triangular to ligulate, always
    armed; styles always glabrous; fruiting
    pedicels stout, strongly recurved, to 3 cm
    long, often armed; seeds rounded not
    flattened (Sect. *Acantophora*) 36
    Plants either armed or unarmed, prickles
    often small and pyramidal; pubescence not
    as above, young stems densely stellate-
    pubescent to glabrescent, long simple hairs
    absent; leaves ovate to elliptic, sometimes
    lanceolate, entire to sinuate-dentate but
    lobes not biserrate (but secondary lobing
    common in *S. aculeastrum, S. dasyphyllum, S.
    thomsonii*); flowering calyx lobes deltate,
    unarmed or armed; styles usually stellate-
    pubescent for part of their length,
    occasionally glabrous; fruiting pedicels not
    usually as above; seeds flattened 38

36. Flowers deep blue or purple with distinct
    yellow star; anthers narrowly bottle-shaped,
    8.5–11 mm long; ovaries conical becoming
    beaked; fruits yellow to orange, obovoid
    berries 4.5–8.5 cm long × 3.5–6.5 cm,
    developing ± 5 large basal protruberances;
    seeds smooth, minutely reticulate to
    punctate . . . .   50. *S. mammosum* (p. 194)
    Flowers white, cream or pale purple with
    purple veining and lobe tips; anthers bottle-
    shaped to sagittate, 5.5–7 mm long: ovaries
    globose to ovoid; fruits *either* veined with
    light to dark green stripes when young
    becoming cream to yellow on maturity *or*
    orange to red, globose to ovoid, 1.2–4 cm
    diameter, basal protruberances absent;
    seeds reticulate-foveolate 37
37. Mature fruits cream to yellow; seeds rounded
    with a thickened margin, usually 2–3.2 mm
    long . . . .   51. *S. aculeatissimum* (p. 196)
    Mature fruits orange to red; seeds without a
    thickened margin bordered by a distinct
    pale wing, 4–5 mm long . . . .   68. *S. capsicoides* (p. 225)
38. Inflorescences with 1–60 long-styled flowers;
    infructescences with 1–40 red (yellow)
    mature berries 0.6–1.5(–5.9) cm diameter
    (Sect. *Oliganthes*) 39
    Inflorescences with 1–3(–6) long-styled flowers;
    infructescences with 1–2(–10) yellow mature
    berries 1.5–6 cm diameter (Sect. *Melongena*) 70
39. No prickles or bristles visible on stems or leaves 40
    Some prickles or bristles clearly visible on stems
    or leaves 48
40. Flowers white; either fruit large and edible
    with a soft pericarp or leaves glabrous and
    used as leaf vegetable; cultivated plants . . .   26. *S. aethiopicum* (p. 164)
    Flowers mauve to purple, sometimes white;
    fruit not large and edible; leaves always with
    stellate trichomes; wild plants 41
41. Leaves orbicular (ovate), usually as long as wide,
    0.7–1.5(–3) × 0.7–1.3(–3.5) cm, almost
    glabrous; flowers 1(–2) per inflorescence; **K** 1   29. *S. cordatum* (p. 171)
    Leaves ovate to obovate or lanceolate, 1.5–8
    times longer than wide, 1.2–25 × 0.6–17 cm,
    sparsely to densely pubescent; flowers 1–22
    per inflorescence 42
42. Leaves 4–8 times longer than wide, entire;
    petiole less than $\frac{1}{6}$ of the leaf blade
    length; corolla lobed for ± $\frac{1}{2}$ of its length   35. *S. lanzae* (p. 176)
    Leaves 1.5–3 times longer than wide, lobed or
    entire; petiole $\frac{1}{3}$–$\frac{1}{6}$ of the leaf blade
    length; corolla lobed for $\frac{2}{3}$–$\frac{5}{6}$ of its length 43
43. Fruit 13–20 mm in diameter, 1(–2) per
    infructescence; calyx 8–16 mm long, long-
    acuminate; 1800–2200 m, rare . . . .   38. *S. polhillii* (p. 178)
    Fruits 6–13 mm in diameter, 1–22 per
    infructescence, calyx 3–6 mm long, obtuse
    to acuminate 44

44. Calyx lobes oblong, the apices obtuse; 0–300 m;
     T 6, 8 . . . .   34. *S. lamprocarpum* (p. 176)
     Calyx lobes deltate (ovate or narrowly oblong),
     the apices acute to long-acuminate 45
45. Leaves 1.2–5.5 × 0.6–2.5 cm, entire to subentire 46
     Leaves (5–)6–25 × 3–17 cm, subentire to lobed 47
46. Leaf apex rounded; flowers 1–2(–3) per
     inflorescence; 0–1500 m . . . .   42. *S. taitense* (p. 183)
     Leaf apex acute or sometimes obtuse; flowers
     3–10 per inflorescence; inland areas
     800–1500 m . . . .   30. *S. cyaneopurpureum* (p. 172)
47. Erect shrub; flowers 5–22 per inflorescence;
     corolla 0.8–1.5 cm in diameter; fruits 6–9 mm
     in diameter; common above 1000 m . . . . .   28. *S. anguivi* (p. 167)
     Scandent shrub; flowers 3–10 per inflorescence;
     corolla 1.8–3 cm in diameter; fruits
     10–13 mm in diameter; 700–2000 m . . . . . .   41. *S. stipitatostellatum* (p. 182)
48. Young stems and inflorescences with numerous
     soft thin bristles (2.5–)3–6 mm long, less
     than 1 mm wide at base 49
     Young stems with hard prickles 0.2–4 mm
     wide at base 50
49. Plant stellate-pubescent; leaves ovate, 1.5–5(–7)
     × 0.7–2(–4) cm; flowers 1–3(–4) per
     inflorescence; *Acacia* bushland, 1000–1500 m   40. *S. setaceum* (p. 180)
     Plant almost glabrous; leaves elliptic 9–12 ×
     2.5–4.5 cm; flowers 10–15 per inflorescence;
     only known from Ruvu forest, **T** 6 . . . .   39. *S. ruvu* (p. 179)
50. Prickles on young branches straight (sometimes
     with a few gently curved prickles also visible) 51
     Prickles on young branches curved (sometimes
     with a few straight prickles also visible) 59
51. Calyx lobes oblong, the apices obtuse; 0–300 m;
     T 6, 8 . . . .   34. *S. lamprocarpum* (p. 176)
     Calyx lobes deltate (ovate or narrowly oblong),
     the apices acute to long-acuminate (obtuse) 52
52. Inflorescences branched more than once;
     flowers 20–60 per inflorescence; 1800–2200 m   43. *S. usambarense* (p. 184)
     Inflorescences simple or branched once;
     flowers 1–22 per inflorescence 53
53. Leaves orbicular (ovate), usually as long as
     wide, 0.7–1.5(–3) × 0.7–1.3(–3.5) cm, almost
     glabrous; flowers 1(–2) per inflorescence;
     K 1 . . . .   29. *S. cordatum* (p. 171)
     Leaves ovate to lanceolate, 1.5–4 times longer
     than wide, 1–25 × 1–17 cm, sparsely to
     densely pubescent; flowers 1–22 per
     inflorescence 54
54. Fruit 13–20 mm in diameter, 1(–2) per
     infructescence; calyx 8–16 mm long, long-
     acuminate; 1800–2200 m, rare . . . .   38. *S. polhillii* (p. 178)
     Fruit 6–14 mm in diameter, 1–22 per
     infructescence; calyx 2–6 mm long, acute to
     shortly acuminate 55

63. Corolla 1.2–2 cm in diameter; anthers 4–6.5 mm
    long; leaves 2–3 times longer than wide 64
    Corolla 1.8–3 cm in diameter; anthers 6–10 mm
    long; leaves ± 2 times longer than wide 65
64. Prickles 2–4 mm long; leaf margin usually
    lobed; fruit 8–14 mm in diameter; seeds
    1.8–2.5 mm long; 0–700 m near the coast   45. *S. zanzibarense* (p. 185)
    Prickles 1.5–2(–3) mm long; leaf margin
    usually entire; fruit 6–10 mm in diameter;
    seeds 2.5–3.5 mm long; inland areas
    800–1500 m . . . .   30. *S. cyaneopurpureum* (p. 172)
65. Leaves 6–13 × 3–7 cm, drying reddish;
    peduncle (2–)6–30 mm long; N and E
    Tanzania montane forest understorey,
    700–2000 m elevation . . . .   41. *S. stipitatostellatum* (p. 182)
    Leaves 3–8 × 1.5–4 cm, drying yellow-green or
    green-brown; peduncle 0–5(10) mm long;
    coastal lowlands . . . .   44. *S. usaramense* (p. 185)
66. Leaves orbicular (ovate), usually as long as
    wide, 0.7–1.5(–3) × 0.7–1.3(–3.5) cm,
    almost glabrous; flowers 1(–2) per
    inflorescence; **K** 1 . . . .   29. *S. cordatum* (p. 171)
    Leaves ovate to lanceolate, 1.5–4 times longer
    than wide, 1.2–25 × 0.6–17 cm, sparsely to
    densely pubescent; flowers 1–22 per
    inflorescence 67
67. Leaves 1.2–3(–3.5) × 0.6–1.3 cm; flowers
    1–2(–3) per inflorescence; 0–1500 m . . . .   42. *S. taitense* (p. 183)
    Leaves 4–25 × 0.8–17 cm; flowers 3–22 per
    inflorescence 68
68. Corolla 1.4–2 cm in diameter, lobes 6–10 mm
    long; anthers 5–8 mm long; berries 1–4 per
    infructescence; inland areas . . . .   32. *S. hastifolium* (p. 174)
    Corolla 0.8–1.5 cm in diameter, lobes 3–6 mm
    long; anthers 3.5–5 mm long; berries 6–22
    per infructescence 69
69. Leaves entire to subentire; prickles always
    hooked and flattened; trichomes with
    stalks 0.15–0.4 mm long; **K** 3, 4, 6, **T** 2;
    1800–3000 m . . . .   37. *S. mauense* (p. 178)
    Leaves lobed (subentire); prickles straight or
    curved, round or flattened; trichomes with
    stalks less than 0.1 mm; common above
    1000 m . . . .   28. *S. anguivi* (p. 167)
70. Petiole usually decurrent, leaf bases attenuate
    (cuneate); leaves with deep lobes and
    frequent secondary lobing; trichomes on
    abaxial leaf surface (if present) stalked with
    4(–5) rays 71
    Petiole never decurrent, leaf bases cordate to
    cuneate; leaves entire or with shallow to
    deep lobes, secondary lobing occasional;
    trichomes on abaxial leaf surface (if
    present) sessile or stalked with 6–16 rays 72

71. Plant clearly stellate-pubescent and armed,
    drying yellow-green to red-brown; wild plant;
    (0–)600–1600 m .... 54. *S. dasyphyllum* (p. 206)
    Plant usually glabrous and unarmed, drying
    a distinctive red-brown colour; cultivated
    plant .... 57. *S. macrocarpon* (p. 210)
72. Plant prostrate or climbing; inflorescence
    6–12 cm long; anthers of long-styled flowers
    7–11.5 mm 73
    Plant erect; inflorescence 2–8 cm long;
    anthers of long-styled flowers 4–9 mm 74
73. Prickles straight; pedicel of long-styled flower
    3–5.5 cm long; anthers of long-styled flowers
    7–9 mm long; fruits 2.5–3 cm diameter, 1 per
    infructescence; **K** 3–6; 2500–3000 m ..... 59. *S. nigriviolaceum* (p. 212)
    Prickles curved; pedicel of long-styled flower
    1–2.5 cm long; anthers of long-styled flowers
    8.5–11.5 mm long; fruits 3–5 cm diameter,
    2–6 per infructescence; Tanzania; 0–1300 m 61. *S. richardii* (p. 214)
74. Shrubs to trees 1–6 m tall; corolla lobed for
    more than $\frac{1}{2}$ of its length; young fruits plain
    green; wet habitats 1200–3000(–3200) m
    altitude 75
    Shrubs 0.2–2(–4) m tall; corolla lobed for less
    than $\frac{1}{2}$ of its length; young fruits striped in
    different shades of green; mostly arid
    habitats 0–2000 m altitude 77
75. Fruit globose, 1.4–1.7 cm in diameter, 4–10 per
    infructescence; leaf lobes deltate to oblong
    or obovate with frequent secondary lobing;
    young stems with yellowish long(1–3 mm)-
    stalked trichomes; **T** 7 .... 62. *S. thomsonii* (p. 215)
    Fruit globose or cylindrical, 2.8–5 × 1.8–4.5 cm,
    1–3(–5) per infructescence; leaves entire or
    leaf lobes rounded to broadly deltate,
    sometimes with secondary leaf lobing; young
    stems usually lacking long-stalked yellowish
    trichomes 76
76. Fruit distinctively cone-shaped, 2.8–3.7 ×
    1.8–2.2 cm; leaves on fertile branches
    elliptic and subentire, 6–8 × 2.5–4 cm, ± 2.5
    times longer than wide; 2100–3000 m .... 60. *S. phoxocarpum* (p. 213)
    Fruit globose, usually apiculate, 3–5 × 2–4.5 cm;
    leaves on fertile branches ovate (elliptic) and
    lobed (subentire), 8–15 × 6–12 cm, 1.5–2
    times longer than wide; 1200–2100(–3200) m 52. *S. aculeastrum* (p. 198)
77. Fruit with soft pericarp, in a variety of shapes
    in colours, edible; common fasciation in the
    flowers such as increase in the number of
    flower parts up to 8, inflated ovaries, and
    straight thick styles not exserted more than
    2 mm above the anthers; cultivated species
    (eggplant) .... 58. *S. melongena* (p. 211)
    Fruit spherical, yellow, with comparatively
    hard pericarp, not edible; flowers 5-merous;
    styles gently curved exserted 2–5 mm above
    the anthers; wild 78

1. **Solanum betaceum** *Cav.* in Anal. Hist. Nat. 1: 44 (1799) & Icon. 6: 15, t. 524 (1800); Aiton in Hort. Kew, ed. 2(1): 400 (1810); Dunal, Hist. *Solanum*: 169 (1813) & Synopsis: 7 (1816); Don in Gen. Hist. Dichl. Pl. 4: 108 (1837); Alliaume in Rev. Hort.: 150 (1880); Bohs in Taxon 44: 584 (1995); Nee in Solanaceae IV: 293 (1999); Gonçalves in F.Z. 8(4): 75 (2005). Type: cultivated in Hortus Madrid from seed of unknown origin in 1798–9, *Cavanilles* s.n. (MA, lecto. designated by Bohs in Flora Neotropica, 63: 51 (1994))

Shrub or small tree to 7 m high, rarely herbaceous, often malodorous; younger parts hispid, with a mixture of small spreading simple eglandular- and glandular-headed hairs, older parts glabrescent and often with small whitish spots. Leaves usually solitary and alternate, rarely opposite, simple, coriaceous, often dark green, often foetid when bruised, broadly ovate to cordate, 10.8–40 × 5.6–26 cm, bases usually cordate occasionally unequal sometimes auriculate with overlapping or clasping lobes, margins entire, apices acuminate, surfaces prominently veined and softly hispid with hairs denser on lower surfaces, midribs and veins; petioles 2.4–13 cm long. Inflorescences terminal pendulous lax long (–60 cm) cymes, simple, forked or branched arising from stem fork or leaf axil and often appearing axillary, (5–)10–50(–100)-flowered, up to 15 cm long; flowers often fragrant, actinomorphic; peduncles 1.5–2.2 cm long in flower and 1.8–5 cm in fruit; pedicels always pendulous, 6–16 mm long in flower, ± hispid, 1.6–5.2 cm long in fruit when woody and thickened, articulate at or near the base leaving rachis scars. Calyx fleshy, cyathiform, 2.5–4.5 mm long, sparsely hirsute with glandular hairs externally, with five broadly triangular acute or apiculate lobes 1–2.5 × 1.9–3 mm, slightly accrescent and persistent in fruit when 1.5–4 × 2.5–5 mm. Corolla white, pink or pale purple, stellate, fleshy, 1.3–2 cm diameter, tube 1.5–3 mm long; lobes narrowly triangular, usually spreading after anthesis with apices recurved, 7–11 × 2–4 mm, acute to acuminate, lobe margins densely pilose internally, otherwise glabrous. Stamens usually equal, connivent; filaments pale pink, free for 1.5–2 mm, glabrous; anthers bright yellow or orange, 4.5–6 × 1.3–2.9 mm, dehiscing by small apical pores, with broad connective forming darker gibbose thickened band (0.7–1.2 mm) along backs. Ovary greenish-white, 4 × 2–3 mm, glabrous, bilocular; style greenish-white, 5–6(–7) × 0.7–0.8 mm, glabrous, exserted up to 2 mm; stigma greenish-white, truncate to subcapitate, sometimes bilobed, 0.6–0.7 mm diameter. Fruit pendulous, sub-fleshy, orange to dark red or purple, dull, ovoid to ellipsoidal, (2.2–)5–10 × 2–5 cm, with acute or acuminate apex, pericarp thick and smooth, softly pubescent to glabrous, bilocular; fruiting calyx lobes adherent becoming reflexed basally. Seeds numerous, pale reddish-brown, discoid, (3–)4–4.2 × (2.5) 3–3.8 mm, rugose, shallowly reticulate-foveate with narrow winged margin; sclerotic granules sometimes present.

UGANDA. Toro District: S Kibale Forest, 9 Dec. 1938, *Loveridge* 207! & 14 Dec. 1938, *Loveridge* 243!; Mengo District: Mawokota near Mpigi Mpanga Forest, about 3 km W Mpigi, 27 Mar. 1951, *Dawkins* 721!

TANZANIA. Lushoto District: W Usambaras, Baga–Bumbuli road, NE of Sakarani, 6 May 1953, *Drummond & Hemsley* 2408! & Kitivo Village, 3 Oct. 1989, *Sigara* 297!; ?Mbeya District: near Tewi village in Bundali Hills, 6 Nov. 1966, *Gillett* 17583!

DISTR. U 2, 4; T 2, 3, 7; probably native to Bolivia or NW Argentina; cultivated, often naturalised in Ghana, Congo, Rwanda, Ethiopia, Angola, Zimbabwe, South Africa (Natal) and Madagascar; tropical and subtropical South and Central America, the West Indies, Spain, India, China, Papua New Guinea, Australia

HAB. Introduced and naturalised (following bird dispersal) on roadsides, drier and moist forest types, often in deep shade, clearings or on wet ground; 1050–2050 m

SYN. *Solanum crassifolium* Ortega, Nov. Pl. Descr. Dec. 9: 117 (1800), *nom. illegit.*, *non* Salisb. (1796) *nec* Lam.(1797). Type: Spain, cultivated at Madrid (no type extant fide Bohs in Fl. Neotropica, Mon. 43 (1994))

Cyphomandra betacea (Cav.) Sendtn. in Flora 28: 172, t. 6 (1845); Mart., Fl Bras. 10: 119 (1846); Dunal in DC., Prodr. 13(1): 393 (1852); Hooker, Bot. Mag. 125: t. 7682 (1899); E.P.A.: 881 (1963); Bailey, Man. Cult. Plants: 870 (1966); Purdie, Symon & Haegi, Fl. Austral. 29: 68 (1982); Symon in Journ. Adelaide Bot. Gard. 3(2): 142 (1981) & 8: 12 (1985); Child in F.R. 95: 289 (1984); Nee, Fl. Veracruz 49: 64 (1986); Bohs in Econ. Bot. 43: 143 (1989); Huxley *et al.*, New RHS Dict. Gard. 1: 808 (1992); D'Arcy & Rakotozafy, Fl. Madagascar, Solanaceae: 20 (1994); Bohs in Fl. Neotrop. Mon. 63: 50 (1994); Hepper in Fl. Egypt 6: 62 (1998); Gonçalves in Fl. Cabo Verde 71: 27 (2002)

Solanum insigne Lowe, Man. Fl. Madeira 2: 84 (1868). Type: cultivated at Caminho do Torriao, Funchal, Madeira, *Lowe* s.n. (BM!, lecto. designated by Bohs in Fl. Neotropica, Mon. 63: 51 (1994))

Cyphomandra crassifolia (Ortega) Kuntze, Rev. Gen. Pl., 3(2): 220 (1898); Huxley *et al.*, New RHS Dist. Gard., 1: 808 (1992), *orth. error.* Type as for *Solanum crassifolium*

NOTE. Commonly known as the tree tomato or tamarillo, the fruits are eaten raw, stewed or made into jam which has a pleasant fragrant taste, though the unripe fruits are considered slightly toxic. The fruits have a high vitamin content, with large amounts of ascorbic acid and vitamin D and they are very rich in carotene, making them good sources of pro-vitamin A. Despite the inflorescences bearing large numbers of flowers, only 1–5 fruits mature on each infructescence; these are predominantly bird-dispersed. The natural range and place of origin of *S. betaceum* remains conjectural (cf. Bohs, 1989). Truly wild populations are unknown though there are tentative reports of them occurring in Bolivia and NW Argentina; the plants are almost always associated with human habitation, with its possible depiction on pre-Colombian Peruvian pottery vessels suggesting that this species might have been domesticated by prehistoric inhabitants of the Andes (Bohs, 1989). This, however, has been disputed by some researchers who favour the relatively recent domestication of this species. It is now widely cultivated for both culinary and ornamental purposes in subtropical countries throughout the world including India, SE Asia, New Zealand and Australia as well as Africa.

Hunziker (Genera Solanaceae: 320 (2001)) summarized the alkaloid mixture isolated from '*Cyphomandra betacea*' roots, surmising that this seemed to be the first reported species of a plant with atropine-like alkaloids and edible fruits. Amines and amides have also been found in '*C. betacea*' (Bohs, 1989). Medicinal uses of this species include the use of warmed leaves in poultices for sore throats, fruit pulp in poultices for inflammed tonsils. The leaves have also been used as a source of dyes (cf. Bohs, 1989).

This is the only species of '*Cyphomandra*' found in the FTEA region where it is universally known as *C. betacea* (Cav.) Sendtn. This genus has largely been separated from *Solanum* on the basis of a thickened connective separating the two anther thecae – which appears as a thickened column on the backs of the anthers. Indeed the generic name is derived from the Greek *kyphos* meaning a tumour or curve and *andros* meaning male and referring to this curved thickened anther connective. The volatile perfumes produced in the epidermal cells of this connective attract pollinating male euglossine bees (cf. Hunziker, 2001).

2. **Solanum nakurense** *C.H. Wright* in K.B. 1897: 275 (1897); C.H. Wright in F.T.A. 4, 2: 219 (1906); Bitter in E.J. 54: 448 (1917); F.P.N.A. 2: 209 (1947); T.T.C.L.: 576 (1949); Polhill, *Solanum* in E & NE Africa: 9 (ined., 1961); E.P.A.: 873 (1963); Jaeger, Syst. studies *Solanum* in Africa: 276 (1985, ined.); Blundell, Wild Fl. E. Africa: 190 (1992): U.K.W.F, 2ⁿᵈ ed.: 242 (1994); Bukenya & Carasco in Bothalia 25(1): 49 (1995); Friis in Fl. Eth. 5: 114 (2006). Type: Kenya, Nakuru, *Scott Elliott* 6800 (K!, holo.; BM!, iso.)

Erect, scrambling, straggling, creeping or dwarf perennial herb or subshrub, 0.3–2.7 m tall, sometimes rhizomatous or arising from a woody rootstock; stems usually woody basally, often conspicuously lenticellate, pubescent with short predominantly simple but some branched (never stellate) hairs when young, becoming pilose with lignification, all young parts with brown stalked glands. Leaves usually alternate, occasionally opposite, usually coarse and coriaceous, dark green, obovate or ovate, sometimes lanceolate, 1.2–4.5(–6) × 0.6–2(–3) cm, bases cuneate and decurrent, margins entire, apices acute to obtuse, often appearing mealy, surfaces prominently strigose/pilose, denser on margins, veins, midribs and on lower surfaces with often densely pubescent domatia; petioles 0.3–1(–2) cm. Inflorescences terminal or lateral, leaf-opposed to extra-axillary umbellate cymes, usually simple, rarely branched when a basal pedicel arises from the peduncles below the main cluster, 2–5 cm long, (2–)3–7(–11)-flowered; peduncles erect in flower and in fruit, 1–3.6 cm long, pilose to glabrescent usually with some branched hairs, always with collar of dense short hairs around junction with pedicels; pedicels slender, 0.8–1.5 cm long and erect in flower, 0.6–2 cm and erect in fruit to semi-reflexed, pubescent as peduncles. Calyx cupulate, (2.5–)3–4(–5) mm long, pilose externally with short hairs; lobes triangular, 1.2–3.5(–4) × 1–2(–3) mm, acute, apiculate or extended-ligulate, often with apical tufts of short hairs; persistent and adherent becoming reflexed in fruit when 1–3 × 1.5–1.8 mm. Corolla purple or blue, occasionally white, stellate, 1.2–2(–2.6) cm diameter with tube 1–1.3 mm long, lobes broadly ovate, (3.5–)4.5–8 × (1.5–)2–3.5 mm, densely pilose/papillate with short hairs externally, glabrous internally, often inrolled with densely papillate margins, strongly reflexed to pedicels after anthesis. Stamens usually equal; filaments free for 0.5–1.5 mm, ± glabrous; anthers bright yellow often drying brown especially on the apical dehiscence margin, often unequal, 2.5–3.6(–4.5) × 0.8–1.5 mm, free and often spreading. Ovary green, 1–1.4 × 0.7–1 mm, glabrous; style geniculate becoming straight, 5–9 × 0.2–0.4 mm, always exserted up to 4 mm, glabrous; stigma capitate, often bilobed, 0.3–0.6 mm diameter. Fruit erect to semi-reflexed, red, usually globose, occasionally ovoid, 4.5–10 mm broad and 4–8 mm long, with shiny smooth coriaceous pericarp. Seeds 5–9 per berry, yellow, ovoid to orbicular, 2–2.5 × 1.8–2 mm, foveolate; sclerotic granules absent, rarely one miniscule present. Fig. 16/10 & 11, p. 109.

UGANDA. Mt Elgon, Namasindwa, 25 May 1924, *Snowden* 889/a!
KENYA. Elgeyo District: Kapsowar, Kipkunurr Forest Reserve, E end of Cherangani Hills, 14 Apr. 1975, *Hepper & Field* 4976!; Nakuru/Masai District: Mau summit, (without date details) 1930, *Mettam* 178!; Kiambu District: Muguga, 21 km Naivasha–Nairobi, 6 June 1952, *Verdcourt* 653!
TANZANIA. Masai District: Ngorongoro Conservation Area, Oldeani Mountain, 1 Jan. 1989, *Pócs & Chuwa* 89002/C!; Njombe District: Elton Plateau, Ndumbi River bank, 11 Jan. 1957, *Richards* 7692! & Kitulo Plateau above Matamba, Ndumbi River bank, 22 Nov. 1986, *Brummitt & Congdon* 18112!
DISTR. U 3; **K** 2–6; **T** 2, 3 & 7; Ethiopia
HAB. On poor soils such as in roadsides, heathlands, plateau grassland, volcanic rocks, bare ground, plantation margins, mixed *Acacia* scrub, also found in wetter and often shady habitats including grassland, stream-banks, drier forest with *Juniperus, Olea, Hagenia* and *Rapanea*, woodland, thickets and *Acacia lahai* woodland; 700–3050 m

SYN. *S. lykipiense* C.H. Wright in F.T.A. 4, 2: 220 (1906). Type: Kenya, Laikipia [Lykipia], *Thomson* s.n. (K!, holo.)
  S. mangaschae Pax in E.J. 39: 648 (1907); Bitter in E.J. 54: 451 (1917); E.P.A.: 871 (1963). Type: Ethiopia, Amaniel at the Gazenit river in Damot, *Rosen* s.n. (WRSL, holo., not traced)

S. aculeolatum Dammer in E.J. 48: 237 (1912). Type: Kenya, Masai highlands escarpment, locality not cited, *F. Thomas* s.n. (?B†, holo.; E (photo!), iso.), *non* Martens & Galeotti
S. massaiense Bitter in F.R. 11: 18 (1912). *nom. nov.* for, and type as for, *S. aculeolatum*
S. penduliflorum Dammer in E.J. 48: 255 (1912). Type: Kenya, Masai highlands, Mau Plateau, Molo, Ravine, Sandiani and Ujoro, *S. Baker* 133 (?B†, holo.)
S. stolzii Dammer in E.J. 53: 327 (1915); Bitter in E.J. 54: 450 (1917); T.T.C.L.: 576 (1949). Type: Tanzania, Rungwe District: Kyimbila, Rungwe Crater Lake, *Stolz* 1035 (B!, holo.; K!, M (photo!) iso.)
S. nakurense C.H. Wright var. *lykipiense* (C.H. Wright) Bitter in E.J. 54: 449 (1917)
S. terminale sensu Agnew in U.W.F.K: 526, fig. (1974), *non* Forssk.

NOTE. Bitter (1917) thought the climbing or twining African species, loosely associated with Dunal's (1852) subsection *Dulcamara*, formed a distinct group and described them as the new section *Afrosolanum* Bitter, which he separated into the two series, *Nakurensia* Bitter and *Bifurca* Bitter. He retained the section *Dulcamara* (Dunal) Bitter for the Eurasian taxa closely related to *S. dulcamara* sensu stricto. Although Dammer (in E.J. 38 (1906) & 48 (1912) & 53 (1915) and in Z.A.E. 2 (1914)) had already described many new species associated with this species group, Bitter (1917 & in F.R. 18 (1922)), recognised more than 24 new taxa plus many infraspecific taxa in his new section. Most of these were based on minor vegetative characters, such as leaf variation and topological hair morphology, characters which are notoriously variable in *Solanum* species and usually of little value in species delimitation. Heine (in K.B. 14 (1960)) thought that his excessive splitting of these species was partly due to his revision being based on Berlin herbarium material and his inability to consult that from other European herbaria. Bitter (1917) allocated *S. nakurense* together with *S. stolzii* and *S. mangaschae* to his series *Nakurensia*.

Though often confused with *S. terminale*, typical *S. nakurense* is a low shrubby non-scandent small-leaved species usually with simple few-flowered inflorescences. However it is largely confined to tropical East Africa with the few specimens found in Ethiopia being small individuals and with some being morphologically intermediate with *S. terminale* (e.g. *Purseglove* 3025 (**U** 2)). A few specimens of this species have been decribed as either being parasitic or epiphytic in that they were found growing on tree trunks in **K** 2, **T** 2 and **T** 7. Sepal shapes seem to be extremely variable in this taxon, sometimes varying from broadly triangular to apiculate to ligulate often between inflorescences on the same plant. Very few specimens are fruiting, and those that exhibit mature berries contain few seeds, perhaps indicating that the plants are only partially fertile. It is browsed by all domestic animals and especially by goats (**K** 3 & 6).

The protologue of *S. aculeolatum* Dammer is brief and fails to mention any inflorescence details. Though Dammer did not give a locality for the holotype of this species, Bitter later gave it the new name *S. massaiense* so it is possible that he saw Thomas' specimen in the Berlin herbarium. He later cited both specific names as synonyms of *S. nakurense* (Bitter 1917). The terminal few-flowered inflorescence and floral dimensions visible on the Edinburgh specimen are characteristic of those found in *S. nakurense*.

The protologue of *S. mangaschae* points to the synonymy of this species with *S. nakurense*. However, Bitter (1917) later described the inflorescences as being terminal, forked and 20-flowered, which would indicate more affinity with *S. terminale*. Unfortunately his description did not include any calyx dimensions. The resolution of the correct placement of this name is dependent on the discovery of duplicate type material. The locality of Rosen's type was not cited by Pax in his very general protologue of this species, though Bitter later added WRSL when he redescribed it in 1917.

These is little doubt from the extant holotype that *S. stolzii* is synonymous with *S. nakurense*. Agnew's (1994) illustration of *S. terminale* clearly portrays *S. nakurense*. However, his descriptions confirm that both species are locally common in upland Kenya, with *S. nakurense* occurring in evergreen upland bushland, and *S. terminale* favouring wet lowland forest edges.

3. **Solanum terminale** *Forssk.*, Fl. Aegypt.-Arab.: 45 (1775); Dunal, Synopsis: 17 (1816); G. Don, Gen. Hist. Dichlam. Pl.: 415 (1837); Dunal in DC., Prodr. 13(1): 105 (1852); Bitter in F.R. 18: 301 (1922); Heine in K.B. 14: 247 (1960); Polhill, *Solanum in E & NE Africa*: 9 (ined., 1961); F.F.N.R.: 436 (1962); Heine in F.W.T.A. 2nd ed., 2:331 (1963); E.P.A.: 879 (1963): Fl. **Rwanda** 3: 382 (1985); Jaeger, Syst. studies *Solanum* in Africa: 277 (1985, ined.); Abedin *et al.* in Pak. J. Bot., 23: 273 (1991); Hepper & Friis, Pl. Forssk. Fl. Aegypt.-Arab.: 233 (1994); U.W.F.K., 2nd ed.: 242 (1994); K.T.S.L.: 583 (1994); D'Arcy & Ratotozafy in Fl. Madagascar, Solanaceae: 126

(1994); Bukenya & Carasco in Bothalia 25(1): 49 (1995); Hepper in Fl. Egypt, Solanaceae, Family 159: 45 (1998); Gonçalves in F.Z. 8(4): 77 (2005); Friis in Fl. Eth. 5: 114 (2006). Type: Yemen, ? Mukham [Mochham fide Forsskål (1775) but Mokhaja is given as the locality on the original label which, according to Hepper & Friis (1994) refers to Mukhajah, Jebel Barad], *Forsskål* 406 (C!, syn; microfiche 102: II, 5,6 !); 419 (C!, syn; microfiche 102: II, 3,4 !).

Climbing or straggling shrub, perennial vine or liana, 1–7 m, usually scandent over dominant vegetation, sometimes with woody rootstock or rhizomatous; stems woody basally, sometimes herbaceous and slender, often prominently lenticellate, pubescent with mixture of simple and occasionally branched (never stellate) hairs when young, often becoming glabrescent; branches often pendent. Leaves usually alternate, membranaceous to coriaceous, light to dark green, ovate to lanceolate, rarely linear-lanceolate, 4.8–9.5(–12) × (1.7–)2.2–7.3 cm, becoming smaller towards stem apices, bases cordate to cuneate and decurrent, margins entire, apices acute to acuminate; venation often prominent, surfaces pilose to glabrescent, hairs as on stems, denser and more branched on veins, midribs and on lower surfaces, pubescent domatia usually present; petioles 0.8–3(–5) cm long, often curved to facilitate climbing. Inflorescences terminal, forked, often multiply so, racemose or corymbose cymes with umbellate fascicles of flowers, either broadening towards the apex with short central rachides or narrowing towards the apex with long central rachides, sometimes pendent, sometimes sweetly fragrant, 4–16(–25) × 4–10(–15) cm, 15–70(> 100)-flowered; peduncles (1–)2–7(–9) cm long though sometimes appearing as a continuation of the main stem; pedicels in umbellate clusters but sometimes lateral, 5–12(–19) mm long in flower, 10–22 mm in fruit; inflorescence axes pilose to glabrescent, rarely villous, hairs short or long, simple and occasionally branched; fuscous papillate collar around junctions with branches or pedicel bases. Calyx cupulate, 1–2(–3) mm long, pilose to glabrescent externally with short hairs; calyx lobes broadly triangular, 0.5–1.5(–2) × 0.9–1.8 mm, acute to apiculate, usually with apical tufts of hairs; reflexed becoming adherent in fruit when 1–2(–3) × 1.2–2.5 mm. Corolla white, whitish-green, cream, mauve or blue with yellowish basal star, stellate, 1–2(–2.8) cm diameter, tube 1–1.6 mm long; lobes broadly ovate to lanceolate, rarely ligulate, 4.5–9 × 1.5–3 mm, densely pilose/papillate externally with short often branched hairs, glabrous internally, strongly reflexed after anthesis. Stamens usually equal; filaments free for 0.3–1.3 mm, glabrous; anthers yellow to orange-yellow, 2.5–3.5(–4) × 0.9–1.2(–1.6) mm, papillate or not internally, free and often spreading. Ovary whitish, 0.9–2.5 × 0.8–2.2 mm, glabrous; style geniculate becoming straight, white to green, often twisted, 5–9 mm, glabrous, exserted 2–3(–4) mm; stigma pale green, capitate, 0.3–0.5 mm diameter. Fruit smooth, orange to red, globose to ovoid with coriaceous pericarp, 5–10 × 5–8 mm. Seeds 3–26 per berry, dark brown, ovoid, discoid or reniform, 2–3 × 1.8–2.5 mm, winged with flattened margins, foveolate; sclerotic granules absent. Fig. 16/1–9, p. 109.

UGANDA. Kigezi/Rukungiri District: Bwindi Impenetrable Forest, Kayonza Sector, Ishasha Gorge, 10 Feb. 1998, *Eilu* 247!; Busoga District: Bunya, Nov. 1937, *Webb* 62!; Mengo District: Kajansi Forest, Entebbe Road, June 1935, *Chandler* 1247!
KENYA. Kiambu District: Muguga, 1 Aug. 1960, *Greenway* 9700!; Teita District: Taita Hills, Chawia Forest, 18 May, 1931, *Napier* 1132!; "Eastern Kenya Forests", 5 Aug. 1914, *Battiscombe* 854!
TANZANIA. Moshi District: Marangu, Kibo Hotel, 7 May 1962, *Wright* 6!; Arusha District: Mt Meru, Kilinga Forest, 6 Apr. 1971, *Richards & Arasululu* 26922!; Lushoto District: Kwamshemshi–Sakare Road, W Usambaras, 4 July 1953, *Drummond & Hemsley* 3155! & B!
DISTR. U 2, 3 & 4; K 1, 3–7; T 1–8; from Nigeria, Togo, Gabon, Congo-Kinshasa, Cameroon, Burundi, Rwanda, Congo-Kinshasa, Sudan, to Ethiopia, and south to Angola, Zambia, Malawi, Mozambique, Zimbabwe and South Africa; Mascarenes
HAB. Forest, riverine and scrub forest, secondary forest and scrub, forest margins and clearings, thicket, scattered tree grassland, bushland, grassland; 400–3000 m

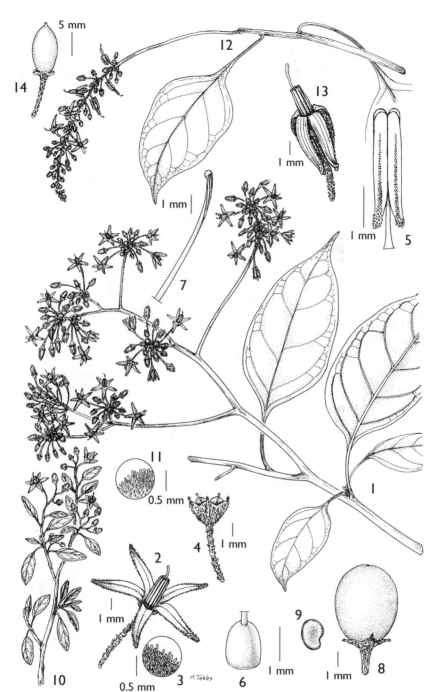

FIG. 16. *SOLANUM TERMINALE* — **1**, habit in flower; **2**, complete flower; **3**, external corolla lobe surface; **4**, calyx and pedicel; **5**, stamen; **6**, ovary; **7**, style and stigma; **8**, berry; **9**, seed. *S. NAKURENSE* — **10**, habit in flower; **11**, external corolla lobe surface. *S. WELWITSCHII* — **12**, inflorescence; **13**, flower; **14**, berry. 1–7 from *Drummond & Hemsley* 3155; 8–9 from *Richards* 26922; 10–11 from *Dummer* 5127; 12–14 from *Mbago* 1109. Drawn by Margaret Tebbs.

Syn. *S. bifurcatum* A. Rich., Tent. Fl. Abyss. 2: 98 (1851); C.H. Wright in F.T.A.: 213 (1906); Exell, Cat. Vasc. Pl. S. Tome: 252 (1944); Milne-Redhead in K.B. 3: 467 (1949); F.F.N.R.: 376 (1962); F.P.U.: 129 (1962). Type: Ethiopia, Adoa, *Schimper* 201 (P, lecto. (photo!) designated by Lester in J.L.S. 125: 282 (1997); BM!, G-DC, E (photo!), M (photo!), NY (photo!), P (x 5), TCD!, U (photo!), W (photo!) all isolecto.)

    S. bifurcum Dunal in DC., Prodr. 13 (1): 77 (1852); Engl., Hochgebirgsfl. Trop. Afr.: 372 (1892); P.O.A. C: 352 (1895); Hiern in Cat. Afr. Pl. Welw. 3: 746 (1898); C.H. Wright in Fl. Cap.: 94 (1904) & in F.T.A. 4, 2: 213 (1906); Dammer in E.J. 38: 180 (1906); Bitter in E.J. 54: 453 (1917); Milne-Redhead in K.B. 3: 467 (1949), ?*nom. illegit.* Type: as for *S. bifurcatum*

    S. phytolaccoides C.H. Wright in K.B. 1894: 126 (1894); Dammer in E.J. 38: 181 (1905). Type: Ethiopia, Tigrè to Begemder, *Schimper* 310 (BM!, K!, syn.); Tanzania, Kilimanjaro, *Johnston* s.n. (BM!, K!, syn.)

    S. togoense Dammer in Schlechter, Westafr. Kautsch.-Exped.: 312 (1900), *nomen nudum*, & in E.J. 38: 183 (1906) descr.; C.H. Wright in F.T.A. 4, 2: 246 (1906); Bitter in E.J. 54: 478 (1917). Type: Togo, around Badja, *Schlechter* 12974 (?B†, holo.; K!, lecto. designated here; BM!, P (photo!), isolecto.)

    S. plousianthemum Dammer in E.J. 38: 180 (1906); Dammer in Z.A.E.: 285 (1914); Bitter in E.J. 54: 456 (1917) & in F.R. 18: 301 (1922); F.P.N.A.: 209 (1947); T.T.C.L.: 576 (1949). Type: Tanzania, Lushoto District: Usambara, *Holst* 232 (?B†, syn.) & *Holst* 3731 (?B†, syn.) & Handei, *Holst* 8927 (?B†, syn.; K!, lecto. designated here, M (photo!), W (photo!) all syn.)

    ?*S. buchwaldii* Dammer in E.J. 38: 180 (1906). Type: Uganda, *Stuhlmann* 1254, 1329 (?B†, syn.); Tanzania, Lushoto District: Usambara, Muoso, *Buchwald* 91, 542, 639, (?B†, syn.)

    ?*S. comorense* Dammer, in E.J. 38: 181 (1906); Bitter in E.J. 54: 474 (1917). Type: Comoro Islands, [Mayotte], *Schmidt* 263 (?B†, syn.); *Schmidt* 284 (?B†, syn.; K!, isosyn. as *Humblot* 284; P (2 photos!) isosyn., see Notes)

    S. bilabiatum Dammer in E.J. 38: 181 (1906). Type: São Tomé, *Moller* 146 (?B†, holo.)

    S. bansoense Dammer in E.J. 48: 237 (1912); Bitter in E.J. 54: 472 (1917) & F.R. 18: 303 (1922). Type: E. Cameroon, Bansso Mts, *Ledermann* 5778 (?B†, holo.)

    S. leucanthum Dammer in Z.A.E. 2: 284 (1914); Bitter in E.J. 54: 456 (1917), as *S. leucanthum* Bitter & Dammer. Type: Rwanda, Rukarara, Rugege forest, *Mildbraed* 894 (?B†, syn.) & Kissenye, Bugoyer, *Mildbraed* 1431a (?B†, syn.). See *S. ruandae* Bitter

    S. lateritium Dammer in E.J. 53: 325 (1915); T.T.C.L.: 576 (1949). Type: Tanzania, Rungwe District: Kyimbila between Uboka and Rufilyo Rivers, *Stolz* 1514 (B!, holo.; K!, U (photo!), W (photo!), WAG (photo!), iso.)

    S. rhodesianum Dammer in E.J. 53: 326 (1915). Type: Zimbabwe, Chirinda Forest, *Swynnerton* 86 (?B†, holo.; K!, BM!, Z, iso.)

    S. meyeri-johannis Dammer in E.J. 53: 328 (1915); T.T.C.L.: 576 (1949). Type: Tanzania, Kilosa District: Ussagara, Buga Mts, *Houy* 1242 (?B†, holo.)

    S. holtzii Dammer in E.J. 53: 329 (1915). Tanzania, Morogoro District: Bauduki Forest Reserve, *Holtz* 3148 (?B†, holo.)

    S. plousianthemum Dammer var. *gracilifilum* Bitter in E.J. 54: 463 (1916); T.T.C.L.: 577 (1949). Type: Tanzania, Kilimanjaro, Marangu, *Volkens* 2265 (?B†, holo.)

    ?*S. plousianthemum* Dammer var. *buchwaldii* (Dammer) Bitter in E.J. 54: 459 (1917); T.T.C.L.: 576 (1949)

    S. plousianthemum Dammer var. *ugandae* Bitter in E.J. 54: 460 (1917). Type: Uganda, *Stuhlmann* 1329 (B†, holo.; K!, iso.)

    S. plousianthemum Dammer var. *rhodesianum* (Dammer) Bitter in E.J. 54: 461 (1917)

    S. plousianthemum Dammer var. *microstelidium* Bitter in E.J. 54: 461 (1917). Type: Rwanda, Kiwu Lake, Lubengera [Rubengera] and Mugarura [Bugarura] Island, *Meyer* 909 (B†, holo.)

    S. plousianthemum Dammer var. *conglutinans* Bitter in E.J. 54: 461 (1917); T.T.C.L.: 577 (1949). Type: Tanzania, Kwa Mshuza near Handei, *Holst* 8927 (?B†, holo.; K!, M (photo!), W (photo!) all probable iso.)

    S. plousianthemum Dammer var. *angustifrons* Bitter in E.J. 54: 462 (1917); T.T.C.L.: 576 (1949). Type: Tanzania, Kilimanjaro, Marangu, *Volkens* 2109 (?B†, holo.)

    S. plousianthemum Dammer var. *epapillosum* Bitter in E.J. 54: 464 (1917); T.T.C.L.: 577 (1949). Type: Tanzania, Kilimanjaro, Useri, *Volkens* 1991 (?B†, holo.)

    S. plousianthemum Dammer var. *commixtum* Bitter in E.J. 54: 464 (1917); T.T.C.L.: 577 (1949). Type: Tanzania, Kilimanjaro, *Johnston* s.n. (B†, holo.)

    S. plousianthemum Dammer var. *endosiphonotrichum* Bitter in E.J. 54: 465 (1917). Type: Rwanda, Niansa Mt, *Kandt* 147 (B†, holo.)

    S. plousianthemum Dammer var. *devians* Bitter in E.J. 54: 466 (1917). Type: Rwanda, Mpororo, Issenge, *Mildbraed* 336 (?B†, holo.)

S. plousianthemum Dammer subsp. *holtzii* (Dammer) Bitter in E.J. 54: 467 (1917); T.T.C.L.: 577 (1949)

S. plousianthemum Dammer var. *kundelunguense* Bitter in E.J. 54: 467 (1917). Type: Congo-Kinshasa, Kundulungu, *Kassner* 2760 (B†, holo.; BM!, K!, iso.)

S. plousianthemum Dammer var. *subtusbarbellatum* Bitter in E.J. 54: 467 (1917). Type: Kenya, Kitui District: Galunka, *Kassner* 804 (B†, holo.)

?*S. plousianthemum* Dammer subsp. *kasima* Bitter in E.J. 54: 468 (1917); T.T.C.L.: 577 (1949). Type: Tanzania, Rungwe District: Tukuyu [Neu-Langenburg], Kyimbila, *Stolz* 355 (B†, holo.; K!, M (photo!), U (photo!), W (photo!), all iso.)

S. ruandae Bitter in E.J. 54: 471 (1917). Type: Rwanda, Kissenye, Bugoyer Highlands, *Mildbraed* 1431a (B†, holo.) [See notes]

S. sychnoteranthum Bitter in E.J. 54: 472 (1917); F.P.N.A.: 210 (1947). Type: Congo-Kinshasa, *Kassner* 3250 (B†, holo.; E (photo!), iso.)

S. butaguense De Wild., Pl. Bequaert. 1: 424 (1922). Type: Congo-Kinshasa, Ruwenzori, Butagu valley, *Bequaert* 3616 (BR!, holo.; BR!, iso.)

S. plousianthemum Dammer var. *angustatum* Bitter in F.R. 18: 302 (1922); T.T.C.L.: 576 (1949). Type: Tanzania, Morogoro District: Uluguru Mountains, *von Brehmer* 891 (?B†, holo.)

S. plousianthemum Dammer var. *kyimbilense* Bitter in F.R. 18: 302 (1922); T.T.C.L.: 577 (1949). Type: Tanzania, Rungwe District: N Lake Nyasa, *Stolz* 2315 (B†, holo.; BM!, K!, iso.)

S. bansoense Dammer var. *episporadotrichum* Bitter in F.R. 18: 303 (1922). Type: Cameroon, Lom, Sanaga, *Mildbraed* 8555 (B†, holo.)

S. bansoense Dammer subsp. *sanaganum* Bitter in F.R. 18: 304 (1922). Type: Cameroon, Lom, Sanaga, Dengdeng, *Mildbraed* 8619 (B†, holo.; K!, iso.)

S. balboanum Chiov., Racc. Bot. Miss. Consol. Kenya: 87 (1935). Type: Kenya, Mt Aberdare E, Tuso [Tusu], *Balbo* 579 (TO, syn.); Mt Kenya SW, Nyeri, *Balbo* 425 (TO, syn.)

S. terminale Forssk. subsp. *terminale*, Heine in K.B. 14: 247 (1960); Polhill, *Solanum* in E & NE Africa: 9 (ined., 1961); Heine in F.W.T.A. 2nd ed., 2:331 (1963); E.P.A.: 879 (1963); Fl. Rwanda 3: 382 (1985)

S. terminale Forssk. subsp. *sanaganum* (Bitter) Heine in K.B. 14: 248 (1960); Polhill, *Solanum* in E & NE Africa: 9 (ined., 1961); Heine in F.W.T.A. 2nd ed., 2: 331, 2nd ed. (1963); Gbile in Biol. & Tax. Solanaceae: 118 (1979); Fl. Rwanda 3: 382 (1985)

NOTE. This is an extremely variable polymorphic and widespread species which is common, especially in montane forests, throughout tropical Africa. Two major variants are apparent in the inflorescence structure of herbarium material. All inflorescences are pyramidal, but the majority are corymbose cymes which broaden towards the apex, have short central rachides and irregular ascending branches and are often referred to as subsp. *terminale*. The less common type has paniculate cymes which narrow towards the apex, have long central rachides (4–14 cm) and ascending lateral branches. This is the form which is often referred to as *S. plousianthemum*. Moreover, though inflorescences in this species are typically forked and many-flowered, some are few-flowered and occasionally simple. These are often identified as *S. inconstans*. There are, however, a number of specimens in which the inflorescence structure is intermediate between all these variants, indicating that the formal recognition of such variation is unwise pending further work with living material.

The variability in inflorescence structure is clearly portrayed by Bitter's treatment of species belonging to his new section when he described 14 infraspecific variants of the "polymorphic species" *S. plousianthemum* Dammer, most of which are minor morphological variants and all synonymous with *S. terminale*. Heine (1960, 1963) too thought that the variation in *S. terminale* should be recognised infraspecifically, and he described the four subspecies: *terminale* itself; *inconstans* (C.H. Wright) Heine characterised by very small, long-peduncled and usually lateral inflorescences with ellipsoid fruits; *sanaganum* (Bitter) Heine delimited by large terminal inflorescences and ellipsoid fruits; and *welwitschii* (C.H. Wright) Heine with spicate, terminal inflorescences and globose fruits. Gonçalves (2005) later elaborated on these differentiating characters, adding discriminatory anther morphology, and describing the inflorescences as varying from racemose to spiciform in subsp. *terminale*, but being shorter than those found in subsp. *welwitschii*. Nevertheless because of intermediate forms he decided not to recognize these subspecific taxa in F.Z. Heine's earlier conclusion that *S. plousianthemum* should also be reduced to subspecific rank was apparently never published (cf. Cameroonian specimen *Ujor* 30026 determined as *S. terminale* Forssk. subsp. *plousianthemum* (Dammer) Heine in 1958).

Solanum terminale is described as a forest species in Uganda where the roots are used medicinally for the relief of stomach pains (Bukenya & Carasco in Nee *et al.* (eds), Solanaceae IV: 356 (1999)). It is also listed as a medicinal plant in PROTA (11: 308 (2002)).

The species can be confused with the Ethiopian species *S. schimperianum* in the herbarium, but Friis (2006) considered the latter to be a somewhat scrambling shrub rather than a true climber, with the stems continuing growth after inflorescence formation thereby not being truly terminal as in *S. terminale*. Hair types too differ in these two species, with those in *S. terminale* being branched but never stellate as are those in *S. schimperianum*.

Superficially *S. terminale* is very similar to *S. madagascariense* Dunal, another climbing Dulcamaroid species which is thought to be endemic to Madagascar, though Bitter did not include it in his Section Afrosolanum (cf. Bitter 1917). While closely related, Madagascan specimens seem to form a comprehensive suite and should probably be recognised specifically – though future field work might suggest subspecific recognition to be more appropriate. The considerable variation in stem indumentum density, the hair type and structure, and the floral and fruiting characters are extremely similar in both species. However, the leaves are invariably glabrous, coriaceous, shining and obovate with prominent veining and mucronate apices and lack domatia in S. *madagascariense*. In addition, while similarly multi-flowered, the inflorescences branches and peduncles do not terminate in such distinct dense umbellate clusters as in *S. terminale*, the anthers are often shorter and broader with distinct and prominent papillae, and the berries seem to be many-more seeded (often > 60/berry). Among the synonyms of *S. madagascariense*, verified from their protologues and as far as possible from the type specimens, are *S. madagascariense* Dammer; *S. nitens* Baker; *S. madagascariense* Dunal var. *nitens* D'Arcy & Rakot.; *S. apocynifolium* Baker; *S. clerodendroides* Hutch. & Dalziel; *S. antalaha* D'Arcy & Rakot.; *S. marojejy* D'Arcy & Rakot., and possibly *S. trichopetiolatum* D'Arcy & Rakot. With the exception of *S. clerodendroides*, these taxa were all described from Madagascar and differentiated on very minor morphological differences which would be expected in such a variable species closely aligned to the extremely diverse *S. terminale*. It is widely thought (cf. Jaeger, 1986 and Heine *in sched.*) that the label on the type specimen of *S. clerodendroides* (*Talbot & Talbot* 3211) citing its provenance as Eket District in Southern Nigeria was misplaced, as this specimen clearly belongs to the Madagascan taxon under discussion, with all other Nigerian specimens examined clearly belonging to *S. terminale* sensu stricto.

The two names *S. bifurcatum* and *S. bifurcum* are apparently orthographic variants, with the Paris specimens suggesting that although Hochstetter originally spelt the name of this taxon as *bifurcum* Richard later changed the spelling to *bifurcatum;* see Lester l.c. (1997) for a full discussion.

The lecto- (K) and isolecto- (BM) type specimens of *S. togoense* are poor with the inflorescence bearing only two and seven buds respectively. The Paris duplicate is similarly poor and bears a note saying that Bitter (1917), who retained Dammer's species (first published without a description), cited *Schlechter* 12974 as the only specimen of this species. It's vine-like stems certainly indicate its affinity with *S. terminale sensu lato,* though the potentially few-flowered inflorescences led Heine (1960) to identify it as *S. terminale* subsp. *inconstans,* and place *S. togoense* in synonymy with this subspecies.

Bitter (1917) based his *S. plousianthemum* Dammer var. *conglutinans* Bitter (see below) on the *Holst* 8927 specimen cited as a syntype of Dammer's species.

Though Dammer cited *Schmidt* 263 and 284 as syntypes of *S. comorense* without mentioning their location, Bitter later redescribed it (in Latin), citing the two specimens *Schmidt* 263 (with the same collection details as those given by Dammer) and *Humblot* 284 (where the collection label simply stated Comoro Islands). Both of these specimens were in the Berlin Herbarium, and it is likely that Dammer mistakenly assigned the number 284 to Schmidt instead of Humblot. A duplicate of *Humblot* 284 is extant at Kew. Unfortunately, this consists of two apical fragments in which the inflorescences appear broken. Superficially these fragments indicate conspecificity with *S. welwitschii,* but the anthers are free and papillate internally, the buds ovoid and one peduncle has one short branch. The affinity of this species is difficult to discern from Dammer's protologue, while Bitter's later elaboration on this so-called endemic species allied to '*S. plousianthemum*' provided little in the way of additional morphological features to differentiate it from *S. terminale sensu lato*. It seems sensible to regard the extant *Humblot* 284 specimens as isosyntypes, though the two P specimens have been labelled neotype and isoneotype.

Bitter (1917) must have decided that the two syntypes cited by Dammer (1914) for *S. leucanthum* represented different species. He redescribed Dammer's species as *S. leucanthum* Bitter & Dammer, basing it on *Mildbraed* 894. He then went on to describe the new species *S. ruandae,* basing it on *Mildbraed* 1431a, the second of Dammer's syntypes. Both of his species are described from high altitudes and from their protologues clearly belong to *S. terminale*.

The locality Handei (**T** 3, Lushoto) cited in the protologue by both Dammer for his *S. plousianthemum* (for which *Holst* 8927 is a syntype) and later by Bitter for his var. *conglutinans* (for which Holst 8927 becomes an isotype) is not given on the labels of any of the duplicate specimens. All plant parts on these specimens are glabrescent, with the inflorescences being multi-(± 65)-flowered and typical of *S. terminale.*

The isotype of *S. plousianthemum* Dammer var. *ugandae* is another poor fragmentary specimen; though most of the flowers are only in bud their floral morphology suggests that this taxon is a synonym of *S. terminale.*

4. **Solanum welwitschii** *C.H. Wright* in K.B. 1894: 126 (1894); P.O.A. C: 352 (1895); Hiern in Cat. Welwitsch Afr. Pl. 3: 747 (1898); C.H. Wright in F.T.A. 4, 2: 231 (1906); De Wild., Miss. Laurent: 440 (1907); Durand & Durand, Syll. Fl. Cong.: 394 (1909); Bitter in E.J. 54: 478 (1917) & in Fl. Rwanda 18: 305 (1922); Friis in Fl. Eth. 5: 115 (2006). Type: Angola, Cazengo, Muxaulo Mts, *Welwitsch* 6081 (BM!, K!, P (photo!), syn.); Golungo Alto, *Welwitsch* 6098 (BM!, K!, P! (photo), syn.)

Scandent or twining subshrub or perennial vine, 5–8 m, sometimes creeping over ground, often deciduous; stems usually basally woody, often corky, sparsely pubescent with short simple and occasionally branched (never stellate) hairs when young, becoming glabrescent. Leaves usually alternate, occasionally opposite, often membranaceous, dark green above, ochraceous below, often decreasing in size towards the inflorescence, ovate to lanceolate, 5.8–11 × 2.8–5.5 cm, bases cordate to cuneate and decurrent, margins entire, apices acute to acuminate; surfaces ± pilose and glabrescent, hairs denser on veins, midribs and on lower surfaces, domatia absent; petioles usually long (0.5–)2.5–10.5 cm, curved. Inflorescences terminal simple spicate cymes, often pendulous, with dense fascicles of flowers intermixed with densely pilose bracts arising at intervals along the whole length of the central rhachis, 6–16(–20) cm long, many (often 100+)-flowered; peduncles densely to moderately pilose/strigose, hairs short and usually simple, becoming glabrous, erect in flower and in fruit, dense papillate ochraceous collar around junctions with pedicels; pedicels (3–)5–9 mm long and erect in flower, 8–12 mm and erect in fruit, spreading or strongly reflexed, densely pilose/papillate. Calyx shallowly cupulate, 1–2 mm long, sparsely to densely pilose externally with short hairs; lobes broadly triangular, 0.5–1 × and 1–1.5 mm, acute to apiculate; adherent in fruit when 1–1.5 × 1–1.5 mm. Corolla purple or blue with yellow star, occasionally white, deeply stellate, 1.4–2 cm diameter, tube 0.9–1.3 mm long; lobes ligulate, 5.8–8 × 0.9–1.5 mm, densely pilose/papillate externally with short often branched hairs, glabrous internally, strongly reflexed to pedicels after anthesis. Stamens usually equal; filaments free for 0.8–1.2 mm, glabrous (in FTEA area); anthers yellow, often unequal, 2.5–4.7 × 0.7–1.4 mm, connate. Ovary green, 1–1.3 × 0.7–0.8 mm, glabrous; style geniculate becoming straight, greenish-white, 5.2–7 × 0.2–0.3 mm, glabrous, exserted up to 2.5 mm; stigma capitate, often bilobed, 0.3–0.5 mm diameter. Fruit orange to red, ovoid (globose outside our area), occasionally spindle-shaped, 4.5–10 × 4–7 mm, apiculate, smooth, with coriaceous pericarp. Seeds 5–12 per berry, ovoid to orbicular, 2–2.8 × 1.6–2 mm, foveolate; sclerotic granules absent. Fig. 16/12–14, p. 109.

UGANDA. Bunyoro District: Budongo Forest, 26 Nov. 1938, *Loveridge* 108!; Kigezi District: Kasyoha-Kitomi Forest, southern part of reserve, Mzozia river area, 4 June 1998, *Eilu* 293!; Toro District: Semliki Forest, 31 Oct. 1905, *Dawe* 672!
TANZANIA. Kigoma District: Gombe Stream Reserve, middle Kakombe, 5 May 1992, *Mbago & Lyanga* 1109!
DISTR. U 2, 4; T 4; Guinea Bissau, Sierra Leone, Liberia, Ivory Coast, Ghana, ?Togo, Nigeria, Cameroon, Bioko, Equatorial Guinea, Gabon, Congo, Congo-Kinshasa, and south to Angola and Zambia; sporadic and uncommon in South Sudan and Ethiopia
HAB. Moist evergreen forest, including clearings, swamps and paths; 700–1350 m

SYN. *S. welwitschii* C.H. Wright var. *oblongum* C.H. Wright in K.B. 1894: 127 (1894); Dammer in E.J. 38: 182 (1906); Bitter in E.J. 54: 480 (1917). Type: Cameroon, Ambas Bay, *Mann* s.n. (labelled 'No. X') (K!, lecto., designated here)

S. *welwitschii* C.H. Wright var. *strictum* C.H. Wright in K.B. 1894: 127 (1894); Dammer in E.J. 38: 182 (1906); C.H. Wright in F.T.A. 4, 2: 213 (1906); Durand & Durand, Syll. Fl. Cong.: 394 (1909); Bitter in E.J. 54: 480 (1917) & 18: 303 (1922); Heine in F.W.T.A. 2: 208 (1931). Type: Congo-Kinshasa, Munza, *Schweinfurth* 3498 (K!, syn.); Bioko [Fernando Po], *Mann* 274 (K!, syn.)

?S. *inconstans* C.H. Wright in K.B. 1894: 127 (1894) & in F.T.A. 4, 2: 211 (1906) pro parte; Bitter in E.J. 54: 482 (1917); Heine in F.W.T.A. 2: 207 (1931); Durand & Durand, Syll. Fl. Cong.: 394 (1909); Type: Bioko [Fernando Po], *Mann* 62 (K!, syn.); Cameroon, *Kalbreyer* 172 (K!, syn.)

?S. *lujaei* De Wild. & Durand in Compte Rendu Soc. Bot. Belg. 38: 209 (1899). Type: Congo-Kinshasa, Sona Gunga, Cataractes District, *Luja* s.n. (?BR, holo.)

S. *symphyostemon* De Wild. & Durand, Contrib. Fl. Congo, 1: 44 (1899); Dammer in E.J. 38: 184 (1906); Bitter in E.J. 54: 483 (1917). Type: Congo-Kinshasa, Bolombo near Gali [Ngali], *Thonner* s.n. (BR!, holo.; K!, iso.)

S. *laurentii* Dammer in E.J. 38: 182 (1906); Durand & Durand, Syll. Fl. Cong.: 393 (1909). Type: Congo, *Laurent* s.n. (BR, holo.)

S. *preussii* Dammer in E.J. 38: 183 (1906); Durand & Durand, Syll. Fl. Cong.: 394 (1909). Type: Cameroon, Barombi Station, Kumba River, *Preuss* 397 (?B†, syn.; E (photo!), P (photo!), syn.) & between Victoria and Bimbia, *Preuss* 1167 (?B†, syn.); Congo-Kinshasa, between Lusambo and Lomami, *Laurent* s.n. (BR, syn.)

S. *suberosum* Dammer in E.J. 38: 182 (1906) & in Z.A.E. 2: 285 (1914); Bitter in E.J. 54: 456 (1917) & in F.R. 18: 307 (1922). Type: Cameroon, between Barombi Station and Ninga Town, *Preuss* 18 (?B†, syn.); Barombi stream, *Preuss* 171 (?B†, syn.); SE Buea, *Preuss* 885 & *Lehmbach* 211 (?B†, syn.); Yaounde [Jaunda] forest, *Zenker & Staudt* 328 (?B†, syn.); Yaounde, *Zenker* 268 (?B†, syn.); Bangwe, *Conrau* 200 (?B†, syn.)

S. *subcoriaceum* Th. & H. Durand, Syll. Fl. Cong.: 394 (1909), *nom. nov.*; Bitter in E.J. 54: 481 (1917). Type: as for *S. laurentii* Dammer, *non* De Wild.

S. *hemisymphyes* Bitter in E.J. 54: 477 (1917). Type: Congo-Kinshasa, KwaMuera, Fort Beni, *Mildbraed* 2238 (?B†, holo.).

S. *welwitschii* C.H. Wright var. *laxepaniculatum* Bitter in F.R. 18: 306 (1922). Type: Cameroon, Lomie, *Mildbraed* 5265 (B†, holo.)

S. *terminale* Forssk. subsp. *inconstans* (C.H. Wright) Heine in K.B. 14: 247 (1960); Heine in F.W.T.A. 2nd ed., 2: 331 (1963); Gbile in Biol. & Tax. Solanaceae: 118 (1979); Bukenya & Hall in Bothalia 18: 83 (1988)

S. *terminale* Forssk. subsp. *welwitschii* (C.H. Wright) Heine in K.B. 14: 248 (1960) & in F.W.T.A., 2ⁿᵈ ed.: 331 (1963); Gbile in Biol. & Tax. Solanaceae: 118 (1979); Bukenya & Hall in Bothalia 18: 82 (1988)

NOTE. This species is common, especially in secondary forests, in West Africa, where it forms a well-defined morphological entity. In East Africa and Ethiopia, some authors have considered that it should be considered as part of the polymorphic species *S. terminale* (cf. Gonçalves, 2005). A Ugandan specimen (*Dawkins* 751), recorded as a single individual in Siba Forest, Bunyoro District (U 2) seems to belong to *S. welwitschii*, though it is totally covered with a mealy indumentum of dense multiply branched hairs. Other morphological features are similar to those found in more typical specimens of this species apart from short basal forks on the otherwise spicate inflorescences. Since no other specimens exhibiting these features have been encountered it is provisionally regarded as an extremely pubescent form of *S. welwitschii*.

Authors of recent Floral accounts have varied slightly in their treatment of these Dulcamaroid species, either recognising them as belonging to the one highly polymorphic species *S. terminale* (e.g. Gonçalves, 2005), or to three major species with the recognition that overlapping morphology often makes the delimitation of these difficult (e.g. Friis, 2006). Within the FTEA area, the three major variants are for the most part well-defined, and have been recognised as separate species. Many protologues of the synonyms included here describe the occurrence of stellate hairs. However, stellate hairs are never found in these species – though (dendritic) branched hairs varying in their complexity are commonly mixed with simple ones in these species; these can be scattered and sparse to dense and complexly branched. These references to stellate hairs are therefore thought to be erroneous.

Hunziker (2001) noted from Bitters' detailed descriptions that there is apparently extraordinary variation in androecium morphology in these taxa. Of the 21 species recognised by Bitter in sect. *Afrosolanum*, 13 were described as having free anthers and filaments, with eight showing a) anthers and filaments completely fused between each other (e.g. *S. inconstans* from Togo and Cameroon); b) filaments free and anthers not fused but closely associated due to lateral papillae (*S. plousianthemum* in Tanzania); and c) filaments free but anthers completely fused (*S. togense* and *S. welwitschii*). With the exception of the fused connate anthers found in *S. welwitschii*, the differentiation as described by Bitter for the other species groups is not readily apparent from herbarium material, and they have all been synonymised with *S. terminale*.

The syntypes on which *S. inconstans* (and *S. terminale* subsp. *inconstans*) is based are poor and fragmentary, though both are clearly vines. The inflorescences are simple but few-flowered, though the floral morphology is typical of *S. welwitschii* with the anthers on *Kalbreyer* 172 appearing connate. The berries on *Mann* 62 are ovoid and apiculate to spindle-shaped. Bitter's (1917) redescription of *inconstans* re-cited the Kew syntypes and included a full Latin 'protologue' describing the few-flowered inflorescences, and the floral and fruiting morphology in considerable detail. Some of the features cited, whilst clearly placing this species in *S. welwitschii*, are not visible on the specimens today.

De Wildeman & Durand's illustration of *S. symphyostemon* clearly illustrates that it is a synonym of *S. welwitschii*. Unfortunately the Kew isotype specimen is merely a leaf fragment yielding very little information. Bitter (1917) later included a very detailed description of *S. symphyostemon* but is not clear whether he based this on the *Thonner* specimen or on the De Wildeman & Durand illustration. The excellent BR holotype is, however, completely morphologically conspecific with *S. welwitschii*.

Durand & Durand (Syll. Fl. Cong.: 394 (1909)) later included *S. lujaei* as a synonym of *S. welwitschii* var. *strictum*. This would be consistent with the description of very narrowly lobed petals and connivent anthers in the protologue of *S. lujaei*. However, this protologue also mentions branched racemose inflorescences which are 9 cm broad basally – which would be more indicative of *S. terminale*. Resolution of the correct placement of this synonym therefore awaits location of a type specimen.

The *Preuss* 397 specimens of *S. preussii* are clearly conspecific in all morphological features with *S. welwitschii*; the syntype specimen in Paris is ex Berlin and bears a determination in Bitter's hand.

In his protologue of *S. hemisymphyes* Bitter described the inflorescences as being terminal and almost spiciform, with the stellate flowers having very short pedicels and narrow lobes - features which clearly place it in synonymy with *S. welwitschii*. He went on to describe the narrow ellipsoid anthers as being joined centrally but free basally and apically – features which are also sometimes apparent in the typically connate anthers. He concluded by noting its affinity with both *S. inconstans* and *S. symphyostemon* but differentiating it on very minor androecial characters not considered taxonomically significant in this taxon.

5. **Solanum seaforthianum** *Andr.* in Bot. Repos. 8: t. 504 (1808); Dunal, Synopsis: 7 (1816); G. Don, Gen. Hist. Dichlam. Pl.: 407 (1837); Dunal in DC., Prodr. 13(1): 67 (1852); Urban, Symb. Antill., 6: 167 (1909); T.T.C.L.: 577 (1949); F.F.N.R.: 377 (1962); D'Arcy in Ann. Missouri Bot. Gard. 60: 758 (1973); Symon in Journ. Adelaide Bot. Gard. 4: 67 (1981); Purdie *et al.*, Fl. Australia 29: 105 (1982); Symon in Journ. Adelaide Bot. Gard. 8: 79 (1985); Jaeger, Syst. studies *Solanum* in Africa: 319 (1985, ined.); Huxley *et al.*, New RHS Dict. Gard. 4: 382 (1992); D'Arcy & Rakotozafy in Fl. Madagascar, 176: 125 (1994); Bukenya & Carasco in Bothalia 25(1): 57 (1995); Hepper in Fl. Egypt, Family 159: 45 (1998); Gonçalves in F.Z. 8(4): 79 (2005). Type: cultivated in Britain from seed sent from West Indies by Lord Seaforth; no specimens known to exist, lectotype t. 504 of Andrews (1808) designated by Symon (in Journ. Adelaide Bot. Gard. 4: 67 (1981))

Vine, creeper, or trailing to scrambling shrub reaching 7 m high, climbing by means of twining petioles; stems woody basally, often becoming brittle, light green to brown, sparsely pilose to glabrescent with short eglandular hairs mixed with short glands. Leaves alternate to spirally arranged, membranaceous, light green, 4.5–13.5 × 3.3–11 cm, usually partially or completely imparipinnate with 2–4 pairs of lateral leaflets, upper pairs often confluent with rachis, lowers ones often small; leaflets lanceolate to ovate-lanceolate, 2–5.5 × 1–4 cm, bases cuneate to attenuate, margins entire, apices acute to obtuse, surfaces ± pilose to glabrescent, hairs as on stems when present but denser on rhachides, margins, veins and midribs, sessile or shortly petiolulate (when up to 3 mm long); petioles 1.5–4.5 cm long. Inflorescences terminal or leaf-opposed to extra-axillary when lateral, lax, pendent, many-flowered (to 50+) showy panicles or corymbs, 4.5–13.5 × 4–10 cm, often pyramidal; peduncles erect becoming pendent, 1.5–6 cm long; pedicels erect, with basal abscission layer, 7–12 mm long in flower, 12–30 mm long in fruit when swollen beneath calyx; calyx cupulate, shallow, 1–1.5 mm long, glabrous externally; lobes indistinct, shallowly deltate to broadly triangular, obtuse, 0.5–0.6 × 1.2–2 mm with indistinct apical tufts of hairs, becoming shrivelled and reflexed in fruit. Corolla mauve, purple or blue, sometimes with yellow basal star, showy, stellate, 1.4–2.8 cm diameter, tube 0.5–1 mm long, lobes broadly ovate, 5–11 × 3–5.5 mm, with shortly pilose margins and apical tufts of hairs, spreading to reflexed after anthesis exposing erect androecium. Stamens usually equal; filaments free for 1.6–4 mm, glabrous; anthers yellow, blue, violet to brown, 2.5–3.5 × 1.5–2 mm, with large apical pores, connivent. Ovary brown, 1.3–2 × ± 1.7 mm, glabrous; style coiled or curved above, sometimes straight, 7–9.5 × ± 0.2 mm, glabrous, exserted to 5 mm; stigma globose, inconspicuous, 0.25 × 0.25–0.4 mm. Fruit usually borne erect, red, globose, 6–10 mm diameter, smooth, surrounded basally by small adherent triangular calyx lobes, often deciduous from calyces. Seeds 21–23 per berry, pale yellow to cream, obovoid, elliptic or orbicular, 2.2–3.6 × 1.3–3 mm, usually encircled with distinct rim, densely covered with long silky hair-like strands of thickening; sclerotic granules absent.

KENYA. Nairobi area, Jan. 1949, *Bally* 6565! & Nairobi, National [Coryndon] Museum grounds, 25 June 1952, *Hiza* 3!; Teita District: Taita Hills, 31 July 1998, *Mwachala et al.* 1203!
TANZANIA. Arusha District, Tengeru at S foot of Mount Meru, forest around Duluti Lake, 2 July 1988, *Pócs* 88166/K!; Moshi District: Northern Prov., Northern Stock Farm at Ngare Nairobi, 10 May 1949, *van Rensberg* 495!; Pangani District: Bushiri Estate, 16 June 1950, *Faulkner* 656!
DISTR. U 1 (fide Bukenya & Carasco, 1995); K 4, 7; T 2, 3, 6; native of the West Indies and C America; now widely cultivated and a common and often naturalised escape in Senegal, Sierra Leone, Ghana, Nigeria, Congo-Kinshasa, Zambia, Malawi, Zimbabwe, Namibia, Botswana, South Africa; Egypt, Madagascar, Mauritius, Reunion and the Comoro Islands, Australia, New Guinea
HAB. Semi-deciduous forest, groundwater forest, grassland, farmland, plantations, forest reserves and gardens; 750–1550 m

SYN. *S. seaforthianum* Andr. var. *disjunctum* O.E. Schulz in Urban, Symb. Antill. 6: 169 (1909); Bitter in F.R. 16: 309 (1923); T.T.C.L.: 577 (1949); Polhill, *Solanum* in E & NE Africa: 11 (ined., 1961); Heine in F.W.T.A., ed. 2, 2: 332 (1963) & in Fl. Nouv.-Caled., 7: 143 (1976); Bukenya & Hall in Bothalia 18: 87 (1988). Types: Cuba, near Cieneguita, often cultivated: Havana near Santiago de las Vegas, *R. Combs* 35 (?ISC, syn. ?†): *Van Hermann* Herb. Cuba 5080 (?HAC or NY, syn.); Haiti near Terre Neuve: *Buch* 80, near St Michel: *Buch* 425 (??, syn.); St Jan near Cruz Bay: *Raunkiaer* 3127, 3128 (C, syn.)

NOTE. Commonly known as St Vincent or Brazilian Lilac, Glycine, Italian Jasmine, Seaforth's Nightshade or Potato Creeper, this species has been cultivated worldwide since its introduction from the West Indies. The berries are reportedly poisonous to poultry, pigs, cattle, sheep and children in Australia, and in Zambia. The leaves sometimes emit an offensive odour when crushed, and the species is commonly spread by birds in Africa, where it has become widely established (cf. White, F.F.N.R.: 377).

The punctuation used for the citation of the syntypes of var. *disjunctum* by Schulz is confusing; none of the syntypes cited has been seen. Schulz (1909) thought this variety occurred in Cuba, Guatemala and Costa Rica. He described the lower leaves in *S. seaforthianum* sensu stricto as being tri- or bi-jugate with the upper ones being simple or sub-jugate. His variety *disjunctum* was differentiated through having both the upper and the lower leaves pinnate. Although most African specimens of this species exhibit the latter feature, leaf morphology in species belonging to the section *Dulcamara* is notoriously variable and insufficiently reliable to be used as a basis for formal infraspecific recognition.

6. **Solanum laxum** *Spreng.*, Syst. Veg., 1: 682 (1824); Gonçalves in F.Z. 8(4): 69 (2005). Type: Uruguay, Montevideo, *Sellow* s.n. (?B†, holo.)

Herbaceous evergreen climber to 5 m high, often rooting by runners; stems slender, sparsely pilose with short simple eglandular hairs, denser on young stems and around nodal areas, usually mixed with glands, glabrescent on older parts. Leaves alternate, sometimes coriaceous, dark green, ovate to hastate, 1.5–5.4(–6) × (0.6–)1.4–2.8 cm, bases cordate to truncate, simple to compound - often differing according to stem position, young leaves often deeply divided but some leaves unlobed, margins usually entire with pair of basal lanceolate lobes, apices acute; venation prominent, surfaces glabrescent with sparse short hairs and stalked glands on veins and midribs, denser on lower surfaces with prominent hairy domatia in axils of lowermost veins and midribs; petioles (0.4–)1–2.6 cm, curved to help climbing or twining. Inflorescences forked, terminal or axillary lax panicles or cymes, up to 20-flowered; flowers fragrant; peduncles erect, 1–2.5 cm long in flower, to 2.7 cm long in fruit, sparsely pilose with stalked glands to glabrous; pedicels nodding and 6–16 mm long in flower, sparsely pilose to glabrous but often with dense stalked glands, spreading and 14–17 mm long in fruit, articulated basally. Calyx cupulate/campanulate, 1.5–2.5 mm long; lobes broadly ovate to broadly triangular, 0.5–0.8 × 1.5–1.8 mm, acute/obtuse to apiculate, often with tufted pilose apices and a ciliate fringe to margins; persistent and 1.4–1.5 × 1.5–1.8 mm in fruit. Corolla white or tinged blue, rotate-stellate, 2.4–2.8 cm diameter, tube 1–1.2 mm long; lobes boat-shaped, 6–9 × 4–6 mm basally, with upturned apices and prominent venation, spreading after anthesis, with densely pilose margins and apical tufts externally. Stamens equal; filaments 1–1.8 mm long, glabrous but sometimes with stalked glands; anthers yellow to brownish, 3–4.5 × 0.8–1.2 mm, connivent. Ovary 1.5–2 × 1–1.2 mm, glabrous; style straight, 6.5–8 × 0.3–0.4 mm basally, densely pilose with short hairs in lower part, exserted up to 3 mm; stigma clavate, 0.3–0.6 mm diameter, sometimes clasping the style. Fruit borne erect on spreading pedicels, bluish, purple or blackish (also recorded as red in Australia), globose, 7–10 mm diameter, smooth, surrounded basally by adherent calyx lobes. Seeds 10–32/berry, light to dark brown, ovoid, discoidal or orbicular, 2.6–3.2 × 2–2.8 mm, reticulate; sclerotic granules absent.

KENYA. Naivasha District: South Kinangop, 11 Dec. 1960, Hort. Polhill, *Verdcourt* 3026! (Polhill 1961 under *S. jasminoides*) also cited a specimen from Nyeri).
TANZANIA. Lushoto Distr.: Kivungilo Mission Garden, 1 Sept. 1982, *Sigara* 251!
DISTR. **K** 3, 4; **T** 3; native of Brazil and Paraguay, Uruguay and Argentina; now widely cultivated in warm temperate and some tropical mountain regions, including Egypt, Ethiopia, South Africa, India, Australia where sparingly naturalised
HAB. Ornamental climber often cultivated in gardens and parks, found as escape in forests; 1600–2650 m

SYN. *S. jasminoides* Paxton in Mag. Bot. 8: t. 5 (1841); Lindley in Bot. Reg. 33: 33 (1847); Dunal in DC., Prodr. 13(1): 82 (1852); Darwin in J.L.S. 9: 42 (1867); Bitter in F.R. 16: 309 (1923); Lawrence in Baileya 8: 24 (1960); Polhill, *Solanum* in E & NE Africa: 11 (ined., 1961); Bailey, Man. Cult. Pl.: 869 (1966); Morton, Rev. Argentine *Solanum*: 64 (1976); Purdie *et al.*, Fl. Australia 29: 104 (1982); Jaeger, Syst. stud. *Solanum* in Africa: 319 (1985, ined.); Huxley (ed.) New RHS Dict. Gard. 4: 318 (1992); D'Arcy & Rakotozafy in Fl. Madagascar, 176: 47 (1994); Hepper in Fl. Egypt, Solanaceae: 33 (1998); Friis in Fl. Eth. 5: 124 (2006). Type: cultivated in London possibly from material collected from Brazil, Rio Grande do Sul, *Paxton* t. 5 in Mag. Bot. 8!, (lecto. designated by Symon in J. Adelaide Bot. Gard. 4: 65 (1981)
    S. boerhaviifolium Sendtn. in Martius, Fl. Brazil, 10: 48, & t. 11 (1846), as *boerhaviaefolium*; Dunal in DC., Prodr. 13(1): 83 (1852); Morton, Rev. Argentine *Solanum*: 65 (1976). Type: Brazil, *Sellow* s.n. (B†, holo.; K! ?iso [*Sellow* 15]; P, iso.)
    S. jasminoides Paxton var. *boerhaviifolium* (Sendtn.) Kuntze in Rev. Gen. Pl., 3: 226 (1898), as *boerhaaviaefolium.* Type: no details given
    S. dietrichiae Domin, Biblioth. Bot., 89: 576 (1929). Type: Australia, Queensland, *Dietrich* 2789 (**PR**, holo.)

NOTE. Few synonyms of *S. laxum* have been included in this account, as this taxon is usually referred to as *S. jasminoides* in African literature. It is commonly known as the Potato Vine or

-Climber and the Jasmine Nightshade. Several varieties and cultivars of "*S. jasminoides*" have been recognised such as the white flowered var. *album*; the more floriferous var. *floribundum*; the larger and more robust inflorescenced var. *grandiflorum* and the white speckled leaved var. *variegatum* (fide Hepper, 1998).

Sendtner's plate of the synonymous *S. boerhaviifolium* illustrates infructescences whereas no fruiting material has been encountered in the herbarium. Fruiting is also apparently rare in cultivated material. The fruiting measurements given above have been taken from specimens of the synonymous *S. boerhaviifolium*. Morton (1976) considered this a distinct species in Argentina, but apart from citing minutely pilose filaments and probable red berries, his distinguishing features are difficult to discern. Kuntze (1898) subdivided Paxton's *S. jasminoides* into a several varieties and forms – all apparently from Bolivia. Morton (1976) synonymised some of these under *S. boerhaviifolium* var. *boerhaviifolium* whilst also describing a new variety var. *calvum* Morton, simultaneously mentioning that this might be conspecific with *S. jasminoides* Pax. var. *boerhaviifolium*. The latter is the only one of Kuntze's varieties to been included here, and it is unclear if these variants occur naturally in Bolivia or were introduced – as were cultivated specimens seen from Colombia.

Darwin (1867, pages 41–43) used "*S. jasminoides*" as an example of a true leaf climber in his paper "On the movement and habits of Climbing Plants"; he illustrated the tendril-like petioles, and experimented on both the time taken for the petioles to clasp sticks or stems and the effect of weights on their revolution.

According to Friis (2006) "*S. jasminoides*" superficially resembles the indigenous forest climber *S. bendirianum* Dammer, but can easily be distinguished by the often coiling petioles, the usually hastate or truncate leaf-bases, and the tufts of hairs in the axils of the primary veins.

The synonymy of *S. boerhaviifolium* (originally spelt with one *a* by Sendtner but with two *aa*'s by later authors including Dunal (1852), Kuntze (1898) and Morton (1976)) is based on Sendtner's protologue and his plate which shows many features typical of "*S. jasminoides*", including the characteristic leaf shape with curved petioles, together with the typical inflorescence and floral morphology. Most of the specimens identified as *S. boerhaviifolium* from Paraguay, Uruguay and Argentina also show these typical features, but some of those from Brazil are atypical with more ovate to rotund coriaceous leaves, without the curved petioles and dense domatia and sometimes with a dense long-haired pubescence. These might simply be indicative of an infraspecific variant, but clearly necessitate more analysis.

7. **Solanum runsoriense** *C.H. Wright* in Johnston, Uganda Protectorate, 1: 362 (1902); C.H. Wright in F.T.A. 4, 2: 212 (1906); Bitter in E.J. 54: 486 (1917); Jaeger, Syst. studies *Solanum* in Africa: 285 (1985, ined.); Hepper & Jaeger in Solanaceae Biol. & Syst.: 45 (1986); Bukenya & Curasco in Bothalia 25: 50 (1995). Type: Uganda, Ruwenzori Mts, *Doggett* s.n. (K!, holo.)

Scrambling or creeping shrub or liana, vine or herb, reaching 10 m in height, often using other vegetation as support; stems woody basally or herbaceous, often purplish, usually angulate, often lenticellate; all parts densely (when ochraceous) to moderately pubescent with mixture of simple and complexely branched (but not stellate) hairs up to 1.5 mm long, or glabrescent with only scattered and usually simple long hairs. Leaves alternate, lanceolate to linear-lanceolate, rarely ovate, membranaceous to coriaceous, dark green above, if pubescent mealy and yellow/green below, (4–)5.6–12 × 2.5–6 cm, bases cordate to cuneate and decurrent, margins entire, apices acute to acuminate; venation often prominent, surfaces pubescent above with pilose margins, glabrescent; pubescent and ochraceous on lower surfaces, with hairs branched and thicker on veins and midribs; domatia usually present; petioles 1–4 cm long, often twisted or curved to facilitate climbing and subtended by two small ovate leaves. Inflorescences subterminal, multiply branched (candelabra-like), lax panicles with umbellate fascicles of 2–3(–5) flowers, often pendulous, (7–)10–28 cm long, 20–80+-flowered; peduncles (1–)2–7.5 cm long often becoming woody; pedicels erect, 0.5–3 cm long in flower, up to 2.5 cm long and expanded beneath calyx in fruit; axes densely tomentose/ochraceous to glabrous, hairs complexely branched and often in clumps. Calyx campanulate, 3.5–8 mm long, often tinged purple, pubescent and ochraceous with short branched hairs or glabrescent externally; lobes narrowly triangular to lanceolate, sometimes unequal, 1.5–5 × 1.5–3 mm, acute, apices often reflexed, usually with apical tufts of hairs; becoming adherent/accrescent in fruit when 3–6 × 2–4 mm. Corolla purple, lilac or blue, rarely whitish, with greenish basal star, rotate/stellate to stellate, 1.8–3(–4) cm diameter, tube 1–2.5(–3) mm long; lobes broadly to narrowly triangular, 3.5–12 × 3.5–8 mm, pubescent externally with short usually branched hairs, semi-reflexed to lateral exposing androecium after anthesis. Stamens often unequal with one longer; filaments free for 1–3.5 mm, pubescent to glabrescent; anthers yellow to yellow-orange with orange-brown apices, 2.5–4 × 1–1.5 mm, free. Ovary whitish, 1–1.5 × 0.8–1 mm, glabrous, bilocular; style straight, white, glabrous, included when 2.5–4.5 mm long or exserted to 3.5 mm when 6–9.5 × 0.2–0.4 mm; stigma green, usually clavate, sometimes capitate, bilobed or clasping, 0.3–0.4 × 0.3–0.8 mm. Fruit smooth, black, globose to ovoid with coriaceous pericarp, 5–9 mm diameter. Seeds up to 8–18 per berry, light to dark brown, orbicular to discoid, 2.2–3.5 × 1.8–2.9 mm, often angled, foveolate; sclerotic granules absent.

NOTE. Several authors including Polhill (1961), Jaeger (1985), Hepper & Jaeger (1986) and Beentje (1994) suggested that this taxon is composed of two subspecies. Specimens from NE Africa confirmed these observations and showed that these can be differentiated by their indumentum and geographical distributions.

a. subsp. **runsoriense**

Plants conspicuously pubescent with simple and complexely branched hairs up to 1.5 mm long; young stems, inflorescence branches and lower leaf surfaces densely tomentose with mealy pubescence appearing ochraceous; upper leaf surfaces moderately to sparsely pubescent, hairs usually simple; floral parts pubescent, hairs especially dense on calyces. Fig. 19/14–19, p. 161.

UGANDA. Ruwenzori, 4 Oct. 1905, *Dawes* 587! & 19 July 1960, *Kendall & Richardson* 25!; Mt Elgon, Apr. 1930, *Liebenberg* 1637!
KENYA. Laikipia District: around Nyeri, Laikipia Plateau and Aberdare Range, (no date details given) 1908, *Routledge* s.n.!; W slopes of Aberdares National Park on N Kinangop–Nyeri Road, 30 July 1960, *Polhill* 250!; Kiambu District: Gatamayu, Aug. 1933, *Napier* 2709!
TANZANIA. Kilimanjaro, track to Shira Plateau, 12 Feb. 1969, *Richards* 24031! (probably)
DISTR. U 2, 3; K 3, 4; T 2?; restricted to Ruwenzori, Mt Elgon, Aberdares and adjacent areas of the western Ruwenzori in the Congo-Kinshasa; a probable specimen has also been collected from Kilimanjaro
HAB. Upper montane forest and -clearings, path-sides, often in or above bamboo forests including rejuvenating bamboo; 2400–3200 m

SYN. *S. keniense* Standley in Smithson. Misc. Coll. 68, 5: 16 (1917), *non* Turrill, *nom. illeg.* Type: Kenya, W slopes of Mt Kenya, along trail to summit, *Mearns* 1416 (US, holo., photo!)
    S. longipedicellatum De Wild., Pl. Bequaert. 1: 428 (1922), *non* Bitter. Type: Congo-Kinshasa, Ruwenzori, valley of Lanuri, *Bequeart* 4676 (BR!, holo.; BM!, BR! iso.)
    S. dewildemanianum Robyns, F. P. N. A. 2: 209 (1947), *nom. nov.*, based on and type as for *S. longipedicellatum* De Wild.

NOTE. Bitter (1917) considered his new section *Bendirianum* to be monospecific when he proposed it for the North East African taxon "*S. bendirianum*". Though he was aware of the closely related species described by C.H. Wright (1902, 1906) as *S. runsoriense* he was unable to examine material of it, and being unsure of its precise affinity, placed this species between his two new sections *Afrosolanum* and *Bendirianum* (cf. Jaeger, 1985). These two species are closely related and indeed conspecific. Unfortunately, *S. runsoriense* was described earlier and so takes precedence over "*S. bendirianum*". The name change now necessary will undoubtedly cause confusion, since the latter is the more widely known and distributed of these two taxa, as well as being reflected in the sectional name. Engler (1892) and Dammer (1906) both spelt the specific name *benderianum* with an **e**, and Bitter (1917) subsequently adopted this spelling both for his sectional name and for the two varieties of this species that he described. However, Schimper (1863) and C.H.Wright (1906) spelt the name with an **i**. Neither Schimpers' nor Engler's names were validly published. Schimper's name was merely in manuscript form on herbarium labels, while Engler repeated Schimper's collection details but altered **i** to an **e** in the spelling of "*benderianum*", and failed to provide a validating description. This was subsequently provided by Wright (in February 1906) who published a description in English. Since this description pre-dates the 1935 nomenclatural rule necessitating a Latin description, Wright's 1906 description was validly published. Clearly Dammer followed Engler, when he re-described Schimper's taxon as *S. "benderianum"* using an **e** instead of an **i**, either as an orthographic error or because he interpreted the use of a capital B for *Benderianum* in Engler's account, intentionally to commemorate the German botanist Bender. Although Dammer also published his protologue in 1906, the date of publication was from August 1906, meaning that Wright's spelling taking precedence. Full bibliographical details of the references cited here are given under subsp. *bendirianum* below.

Wright (1902, 1906) described the leaf surfaces of *S. runsoriense* as being covered with scattered deciduous simple hairs above and dense stellate hairs below; he further described most parts of the inflorescences as being covered with stalked stellate hairs. These characteristics together with the fact that he was unable to examine representative material of this species resulted in Bitter deciding not to allocate it to his new section (see above). The multicellular hairs found in this section have complex branching, and are often pectinate with the basal part appearing stalk-like, but never stellate despite being described as such by several early authors.

De Wildeman described *S. longipedicellatum* as being endemic to the subalpine regions of Ruwenzori in the Congo-Kinshasa, and also mentioned its affinity to *S. runsoriense*; his protologue and the type specimens confirm the conspecificity of these two taxa.

Both Standley's (1916) protologue of *S. keniense* and the photograph of his holotype leave little doubt that the taxon he described is synonymous with *S. runsoriense* subsp. *runsoriense*. This is further supported by his observation that *S. keniense* was a very distinct species growing at an unusually high altitude (3630 m). Unfortunately, Turrill had previously used this epithet to describe a taxon belonging to the *S. indicum* L. Group, making Standley's name illegitimate.

b. subsp. **bendirianum** (*C.H. Wright*) *Edmonds*, **comb. nov.** Type: Ethiopia, near Gaffat, *Schimper* 1227 (?B†, syn.; BM!, CGE!, K!, E, isosyn.); Uganda, Ruwenzori, Kivala (not traced), *Scott Elliott* 7733 (K!, syn.; BM!, isosyn.)

Plants moderately pilose to glabrescent with simple and scattered short branched (dendritic) hairs; young stems and inflorescence branches and lower leaf surfaces glabrescent, non-ochraceous; upper leaf surfaces sparsely pilose to glabrescent, hairs simple; calyces sparsely pubescent to glabrescent. Fig. 19/12 & 13, p. 161.

UGANDA. Toro District: Ibonde Pass, Bwamba, 11 Nov. 1933, *A.S. Thomas* 767! & Ruwenzori range, 1915, *Fyffe* 20!
KENYA. Northern Frontier District: Mt Nyiro, S end of Lake Turkana [Rudolph], 28 June 1936, *Jex-Blake* 28! & Mt Nyiro, 30 Dec. 1955, *T. Adamson* 543! & Mt Nyiru, Mbarta Forest zone, 29 Mar. 1995, *Bytebier et al.* 28!
DISTR. U 2; K 1; restricted to Mt Nyiro, Elgon, Ruwenzori, widespread in Ethiopia; possibly also occurring in the Comoro Islands
HAB. Moist montane forest e.g. Juniper or *Podocarpus*, bamboo and *Hagenia- Hypericum* zone, in undergrowth and clearings, on forest margins and on tracks, above tree heath zone with *Echinops*, moorlands; 2400–2750 m

SYN. *S. bendirianum* Schimper in Pl. Abyss., nr. 1227 (1863); Engler in Hochgebirgsfl. Trop. Afr.: 372 (1892), *nom. nud.*
  S. bendirianum C.H. Wright in F.T.A. 4, 2: 212 (1906); Dammer in E.J. 38: 184 (1906); Bitter in E.J. 54: 487 (1917); Polhill, *Solanum* in E & NE Africa: 10 (ined., 1961); E.P.A.: 863 (1963); U.K.W.F.: 526 (1974); Jaeger, Syst. studies *Solanum* in Africa: 285 (1985, ined.); Hepper & Jaeger in Solanaceae Biol. & Syst.: 45 (1986); U.K.W.F. ed. 2: 242 (1994); K.T.S.L.: 580 (1994); Bukenya & Carasco in Bothalia 25(1): 50 (1995); Friis in Fl. Eth. 5: 115, fig. 158.5–7 (2006)
  ?*S. macrothyrsum* Dammer in E.J. 38: 185 (1906). Type: Comoro Islands, *Humblot* 387 (?B†, holo.; K!, lecto selected here; P, (photo!), W (photo!), isolecto.)
  S. bendirianum C.H. Wright var. *lanceolatum* Bitter in E.J. 54: 488 (1917). Type: Ethiopia, ?Sidamo, Kiritscha (Utadera), *Neumann* 62 (B†, syn.) & Sidamo, Jem-Jem, *Ellenbeck* 1761 (?B†, syn.)
  S. bendirianum C.H. Wright var. *ruwenzoriense* Bitter in E.J. 54: 489 (1917). Type: Uganda, Ruwenzori, Kivata, *Scott Elliott* 7733 (K!, holo.; BM!, iso.) *nom. illegit.* (based on one of syntypes of *S. bendirianum* C.H. Wright)

NOTE. The flowers of Ethiopian specimens of this subspecies are generally slightly smaller and appear more delicate than those found in the FTEA region. Both subspecies exhibit stylar heteromorphism with styles either being short and included, or long and exserted up to 4 mm beyond the anthers. Bitter (1917) first noted this phenemenon in *S. bendirianum*. Short-included styles seem to predominate in the flowers examined, with all flowers in a single inflorescence exhibiting styles of a uniform length. Jaeger (1985) noted that the short-styled form exceeded the long-styled form in a ratio of around 5:2. Jaeger's observation that the species exhibited unequal stamens, with one filament being shorter than the other four was not easily discernable. The floral situation is further complicated in this taxon by reports of the occurrence of andromonoecy; both Symon (in Biol. & Tax. Solanaceae: 385–397, 1979) and Whalen & Costich (in D'Arcy (ed.), Solanaceae: Biology and Systematics: 284–302, 1986) described inflorescences in the section *Bendirianum* as bearing a high proportion of hermaphrodite and staminate flowers. Symon discussed the evolution of this sexual dimorphism proposing that it occurred in response to the provision of pollen for specialised pollen vectors in these Solanums. Whalen & Costich's studies supported this, and they suggested that excess staminate flowers aid pollinator attraction, with gynoecial suppression allowing a flexible mechanism for controlling fruit set. The relevance of such observations to the subspecies of *S. runsoriense* requires extensive analysis of living inflorescences collected throughout their eco-geographical ranges together with genetic analyses, and is outside the scope of this Flora account. However, in those flowers examined, although there was no evidence of gynoecial reduction, there were few fruits on the specimens, with those occurring being unexpectedly small.

Superficially the lectotype of *S. macrothyrsum* is identical to the glabrescent subspecies of *S. runsoriense*, especially with regard to the form of the inflorescence, its vegetative and most floral features. It only differs in the shape of the corolla, which is strikingly stellate rather than rotate/stellate, and has ligulate lobes. Dammer (1906) mentioned the similarity of *S. macrothyrsum* to "*S. benderianum*" and Jaeger (1985) too thought that this species was probably synonymous with "*S. benderianum*". However, *S. runsoriense* was thought to be confined high altitude habitats in Uganda, Kenya and southern Ethiopia. The inclusion of *S. macrothyrsum* as a synonym would extend its distribution not only into the Comoros but also to very low altitudes.

8. **Solanum lycopersicum** *L*., Sp. Pl. 1: 185 (1753); Engler in P.O.A. C: 356 (1895); Bitter in E.J. 54: 500 (1917); Gonçalves in F.Z. 8(4): 69 (2005); Peralta, Spooner & Knapp, Taxonomy of Wild Tomatoes and their Relatives, in Syst. Bot. Monogr. 84; 133 (2008). Type: "Habitat in America calidiore", *Herb. Linnaeus* 248.16 (LINN!, lecto. designated by Deb in Journ. Econ. Bot., 1: 41 (1980)) [see also Knapp & Jarvis in J.L.S. 104: 342, f.12 (1990) & Jarvis, Order out of Chaos: 860 (2007)]

Annual herb, erect, procumbent, trailing or straggling up to 2 m in height, sometimes using associated vegetation for support; branches spreading or ascending, stems fleshy or herbaceous, light to dark green; all parts viscid-pubescent with eglandular and glandular hairs, denser on young vegetative and floral stems. Leaves alternate, imparipinnate, usually strongly aromatic, 9–31 × 6–14 cm with rachides 4.2–14 cm long, usually interrupted imparipinnate with 2–8 pairs of leaflets which light green above, ovate to ovate-lanceolate, 3–5(–7) × 1–3.8(–5) cm, bases cordate to cuneate, usually oblique, irregularly and usually deeply incised and serrate to sinuate, apices obtuse, acute or acuminate; surfaces viscid, pubescent below, sparser above, often stipulate and alternating with small entire ovate interstitial leaflets 0.8–1.2 × 0.6–0.7 cm; petioles 1.2–7 cm long, petiolules 0.4–2 cm, often with small pairs of leaflets. Inflorescences leaf-opposed to extra-axillary racemose simple or forked cymes, usually 5–12-flowered; flowers pendent, usually 5-partite but often 6–10-partite in cultivars; peduncles 0.5–3.5 cm long in flower, 2–4.8 cm long in fruit; pedicels 4–15 mm long in flower, 1–1.6 cm long in fruit, articulated at or above the middle from where deflexed in flower and in fruit; axes pubescent with short appressed and long spreading glandular-headed hairs. Calyx campanulate/stellate, divided almost to base, 3.5–11 mm long, densely pubescent externally; lobes narrowly triangular to lanceolate, 3–9 × 0.7–2 mm, acute, strongly reflexed in fruit when 7–14 × 1.5–2.8 mm. Corolla yellow, with greenish basal star, stellate, 1.4–2.4 cm diameter, tube 0.5–1.3 mm long; lobes narrowly triangular to lanceolate, 5–9 × 1.3–3(–4) mm, glandular-pubescent externally, glabrous internally, strongly reflexed exposing androecium after anthesis. Stamens equal; filaments flat, fused to each other forming a ring, 0.3–0.8 mm long, pubescent to glabrescent; anthers yellow, 5.5–7 × 0.6–1.3 mm, connivent to form a bottle-shaped cone 5–8 × 1.8–3.6 mm with sterile beak, dehiscing by lateral fissures from apex of fertile section. Ovary usually bilocular but plurilocular with false septa in cultivated varieties, 1.3–1.6 × 1.8–2 mm, ridged, glabrous; style straight, 5.5–7.3 × 0.2–0.4 mm, hairy in lower part, usually included, occasionally exserted to 1 mm; stigma green, capitate, sometimes bilobed, 0.2–0.8 × 0.3–0.5 mm. Fruit red, orange or yellow, globose, ovoid to pyriform berries, 1–10 cm diameter, often depressed or irregularly lobed, glabrous and smooth, pericarp thick, sweet-tasting. Seeds usually > 100, yellow to light brown, elliptic-ovoid, 3–3.8 × 1.7–2.4 mm, flattened, verrucate, often covered with the remnants of strands of thickening appearing as pseudo-hairs, enclosed in mucilage; sclerotic granules absent.

UGANDA. Mengo District: Entebbe, Xahu Ebone (?), Apr. 1923, *Maitland* 702!
KENYA. Marsabit District: Marsabit, Songa, 12 km KWS Gate–Songa, Feb. 2005, *Muasya* 490!; Nairobi, 22 Sep. 1916, *Dowson* 465!; Machakos District: Kilungu, Nduu village ± 1 km E Muindi Mutio's Garden, 23 Aug. 1971, *Mwangangi* 1710!
TANZANIA. Arusha District: Arusha National Park, near Longil Lake, 7 Apr. 1968, *Greenway & Kanuri* 13437!; Moshi District: Kilimanjaro Timbers, 3 Aug. 1994, *Grimshaw* 94674!; Mpanda District: Kungwe Mountain, Kasoje, 17 July 1959, *Newbould & Harley* 4411!
DISTR. U 4; K 4; T 2–4; native to the Andes of S America, but now cultivated and escaped throughout the world including Cape Verde Islands, Sierra Leone, Nigeria, Cameroon, Bioko, Central African Republic, Congo-Kinshasa, Sudan, Ethiopia, Angola, Zambia, Malawi, Mozambique, Zimbabwe, Namibia, South Africa; Madagascar, Comoro Islands, St Helena, Seychelles & Ascension Islands, Rodrigues and Mauritius, S, C & E Europe, the Middle East, India, China, N, C and S America but nowhere truly naturalised
HAB. Widely cultivated, and a common to occasional escape on roadsides and in waste places, frequently near habitation; often naturalised on river banks, in marshy areas, woodland, forest areas and open places; spread by birds and baboons; 750–2000 m

SYN. *Lycopersicon esculentum* Mill., Gard. Dict. ed. 8: no. 2 (1768); Dunal in DC., Prodr. 13(1): 26 (1852); Hiern in Cat. Welw. Afr. Pl. 1: 744 (1898); De Wild., Miss. Laurent: 441 (1907); Durand & Durand, Syll. Fl. Cong.: 395 (1909); Exell, Cat. Vasc. Pl. S Tome: 252 (1944); Polhill, *Solanum* in E & NE Africa: 8 (1961, ined.); E.P.A.: 881 (1963); Heine in F.W.T.A. 2nd ed., 2: 335 (1963); Heiser, Nightshades: 52–61 (1969); Hawkes in Fl. Europaea 3: 199 (1972); D'Arcy in Ann. Missouri Bot. Gard. 60: 649 (1973); Symon in Journ. Adelaide Bot. Gard. 3: 144 (1981) & 8: 14 (1985); Abedin *et al.*, in Pak. J. Bot. 23: 267 (1991); Huxley in New RHS Gard. Dict. 3: 138 (1992); D'Arcy & Ratotozafy, Fl. Madagascar, Solanaceae: 28 (1994); Hepper in Fl. Egypt Family 159, Solanaceae: 55 (1998); Mansfeld, Encycl. Ag. & Hort. Crops 4: 1833 (2001); Friis in Fl. Eth. 5: 146 (2006); Thulin in Fl. Somalia 3: 219 (2006). Type: cultivated in the Chelsea Physics Garden, *Miller* s.n. (BM, not found)
    L. lycopersicum (L.) Karsten & Farwell in Ann. Rep. Comm. Parks Boul., Detroit 11: 83 (1900); Purdie *et al.*, Fl. Australia 29: 175 (1994), *nomen rejicendum* based on *Solanum lycopersicum* L.
    L. cerasiforme Dunal in Hist. Solan.: 113 (1813), *nom. illeg.* (Dunal cited *Solanum lycopersicum* in synonymy)

NOTE. Commonly known as the Tomato, Golden- or Love-apple, this species is one of the most important global vegetables which is cultivated in most tropical and temperate countries, but which is now only known in cultivation. Over 500 cultivars have been bred resulting in enormous variations in fruit size, shape and colour. The small-fruited *L. esculentum* Mill. var. *cerasiforme* (Dunal) Alef., for example, is commonly known as the cherry tomato. The fruits are eaten raw or cooked, with commercial tomato sauce being made from the berry flesh, while the leaves are used as a vegetable. The species probably originated in Peru or Ecuador from where, in pre-Colombian times, it could have spread north as a weed (cf. Mansfeld, 2001). The considerable morphological diversity found in Mexico suggests that it was first domesticated there, and this may have been the source of the first tomatoes to reach Europe (cf. Heiser, 1969). It is widely grown throughout Africa; in the floral area it is cultivated for home consumption, and for sale in both local markets and the cities. Nevertheless, African herbarium specimens of this species are rare even though the tomato is considered to be the most important vegetable crop in most West African countries (Rouamba in van den Berg *et al.* (eds), Solanaceae V: 245–250 (2001)).

An extensive list of synonyms of this species – as *Lycopersicon esculentum* - is given in Mansfeld (2001) together with a bibliography of relevant references and in the more recent publication by Peralta *et al.* (2008). This species and related taxa contain steroidal glycoalkaloids, with the major component being α-tomatine.

9. **Solanum tuberosum** *L.*, Sp. Pl. 1: 185 (1753); Dunal, Hist. *Solanum*:117 (1813) & in Synopsis: 5 (1816) & in D.C. Prodr., 13(1): 31 (1852); Dumortier, Fl. Belg.: 382 (1827); G. Don, Gen. Hist. Dichlam. Pl. 4: 400 (1837); Richard in Tent. Fl. Abyss. 2: 97 (1851); Bitter in E.J. 54: 499 (1917); U.O.P.Z.: 448 (1949); Hawkes in Proc. Linn. Soc. 166: 106 (1956); Polhill, *Solanum* in E & NE Africa: 8 (1961, ined.); Correll, The potato and its wild relatives: 499 (1962); E.P.A.: 879 (1963); Heine in F.W.T.A. 2nd ed., 2: 335 (1963); Heiser, Nightshades: 28 (1969); Hawkes & Hjerting, Potatoes of Argentina, Brazil, Paraguay & Uruguay: 431 (1969); Hawkes & Edmonds in Fl. Europ. 3: 198 (1972); Symon in Journ. Adelaide Bot. Gard. 8: 77 (1985); Jaeger, Syst. studies *Solanum* in Africa: 313 (1985, ined.); Bukenya & Hall in Bothalia 18: 87 (1988); Hawkes, The Potato: 180 (1990); Abedin *et al.*, in Pak. J. Bot., 23: 272 (1991); Hawkes in Solanaceae III: 347 (1991); Huxley in the New RHS Gard. Dict. 4: 320 (1992); Purdie *et al.*, Fl. Australia 29: 70, 74 & 76 (1994); Bukenya & Carasco in Bothalia 25(1): 57 (1995); Hepper in Fl. Egypt 6: 48 (1998); Mansfeld, Encycl. Ag. & Hort. Crops 4: 1815 (2001); Gonçalves in F.Z. 8: 71 (2005); Friis in Fl. Somalia 3: 208 (2006) & in Fl. Eth. 5: 123 (2006). Type: ? Cult in Europe, "Habitat in Peru", *Herb. Linnaeus* 248.12 (LINN!, lecto, designated by Hawkes in Proc. Linn. Soc. London 166: 106 (1956)) [see also Knapp & Jarvis in J.L.S. 104: 358, f. 25 (1990) & Jarvis, Order out of Chaos: 862 (2007)]

Perennial herb, procumbent, creeping or sprawling to 1 m in height with long tuberiferous stolons; tubers extremely variable in colour and shape, small to large; branches spreading or ascending, often flexuous, stems angular or winged, pilose with simple, appressed, eglandular occasionally glandular hairs. Leaves alternate, imparipinnate, 8–28 × 6.5–16 cm, rachides 6–10 cm long, with 3–5 alternate or opposite primary leaflets, ovate to lanceolate and decreasing in size towards the base, 3–6(–10) × 2–4(–7) cm, bases cuneate, entire, apices acute, often oblique, terminal often larger and sessile, usually alternating with small sessile interstitial leaflets 0.4–3 × 0.3–2 mm; surfaces pilose with eglandular hairs, denser below; petioles (and rhachides) often angular or narrowly winged, 2–4(–12) cm long, petiolules absent to 2 mm. Inflorescences terminal becoming lateral with sympodial growth continued by axillary branches, forked cymose panicles, 7–18(–30)-flowered; peduncles 6–12 cm long; pedicels usually erect in flower and in fruit, 1.6–2 cm long, articulated above the middle; axes pilose/villous with long eglandular hairs, dense around point of pedicel articulation. Calyx campanulate/stellate, 6–7 mm long, pilose/villous externally with scattered long eglandular hairs; lobes broadly triangular tapering narrowly above, 3–9 × 2–3 mm, acute. Corolla white, blue or violet, often with white radial stripes between the lobes and greenish-yellow basal star, rotate to rotate-pentagonal, 1.8–3.5 cm diameter, tube 1.8–2 mm long; lobes broadly triangular, 3–5 × 3–5 mm, tips pubescent, glabrous internally. Stamens equal; filaments free for 1–2.2 mm, glabrous; anthers yellow to brown, 4–6.2(–8.5) × 1.1–3 mm. Ovary 2.2–4 × 1.8–3.5 mm, glabrous, usually bilocular; style straight, sometimes twisted or sigmoidal, 6.5–13 × 0.2–0.4(–0.8) mm, glabrous above, minutely papillate below, exserted to 3 mm; stigma capitate to clasping, 0.5–1 × 0.5–1.3 mm. Fruit green, globose to ovoid berries, 1–2(–4) cm diameter, glabrous and smooth, rarely produced in cultivation. Seeds up to 500, ovoid to elliptic-ovoid, ± 2 × 1.7–2.4 mm, flattened, often covered with the remnants of strands of thickening appearing as pseudo-hairs; sclerotic granules absent.

KENYA. Nairobi, July 1961, *Polhill* 12425!; Machakos District: Kilungi, E of Kauti Fall Primary School, 9 Jan. 1972, *Mwangangi* 1956!

DISTR. **K** 4; cultivated in Ghana, Cameroon, Somalia, Ethiopia, Uganda, Tanzania, Malawi, South Africa; Europe, Middle East, India, China, Australia and New Guinea, N, C and S America, sometimes occurring as an escape

HAB. Commonly cultivated in gardens and vegetable plots, but rarely collected; 1650–1950 m

SYN. *Lycopersicon tuberosum* Mill., Gard. Dict. ed. 8: no. 7 (1786). ?Based on and type as in *Solanum tuberosum esculentum*? [none cited apart fom C.B.P. 167 and refers to Linnaeus in discussion]; type specimen not found

[The numerous synonyms of this species may be found in Orchinnikova *et al.* in Bot. J. Linn. Soc. 165: 107–155 (2011)]

NOTE. The potato is probably the best-known and economically important member of the Solanaceae ranking among the top five of the world's most important crop species. It is indigenous to the Andes of Peru, Chile, Bolivia and adjacent parts of South America, with Hawkes (1991) suggesting that its cultivation probably originated in northern Bolivia. The potato was introduced to Europe by the Spaniards in the 1570s, and is commonly known as the White, European or Irish potato. Widely cultivated for its edible tubers or potatoes in temperate and tropical mountain regions both throughout Africa and world, it was introduced into East Africa approximately 100 years ago, though West African introductions occurred much later - after the Second World War. Correll (1962), however cited evidence that potatoes were grown in South Africa in 1833; he thought this may have indicated that some representatives of section *Tuberarium* (Dunal) Bitter occurred naturally in Africa, suggesting that an indigenous species might occur there possibly due to Continental Drift. Jaeger (1985) considered this highly unlikely, stating that no close relatives of section *Petota* are indigenous in Africa. Despite its widespread use as an important vegetable throughout these regions, very few herbarium specimens of this species have been collected. Though frequently planted, this species is rarely naturalised. References to the vast literature dealing with the origins, distribution, inter-relationships with wild species and the cultivars of this tetraploid species may be found in Correll (1962) and Hawkes & Hjerting (1969), while a summary of the evolution and domestication of the potato is given in Hawkes (1991). Mansfeld (2001) similarly gives relevant references to the two subspecies commonly recognized in this species: subsp. *tuberosum* and subsp. *andigena* (Juz. & Buk.) Hawkes together with their differentiating features, origins and probable current distribution ranges. The vegetative parts and immature berries of this species contain high glycoalkaloid contents, resulting in 'green' potatoes being commonly regarded as poisonous.

10. **Solanum nigrum** L., Sp. Pl. 1: 186 (1753); Pers., Syn. Pl., 1: 224 (1805); Dunal, Synopsis: 12 (1816); G. Don, Gen. Hist. Dichlam. Pl. 4: 412 (1837); A. Rich., Tent. Fl. Abyss. 2: 99 (1851); Dunal in DC., Prodr. 13(1): 50 (1852); Lowe, Man. Fl. Madeira 2: 73 (1872); Boissier, Fl. Orient. 4: 284 (1879) *pro parte*; Clarke in Fl. Brit. India: 229 (1885) *sensu lato*; Engler in Hochgebirgsfl. Trop. Afr.: 372 (1892); E. & P. Pf.: 22 (1895); P.O.A. C: 351 (1896); Hiern, Cat. Afr. Pl. Welw. 3: 745 (1898); C.H. Wright in Fl. Cap., 4(1): 89 (1904) *sensu lato*; Dammer in E.J. 38: 178 (1907); C.H. Wright in F.T.A. 4, 2: 216 (1906) *sensu lato*; Durand & Durand, Syll. Fl. Cong.: 393 (1909); Post, Fl. Syria, Palestine & Sinai: 257 (1933); F.P.N.A.: 874 (1947) *pro parte*, W.F.K.: 89 (1948); Wessely in F.R. 63: 290 (1960); E.P.A.: 874 (1963); Heine in F.W.T.A. 2nd ed., 2: 335 (1963); Hawkes & Edmonds in Fl. Europ. 3: 197 (1972); Schönbeck-Temesy in Fl. Iran.: 7 (1972) *sensu lato*; Ross, Fl. Natal: 308 (1972); U.K.W.F: 526 (1974) *sensu lato*; Täckholm, Students Fl. Egypt, ed. 2: 473 (1974) *pro parte*; Henderson in Contrib. Queensland Herb. 16: 19 (1974); Karschon *et al.* in Israel Journ. Bot. 27: 94 (1978); Ormonde in Pl. Cabo Verde 8: 177 (1980): Symon in Journ. Adelaide Bot. Gard. 4: 46 (1981); Purdie *et al.*, Fl. Austral. 29: 99 (1982); Jaeger, Syst. studies *Solanum* in Africa: 299 (1985, ined.); Troupin, Fl. Rwanda: 380 (1985); Karschon & Horowitz in Phytoparasitica 13(1): 63 (1985); Bukenya & Hall in Bothalia 18(1): 82 (1988); Abedin *et al.* in Pak. Journ. Bot. 23(2): 270 (1991); Blundell, Wild Fl. E. Afr.: 190 (1992) *sensu lato*; Huxley *et al.*, New RHS Dict. Gard. 4: 319 (1992); U.K.W.F. ed. 2: 242 (1994); Bukenya & Carasco in Bothalia 25(1): 46 (1995); Edmonds & Chweya, Black Nightshades: 28 (1997); Hepper in Fl. Egypt, Family 159: 18 (1998); Schippers in Afr.

Indig. Veget.: 188 (2000) & in Légum. Afr. Indig.: 373 (2004); Mansfeld, Encycl. Ag. & Hort. Crops 4: 1807 (2001); Hepper in Fl. Egypt 3: 39 (2002); Edmonds in Fl. Eth. 5: 117 (2006); Olet, Taxonomy *Solanum* section *Solanum* in Uganda: papers i-vi (2004) & Olet *et al.* in African Journal of Ecology, 43:158 (2005); Gonçalves in F.Z. 8(4): 85 (2005); Manoko, Syst. Study African *Solanum* sect. *Solanum*: chapters 1, 5–8 (2007). Type: Probably from Europe, *Herb. Linnaeus* 248.18 (LINN!, lecto. designated by Henderson in Contrib. Queensland Herb., 16:19 (1974))

Annual to perennial herb, erect, decumbent or prostrate, 0.1 to 1.1 m high, main stem sometimes woody, spreading up to 2 m; stems smooth or with inconspicuously dentate ridges, villous to glabrous with simple hairs with glandular or eglandular heads, mixed with glands, often densely leafy. Leaves membranaceous, light to dark green, ovate, ovate-rhomboidal or ovate-lanceolate, rarely lanceolate, 2.5–9.5 × 1.8–6 cm, bases cuneate and decurrent, margins entire, sinuate or sinuate-dentate with a few (< 4) acute or obtuse lobes, apices acute to acuminate, rarely obtuse; surfaces pubescent or glabrescent as stems, hairs usually denser on margins, veins, midribs and on lower surfaces; petioles 0.5–4.2 cm. Inflorescences simple leaf-opposed to extra-axillary cymes, usually lax appearing racemose with fruiting rachis (0–)2–7 mm long, (3–)5–10-flowered; peduncles erect and 6–18 mm in flower, erecto-patent and 11–19(–30) mm long in fruit, pubescent as stems; pedicels erect and 4–8 mm long in flower, erecto-patent often becoming recurved and 8–16 mm in fruit, pubescence as stems. Calyx campanulate, (1.5–)2–3(–4) mm long, pilose externally; lobes broadly ovate, 0.5–1.8 × 0.5–1.6 mm, obtuse, enlarging to 1.4–3 × 1.5–3 mm in fruit. Corolla white or tinged purple, with yellow-translucent basal star, stellate, (0.8–)1.2–1.6 cm diameter, tube 1–1.5 mm long; lobes broadly ovate, (2.7–)4–6 × (2.8–)3–4(–5) mm, spreading or reflexed after anthesis. Filaments free for 1–1.5 mm, densely villous internally; anthers yellow to orange, 1.5–2.5(–2.7) × 0.6–0.8(–1) mm. Ovary brownish, 0.7–2 × 0.6–2 mm, glabrous; style never geniculate, (2.4–)3–4.5 × 0.2–0.4 mm, lower half shortly pilose, not or only the stigma exserted at full anthesis; stigma capitate, 0.2–0.5 mm diameter. Berries dull purple to black and dull or shiny becoming dull with opaque cuticles in Flora area, sometimes greenish to yellowish-green with translucent cuticles in Eurasia, usually broadly ovoid, sometimes globose, 5–10 × 5–11 mm, basal calyx lobes adherent becoming reflexed; falling from calyces or remaining on plant when ripe. Seeds (15–)26–60(–96) per berry, yellow to brown, ovoid, discoid or orbicular, (1.7–)2–2.4 × 1.4–2 mm; sclerotic granules absent. Fig. 17/15 & 16, p. 126.

subsp. **nigrum** *Wessely* in F.R. 63: 306 (1960); Hawkes & Edmonds in Fl. Europ. 3: 197 (1972); Henderson in Contrib. Queensland Herb. 16: 20 (1974); Ormonde, Pl. Cabo Verde 8: 177 (1980); Bukenya & Hall in Bothalia 18(1): 82 (1988); Abedin *et al.,* in Pak. Journ. Bot. 23(2): 271 (1991); Bukenya & Carasco in Bothalia 25(1): 47 (1995); Edmonds & Chweya, Black Nightshades: 31 (1997); Mansfeld, Encycl. Ag. & Hort. Crops 4: 1807 (2001); Edmonds in Fl. Eth. 5: 117 (2006)

Plants moderately pubescent to glabrescent with appressed, eglandular-headed multicellular hairs.

UGANDA. ?West Nile District: Jumbi, 26 Nov. 1941, *A.S. Thomas* 4070!; without locality, May 1880, *Wilson* 77!; Toro District: Kitagwenda, 30 July 1906, *Bagshawe* 1121!
KENYA. Uasin Gishu District: near Kapsaret Forest Reserve, 15 June 1951, *Williams Sangai* 241!; Masai District: Narok, Mara River, Ngerendei, 29 Mar. 1961, *Glover et al.* 180!; Nairobi District: Nairobi, recd. 1916, *Dowson* 482!
TANZANIA. Masai District: Oldonyo Lengai, E Rift Wall, 9 July 1931, *Clair Thompson* 204!; Kondoa District: Great North Road, 11 km S Kondoa, 18 Jan. 1962, *Polhill & Paulo* 1207!; Iringa District: top of Mpululu Mountain, 21 May 1968, *Renvoize & Abdallah* 2312!
DISTR. U 1–3; K 3, 4, 6; T 1–3, 5–8; ?Sierra Leone, Ghana, Nigeria, Cameroon, Bioko, Sudan, Eritrea, Ethiopia, Somalia, Angola, Zambia, Malawi, Mozambique, Zimbabwe, Namibia, Botswana, South Africa; Madagascar, Madeira, Azores, Europe, the Middle East and Asia; introduced and often naturalised in North America, Australia and New Zealand

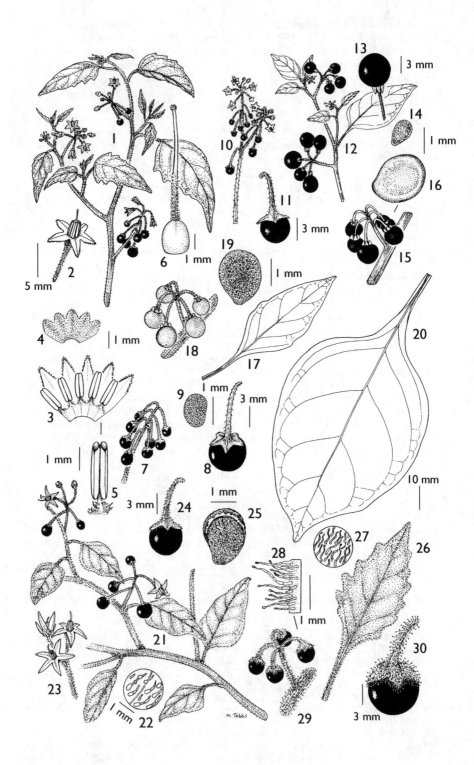

M. Tebbs

HAB. Old or secondary cultivation areas, grassland, forest edges and in clearings or on stream banks in upland forests, on mountain tops and on rocky hillsides with *Brachystegia*, *Acacia* and *Commiphora;* (15–)1200–2200 m

SYN. ?*S. nigrum* L. var. *judaicum* L., Sp. Pl. 1: 186 (1753) & Willd. ed. 4, 1: 1035 (1797); Pers., Syn. Pl. Bot.: 224 (1805). Type: "throughout the world"; specimen not designated and no type material located

S. suffruticosum Willd. in Enum. Hort. Berol: 236 (1809); Dunal in Hist. *Solanum*: 154 (1813) & in Synopsis: 13 (1816); G. Don, Gen. Hist. Dichlam. Pl. 4: 413 (1837); E. & P. Pf.: 22 (1895); Dunal in DC., Prodr. 13(1): 53 (1852); Hiern, Cat. Afr. Pl. Welw. 3: 745 (1898); Dammer in E.J. 38: 178 (1907); E.P.A.: 878 (1963). Type: Morocco [Barbaria], *Schousboe* s.n. (B-W 4363!, holo.)

?*S. judaicum* (L.) Bess., Fl. Gallic., 1: 183: (1809); Roem. & Schultes, Syst. Veg. 4: 589 (1819); G. Don, Gen. Hist. Dichlam. Pl. 4: 413 (1837); Komarov, Flora USSR 22: 28 (1955) [There is a *Besser* s.n. specimen in LE! labelled as an isotype (see below)]

S. moschatum Presl, Delic. Prague: 77 (1822); E. & P. Pf.: 22 (1895). Type: cultivated in Sicily, Panormi; type specimen not cited

?*S. luteo-virescens* Gmel., Fl. Badensis 4: 177 (1826). Based on and type as for *S. humile* Bernh. *non* Lam., (see page 132 under *S. villosum* subsp. *miniatum*)

S. vulgatum L. var. *chlorocarpum* Spenn. in Fl. Frib. 3: 1074: (1829); type specimen not cited

?*S. nigrum* L. subsp. *luteovirescens* (Gmel.) Kirsch., Fl. Alsace 1: 532 (1852)

S. nigrum L. subsp. *humile* (Bernh.) Hartman in Svensk & Norsk Exc.-Flora: 34 (1846)

S. nigrum L. var. *glabrum* Lowe, Man. Fl. Madeira 2: 73 (1872). Type: Portugal, Madeira, Funchal, the Mount & Sta Anna, Madeira, *Lowe* s.n. (BM!, *Lowe* 16, lecto. designated here).

S. nigrum L. subsp. *chlorocarpum* (Spenn.) Arcang. in Comp. Fl. Ital.: 497 (1882)

S. morella Desv. subsp. *nigrum* (L.) Rouy, Fl. France 10: 364 (1908)

S. kifinikense Bitter in F.R 10: 545 (1912). Type: Tanzania, Kilimanjaro, Kifinika volcano, *Volkens* 1909 (B†, holo.; HBG!, lecto., designated here; BR!, isolecto.)

S. nigrum L. var. *incisum* Täckholm & Boulos in Publ. Cairo Univ. Herb., 5: 101, F.16a (1974); Täckholm, Student's Flora Egypt, ed. 2: 473 (1974); Hepper in Flora of Egypt 6: 19 (1998) & in Fl. Egypt 3: 39 (2002). Type: Egypt, Faiyum, Sinnuris, *Boulos* s.n. (CAI, holo.)

NOTE. Possibly originating in Eurasia, this taxon is now widespread in virtually all parts of the world, its natural distribution is difficult to ascertain. It is the generic type species, is hexaploid ($2n=6x=72$) and is one of the most morphologically variable of those in the section *Solanum*. In Australia, Symon (1981) was able to recognise several biotypes of this species though he avoided giving them any formal taxonomic recognition. This is the only subspecies of *S. nigrum* to be found in the Floral area; the subsp. *schultesii* (Opiz) Wess., characterised by spreading glandular-headed hairs, is common in warm, dry areas throughout Europe, the Middle East and Asia.

The species described as *S. nigrum* in most of the cited Floras undoubtely comprise a mixture of several different species belonging to the Section *Solanum* with the name *S. nigrum* being used to embrace any specimen showing an affinity with the true species. This is certainly true in Africa, where most specimens superficially resemble the true *S. nigrum* and have been identified as such. However, increasing knowledge of the morphological variation and distribution of its endemic relatives is demonstrating that *S. nigrum* sensu stricto is relatively rare, especially in East Africa. Though this species has been introduced into many African countries, it is thought to occur only sparingly in some, and is more widespread in temperate rather than tropical regions, and at higher altitudes.

FIG. 17. *SOLANUM TARDEREMOTUM* — **1**, flowering and fruiting habit; **2**, complete flower; **3**, opened flower; **4**, opened calyx; **5**, stamen; **6**, gynoecium; **7**, infructescence; **8**, berry; **9**, seed. *S. FLORULENTUM* — **10**, infrutescence; **11**, berry. *S. AMERICANUM* — **12**, flowering and fruiting habit; **13**, berry; **14**, seed. *S. NIGRUM* — **15**, infructescence; **16**, seed. *S. VILLOSUM* — **17**, leaf; **18**, infructescence; **19**, seed. *S. SCABRUM* — **20**, leaf. *S. PSEUDOSPINOSUM* — **21**, flowering and fruiting habit; **22**, upper leaf indument; **23**, inflorescence; **24**, berry; **25**, seed. *S. MEMPHITICUM* — **26**, leaf; **27**, leaf indument; **28**, stem indument; **29**, infructescence; **30**, berry. 1–6 from *Richards* 13004; 7–9 from *Richards* 21308; 10–11 from *Laboratory Staff* 2153; 12–14 from *Faulkner* 3120; 15–16 from *Thomas* 4070; 17 from *Greenway* 10635; 18–19 from *Hepper & Field* 5500; 20 from *Olof* 048; 21–22 from *Mann* 1938; 23–25 from *Lugard & Lugard* 492; 26–28 from *Greenway & Kanuri* 11332; 29–30 from *Kirima* 241. Drawn by Margaret Tebbs.

Foche *et al.* (2002) considered that *S. nigrum* was not grown as a vegetable in Africa, and probably only occurred in cooler parts of the continent. Olet (2004, 2005, 2006) later concluded that plants widely known as *S. nigrum* in Uganda have been misidentified and are a wild form of the cultivated *S. scabrum* which she described as subsp. *laevis* (see page 130). Her morphological description of this together with its illustration are very similar to plants usually identified as *S. nigrum* but the Ugandan material cited by Olet has not been seen. She differentiated her new taxon from *S. nigrum* sensu stricto on leaf thickness, margin sinuation and style posture and exsertion, characters which are among those usually found to be variable in this species complex. Whether the specimens thought to be *S. nigrum* sensu stricto are those sold in local markets, and eaten as a vegetable in **K** 3, **U** 1 & 3 and **T** 3, or a different taxon as suggested by Olet, clearly requires further investigation.

Cufodontis (in E.P.A.,1963) regarded *S. suffruticosum* as a distinct species, citing the *Schousboe* specimen collected in "Barbaria" as the type. However, the *Willdenow* holotype is clearly conspecific with the eglandular subspecies of *S. nigrum* (including having pollen in the hexaploid range of (25.1–)27.4(–29.5) µ). *Solanum suffruticosum* is therefore considered a synonym of the subsp. *nigrum.*

With the exception of the var. *judaicum,* Linnaeus' other four varieties of *S. nigrum* were based on taxa described by earlier authors. His brief description gave little indication of its true identity, merely describing stems with incurved spines and glabrous, repand leaves. Specimens of this taxon were possibly collected by Hasselquist, a Linnean student, from around the Mediterranean, with *judaicum* being a popular epithet for species from this area. Besser (1809), later elevated this variety to specific status (see below), elaborating on the diagnostic characters and adding that the berries were black. No details of type material or geographical location were given. Several possible authentic Besser specimens have been located in MPU, W, and LE, and appear to belong to two different taxa – *S. nigrum* sensu stricto and *S. villosum* subsp. *miniatum.* That in LE has been labelled [wrongly, ed.] as an isotype; the label appears to be in Besser's hand and it is morphologically conspecific with *S. nigrum* subsp. *nigrum.* Presumably this taxon was described from plants common in northern and central Europe at the time, which were most likely to have been of the ubiquitous *S. nigrum.* However, Schultes (in Osterreichs Flora, 393 (1814)) mentioned that Besser's species was described from "Galizien (unter des Israeliten)", and Don (1837) called it the "Jewish Nightshade", raising the possibility that it might belong to the Middle Eastern *S. sinaicum,* whose berries are yellow turning black. Historically, Galizien refers to Galicia, part of Poland acquired by the Austrians after its partition in 1772. This would eliminate *S. sinaicum* and reinforce the synonymy of *S. nigrum* var. *judiacum* with *S. nigrum* sensu stricto.

Komarov's (1955) description of *S. judaicum* and his illustration clearly refer to *S. nigrum* subsp. *nigrum,* though he thought it might be endemic to the European part of the USSR; he cited types of this species as being in Kiev and Leningrad.

In his very brief description of subsp. *humile,* Hartman (1846) repeated the description of the berry colour as yellowish green, and also included the red-berried subsp. *miniatum* in his floral account, suggesting that his taxon is synonymous with *S. nigrum.*

Lowe described his var. *glabrum* as being one of the commonest weeds in his cited localities. The sheet *Lowe* 16, labelled as this taxon, is composed of four specimens, none of which mentions Funchal. One (b) is labelled "everywhere in roads ...., on walls, June 31 1828, and another (a) as having been collected from "Mr Gordons' kitchen garden at the Mount". They are all young specimens but from their morphology and Lowe's protologue they seem to be conspecific with *S. nigrum* subsp. *nigrum.*

Though the Berlin holotoype of *S. kifinikense* was destroyed, good duplicates survive. The habitat collection details on these differ from those cited in the protologue, but the HBG specimen has been selected as the lectotype of this species, since its morphological features match those given in the protologue. The pollen from this specimen falls within the parameters expected for this hexaploid species.

Several varieties of *S. nigrum* have been recognised in Egypt; the latest Floral account (Hepper 2002) included two of these in addition to the var. *nigrum.* These are characterised by minor differences in leaf shape and ecogeographical distribution; though one is described as endemic, from its protologue and holotype photograph, the third variety *incisum* is synonymous with subsp. *nigrum.*

11. **Solanum scabrum** *Mill.,* Gard. Dict. ed. 8: *Solanum* No. 6 (1768); E.P.A.: 876 (1963); Henderson in Contrib. Queensland Herbarium 16: 61 (1974); Edmonds in J.L.S. 78: 224 (1979); Symon in Journ. Adelaide Bot. Gard. 4: 53 (1981); Jaeger, Syst. studies *Solanum* in Africa: 298 (1985, ined.); Gbile & Adesina in Fitoterapia 56: 11

(1985); Bukenya & Hall in Bothalia 18(1): 82 (1988); D'Arcy & Rakotozafy, Fl. Madag.: 124 (1994); Bukenya & Carasco in Bothalia 25(1): 47 (1995); Edmonds & Chweya, Black Nightshades: 38 (1997); Mansfeld, Encycl. Ag. & Hort. Crops 4: 1809 (2001); Schippers in Afr. Indig. Veget.: 188 (2000) & Légum. Afr. Indig.: 379 (2004); Focho *et al.* in Plant Genetic Resources Newsletter 131: 42 (2002); Gonçalves in F.Z. 8(4): 71 (2005); Edmonds in Fl. Eth. 5: 117 (2006); Olet, Taxonomy *Solanum* section *Solanum* in Uganda: papers i-vi (2004); Olet *et al.* in Novon 16(4): 508 (2006); Manoko, Syst. Study African *Solanum* sect. *Solanum*: chapters 1, 5–8 (2007). Type: cultivated in Chelsea Physic Garden, introduced from North America, *Miller* s.n. (BM!, lecto. designated by Henderson in Contrib. Queensland Herbarium, No. 16: 61 (1974))

Annual or short-lived perennial herb, erect, to 1 m high, base occasionally woody; stems prominently and dentately winged, often purple especially basally; lateral branches sparse and usually spreading horizontally up to 1.5 m, pilose to glabrescent with scattered eglandular hairs on mature parts. Leaves often coarse, light to dark green or purple, broadly ovate, sometimes rotund or ovate-lanceolate, 8–12(–25) × 6–9(–20) cm, bases truncate and decurrent onto stem, margins entire to sinuate, never lobed, apices acute to obtuse, occasionally cordate; surfaces sparsely pilose (especially on lower veins and midribs) to glabrescent; petioles 3.2–8.2(–10) cm. Inflorescences simple or forked (sometimes multiply), umbellate to lax cymes, leaf-opposed to extra-axillary 5–17(–30+)-flowered; peduncles erect, 1–5 cm long in flower, 1.5–6 cm long in fruit; pedicels erect and 8–13 mm in flower, erect or spreading, sometimes recurved, and 8–17 mm long in fruit; axes sparsely pilose to glabrescent. Calyx campanulate, 1.5–3(–4.5) mm long, moderately pilose externally; lobes obovate to broadly triangular, 0.9–2 × 1.2–1.6 mm, obtuse, sometimes irregular, enlarging to 2–2.8 × 2–2.5 mm when in fruit. Corolla white or tinged purple with yellowish-green basal star, stellate, 1.1–1.8 cm diameter, tube to 1.2 mm long; lobes broadly ovate, 5–7 × 2–5.5 mm, pilose externally with ciliate margins, spreading after anthesis. Filaments free for 0.8–1.5 mm, densely villous internally; anthers typically purple to purplish-brown but often orange/yellow in Africa, 2–2.9(–3.3) × 0.8–1 mm. Ovary brown, 0.8–1.5 × 1.5–1.8 mm, glabrous; style 2–4(–4.5) × 0.2–0.4 m, lower third densely pilose, not usually exserted beyond anthers; stigma capitate, 0.3–0.5 mm diameter. Berries usually borne on erect pedicels, deeply purple to black, broadly ovoid with tough opaque cuticles, glossy becoming dull, 1–2(–2.4) × 0.8–1.4 cm, smooth, surrounded basally by adherent ovate to triangular calyx lobes which become reflexed, remaining on plant at maturity. Seeds numerous, up to 144 per berry, yellow, brown or purple, obovoid, discoidal, 2–2.3 × 1.5–1.7 mm, reticulate; sclerotic granules absent. Fig. 17/20, p. 126.

UGANDA. Kigezi District: Bungangari, Ruzhumbura, Feb. 1949, *Purseglove* 2172! & 28 Aug. 1972, *Goode* 3/72! & Kabale, Kachwekano DFI, 25 Sep. 2000, *Olet* 048! & 049!
DISTR. U 2; extensively cultivated in Nigeria and the Cameroon as well as in the Ghana, Congo, Gabon and the Ivory Coast; found sporadically in Sierra Leone, Liberia, Angola, Zambia, Namibia, South Africa, and Madagascar; a casual introduction in Europe, Australia, New Zealand and North America (where it is sometimes cultivated)
HAB. Widely cultivated as a leaf vegetable or garden plant, often occurring as an escape and a weed of crops such as beans or maize; 1200–2150 m

SYN. *S. nigrum* L var. *guineense* L., Sp. Pl.: 186 (1753) – pro *Solanum guineense fructo magno, instar cerasi* [*nigerrimo, umbellato*]. Dillenius, Hort. Eltham. 2: 366, t. 274 (1732); Hiern, Cat. Afr. Pl. Welw. 3: 745 (1898). Type: Dillenius, Hort. Elth., 366; t. 274, f. 354, (1732) ! (lecto. designated by Edmonds in J.L.S. 78: 224 (1979))
    S. guineense (L.) Mill., Gard. Dict., ed. 8: *Solanum* No. 7 (1768), *non* L., *nom. illegit.*; Dammer in E.J. 38: 177 (1906). Type: cultivated in Chelsea Physic Garden (BM!, lecto.)
    S. melanocerasum All., Auct. Syn. Stirp., Horti Taur.: 12 (1773), pro *Solanum guineense, fructo magno, instar cerasi nigerrimo, umbellato,* Dillenius, Hort. Elth., 366 ["336"] (1732); Hawkes & Edmonds in Fl. Europ. 3: 197 (1972); Huxley *et al.*, New RHS Dict. Gard. 4: 319 (1992). Type: *Dillenius* 336 (OXF!, lecto., selected here)

S. guineense Lam., Encycl. Meth. Bot., 2: 18, no. 2339 (1793); G. Don, Gen. Hist. Dichl. Pl., 4: 411 (1837); P.O.A. C: 351 (1895). Type: Guinea, *Lamarck* s.n. (P-LAM!, lecto.)

S. guineense (L.) Lam., Illustr., 2: 18 (1797); P.O.A. C: 351 (1895); Dammer in E.J. 38: 177 (1907), *nom. illegit.*

S. nodiflorum sensu Hiern, Cat. Afr. Pl. Welw. 3: 745 (1898)

S. nigrum L. subsp. *guineense* (L.) Pers., Syn. Pl. 1: 224 (1805)

S. memphiticum Mart. in Pl. Hort. Erlangensis: 63 (1814). Type ? hort. *Erlang* s.n. (not located)

S. tinctorium Welw., Apont. Phytogeo. Angola, 1: 590 (1859). Type: Angola, Golungo Alto, *Welwitsch* 6103 (BM!, ?lecto & ?isolecto., cited by Edmonds in J.L.S. 78: 224 (1979)

S. intrusum Soria in Baileya 7: 33 (1959) *nom. nov.* pro *S. guineense* (L.) Mill., *non* L.

S. nodiflorum [no authority] sensu F.P.U.: 130 (1962)

S. scabrum Mill. subsp. *nigericum* Gbile in Fitoterapia 56(1): 11 (1985), *nom. nud.*

NOTE. The complex synonymy surrounding the correct use of the epithet *guineense* specifically for a species possibly belonging to the section *Cyphomandropsis* is considered by Nee (ined., 1994), while the confusing use of it to describe several variants of *S. nigrum* is discussed by Heine (Kew Bull. 14: 245 (1960)). Though invariably described as being native to Guinea, the origin of this hexaploid species, which is often known as the Garden Huckleberry and which Linnaeus called *S. nigrum* var. *guineense*, is unknown. Heine stated that it does not occur wild in Guinea, where it is cultivated as a potherb and medicinally, and he doubted that it is native to any part of Africa.The species shows its greatest diversity in Nigeria and Cameroon, where it is a popular leafy vegetable - suggesting that it could have originated in the warm humid forests of western Africa. Indeed the work by Focho *et al.* (2002) on the variability exhibited by Cameroonian populations of this species, together with the many different names for *S. scabrum* in West African languages are suggestive of a West African origin for this species. These authors considered that it is probably Africa's most important indigenous leafy vegetable, widely grown in the humid forest zone of West and Central Africa, but also cultivated for its leaves in more than 20 African countries. It is now increasingly cultivated in Uganda, Kenya and Tanzania for its fruits and leaves (*cf.* Schippers, 2000 & 2004) though Olet *et al.* (2005) recorded that this species is only cultivated in Kigezi in Uganda. Focho *et al.* (2002) considered it to be of minor importance in East Africa, where the less bitter *S. villosum* is a more popular vegetable. Very little herbarium material has been collected from these areas.

Schippers (2000) noted that the variation exhibited by this species has resulted in the recognition of many different cultivars in West Africa – mainly of local importance – and this inherent variation has generated the publication of various infra-specific names by some African workers. The subspecies *nigericum* Gbile, for example, was based on the lectotype of *S. scabrum*, and was illegitimate. More recently, Olet (in Novon, 16(4): 510 (2006)) described the subspecies *laevis* Olet for a variant that she considered to be the wild form of *S. scabrum* in Uganda. Basing her conclusions on a variety of systematic analyses, (cf. Olet, 2004) she suggested that these plants are those widely identified as *S. nigrum* (see page 124). Olet further suggested that although this new subspecies was described from Uganda, this area probably represents the eastern edge of its distribution; she considered it to be a West African plant. Among the characters used to differentiate this new taxon from the subspecies *scabrum* were terete edentate stems, simple inflorescences, geniculate styles and smaller leaves and fruits. Type material cited for this subspecies has not been seen (the paratypes cited by Olet as being at K could not be found), so this subspecies has neither been included nor treated as a synonym in this account. However, Manoko's (2007) analyses failed to support the recognition of these two subspecies of *S. scabrum*.

The berry cuticles of *S. scabrum* sensu stricto are extremely tough, and the ripe berries often split open while still in situ leaving a sparsely foliated plant covered with large blackish/purple fruits. The deep anthocynanin pigmentation characterising this species has led to the berries being used as a source of ink in the Congo and Zambia. Three variants are recognised in the Cameroon; the so-called Foumbot type which is grown on a large scale as a monocrop and is characterised by green stems and large dark green leaves; the purple-leaved type Bamenda and and an intermediate-sized leaf type found in Buea. Various regions favour certain types and degrees of bitterness (Focho *et al.*, 2002). Though purplish or brownish anthers characterise this species, African specimens often have yellow/orange anthers. *Solanum scabrum* is another hexaploid species ($2n=6x=72$) (cf. Edmonds 1977).

The complicated arguments concerning the synonymy of Martius' *S. memphiticum* (not to be confused with species 17. *S. memphiticum* Gmel.) are discussed in Edmonds (1979), as is the synonymy of *S. tinctorium* Welw.

12. **Solanum villosum** *Mill.*, Gard. Dict. ed. 8: *Solanum* No. 2 (1768); G. Don, Gen. Hist. Dichlam. Pl. 4: 413 (1837), as *S. villosum* Lam; A. Rich., Tent. Fl. Abyss. 2: 100 (1838–43), as *S. villosum* Lam.; Lowe, Man. Fl. Madeira 2: 76 (1872), as *S. villosum* Lam.; E. & P. Pf.: 22 (1895), as *S. villosum* Lam; Post, Fl. Syria, Palestine & Sinai: 257 (1930), as *S. villosum* Lam.; F.P.N.A.: 211 (1947), as *S. villosum* Lam; Henderson in Contrib. Queensland Herb. 16: 54–58 (1974) & in Austrobaileya 1(1): 18–19 (1977); Edmonds in J.L.S. 75: 141–178 (1977) & in Biol. & Tax. Solanaceae: 529–548 (1979)); Symon in Journ. Adelaide Bot. Gard. 4: 55–56 (1981); Purdie *et al.*, Fl. Austral. 29: 100 (1982); Edmonds in J.L.S. 78: 214 (1979) & 89: 165 (1984); Jaeger, Syst. studies *Solanum* in Africa: 296 (1985, ined.); Karschon & Horowitz in Phytoparasitica, 13(1): 63 (1985); Abedin *et al.* in Pak. Journ. Bot. 23(2): 271 (1991) *pro parte*; Bukenya & Carasco in Bothalia 25(1): 48 (1995); Edmonds & Chweya, Black Nightshades: 42 (1997); Hepper in Fl. Egypt. 159, Solanaceae: 14 (1998); Mansfeld, Encycl. Ag. & Hort. Crops 4: 1809 (2001); Schippers in Afr. Indig. Veget.: 188 (2000) & in Légum. Afr. Indig.: 382 (2004); Hepper in Fl. Egypt 3: 39 (2002); Olet, Taxonomy *Solanum* section *Solanum* in Uganda: papers i-vi (2004); Edmonds in Fl. Eth. 5: 118 (2006) & in Fl. Somalia 3: 207 (2006); Manoko, Syst. Study African *Solanum* sect. *Solanum*: chapters 1, 3, 6–8 (2007). Type: cultivated in Chelsea Physic Garden, origin Barbados, *Miller* s.n. (BM!, lecto., designated by Edmonds, in J.L.S. 78: 214 (1979))

Annual herb, erect to sprawling, decumbent or prostrate, 0.3–0.5(–1) m high, sometimes woody basally, spreading up to 1(–2) m, with woody tap root or fibrous roots; stems much-branched, green to purple, with smooth or dentate ridges, densely pilose to glabrescent with simple glandular- or eglandular-headed hairs mixed with glands in the Floral area, moderately to densely pilose to villous with glandular- and eglandular-headed hairs elsewhere. Leaves coarse to membranaceous, light to dark green or purplish, rhomboidal to ovate-lanceolate, occasionally lanceolate, (2–)3–7(–11) × 1.5–4(–7) cm, bases broadly cuneate to truncate and decurrent, occasionally cordate, margins usually sinuate-dentate with obtuse to acute lobes though sometimes entire to sinuate, apices acute; surfaces pubescent as stems; petioles 0.4–3.2(–5) cm. Inflorescences simple, leaf-opposed to extra-axillary umbellate to lax cymes often appearing racemose, 3–6(–9)-flowered; peduncles erect, 3–14 mm long in flower, 7–12(–26) mm long in fruit; pedicels erect and 3–9 mm long in flower, reflexed and 4.5–12 mm long in fruit; axes pubescent as stems. Calyx campanulate, 1.5–2.5(–3.5) mm long, pilose to villous externally; lobes broadly triangular, rarely ovate, 0.4–1.5 × 0.3–1 mm, acute; persistent and enlarging to 1–2.5 × 1.5–3 mm in fruit. Corolla white to greenish-white, rarely purple, with translucent, yellow or yellowish-green basal star, sometimes with a purple median vein, stellate, 1–1.6 cm diameter, tube 0.5–1.3 mm long; lobes usually triangular, 3.5–6 × 3.5–5.5 mm, spreading to reflexed after anthesis. Filaments free for 1.2–2.5 mm, slender, densely pilose/villous internally; anthers yellow to orange, 1.5–2(–2.2) × 0.6–0.8 mm. Ovary brownish, 0.7–1.5 × 0.5–1.3 mm, glabrous; style 2.4–4.2 mm long, densely pilose for lower three- to one-quarter of length, exserted up to 1 mm but finally included; stigma capitate, 0.3–0.4(–0.5) mm diameter. Berries smooth, red, orange or yellow, cuticles shiny often becoming dull, typically longitudinally ovoid sometimes globose, 4.5–11 × 6–9 mm, with basal calyx lobes adherent becoming fully reflexed; eventually falling from calyces leaving dried pedicels and calyces. Seeds (18–)30–56(–70) per berry, yellow to brown, obovoid, discoidal to orbicular, 1.8–2.2 × 1.3–1.8 mm, often visible through translucent mature cuticle, reticulate; sclerotic granules absent. Fig. 17/17–19, p. 126.

SYN. *S. nigrum* L. var. *villosum* L. (Sp. Pl.: 186 (1753)). Since this was not referred to by Miller (1768), *S. villosum* Mill. (1768) is not considered to be a new combination based on the Linnaean varietal name.

NOTE. This is a tetraploid ($2n=2x=48$) species which might have originated in southern Europe but which is now widespread throughout central and southern Europe, the Middle East and Asia. It is also common in northern and tropical Africa whence it might have been introduced as it was to North America, New Zealand and Australia (where it is occasionally naturalised) and probably also to the UK and northern Europe. Two subspecies have been recognised in this taxon, which are distinguishable on minor morphological differences and ecogeographical preferences. The subsp. *villosum*, characterised by a dense villous indumentum of spreading multicellular hairs with both glandular and eglandular heads and terete stems is predominantly found in hot, dry habitats from Europe to the Middle East and Asia but is rare or absent from our area. The second subspecies *S. villosum* subsp. *miniatum* (Willd.) Edmonds also occurs throughout Europe, the Middle East and India but is a dominant and frequent taxon throughout eastern and southern Africa. These subspecies, their variability and their complete compatibility, are discussed in Edmonds (1977, 1979). Both are analogously morphologically variable, particularly in their leaf margin shape, berry colours and inflorescence structure. Both subspecies are characterised by yellow, orange or red berries, though they are green and opaque when immature. This contrasts with the subspecies of *S. nigrum* where though the ripe berries are typically purplish-black and opaque, some variants have yellowish-green mature berries with tranluscent cuticles. This has led to much confusion in the description of infraspecific taxa of both species (see *S. humile* below, for example).

subsp. **miniatum** (*Willd.*) *Edmonds* in J.L.S. 89: 166 (1984) & 78: 214 (1979) & 89: 165 (1984); Jaeger, Syst. studies *Solanum* in Africa: 296 (1985, ined.); Abedin *et al.* in Pak. Journ. Bot. 23(2): 272 (1991); Edmonds & Chweya, Black Nightshades: 45 (1997); Hepper in Fl. Egypt 159, Solanaceae: 18 (1998); Mansfeld, Encycl. Ag. & Hort. Crops 4: 1810 (2001); Hepper in Fl. Egypt 3: 39 (2002); Olet, Taxonomy *Solanum* section *Solanum* in Uganda: papers i-vi (2004); Edmonds in Fl. Eth. 5: 118 (2006) & in Fl. Somalia 3: 208 (2006); Manoko, Syst. Study African *Solanum* sect. *Solanum*: chapters 1–8 (2007). Type: cultivated Berlin Botanic Garden (B-W 4366!, sheet 3!, lecto., designated by Edmonds in J.L.S. 89: 166; 1984)

Stems angled with dentate ridges. Plants subglabrous to moderately pilose with appressed, eglandular-headed multicellular uniseriate hairs.

UGANDA. Karamoja District: Moroto Mountain, Jan. 1959, *Wilson* 633!; Ankole District: Ruizi River, 3 Apr. 1951, *Jarrett* 454!; Busoga District: Bugagali Falls, River Nile near Jinja, July 1952, *Lind* 82!
KENYA. Masai District: Entasekera, Morijo Loita, 14 July 1961, *Glover et al.* 2252!; Naivasha District: Lake Naivasha, 7 Oct. 1961, *E. Polhill* 139!; Nairobi District: Thika Road House, 21 July 1951, *Verdcourt* 564!
TANZANIA. Ngara District: Kirushya, Bugufi, West Lake, 23 Nov. 1959, *Tanner* 4532!; Mbulu District: Kitingi, 25 Feb. 1965, *Hukui* 26!; Iringa District: Image, 80 km NE Iringa a little N of Morogoro road, 8 Mar. 1962, *Polhill & Paulo* 1686!
DISTR. U 1, 3, 4; K 1–7; T 1, 2, 4, 7; probably introduced from Eurasia, but now occurring from Israel and Egypt, through Somalia, Ethiopia, Eritrea and Burundi to tropical East Africa; also found in Nigeria, Angola and Madeira
HAB. Forest including riverine *Acacia* forest, flooded woodland, bushland and bushed grassland, grassy river banks, cultivated land, abandoned fields and settlements, pasture-land, disturbed and stony ground, a weed of cultivation and gardens; 50–2250 m

SYN. *S. rubrum* Mill., Gard. Dict., ed. 8: *Solanum* No. 4 (1768), *non* L.; G. Don, Gen. Hist. Dichlam. Pl. 4: 412 (1837). Type: cultivated in Chelsea Physic Garden from seed from West Indies, *Miller* s.n. (BM!, lecto.)
  S. alatum Moench., Meth. Pl. Hort.: 494 (1794), pro *S. nigrum* var. *virginicum*; Post, Fl. Syria, Palestine & Sinai: 257 (1930). Type: *Solanum nigrum vulgare simile, caulis exasperatis*, Dillenius Hort. Elth., 2: 368, t. 275, f. 356 (1732) (lecto. designated by Edmonds in J.L.S. 78; 215 (1979); *Dillenius* 443 (OXF!, epi. designated here)
  S. humile Willd., Enum. Pl. Hort. Berol., 1: 236 (1809), *non* Lam.; G. Don, Gen. Hist. Dichlam. Pl. 4: 413 (1837); Dunal in DC., Prodr. 13(1): 56 (1852); E. & P. Pf.: 22 (1895); E.P.A.: 868 (1963). Type: Southern Europe, cult. Berlin Botanic Garden (B-W 4367, sheet 2!, lecto., designated here; B-W 4367 sheet 1! isolecto.)
  S. luteo-virescens Gmel., Fl. Badensis 4: 177 (1826) *superfl. nom. illegit.* pro *S. humile* Bernh.
  S. miniatum Willd. in Enum. Pl. Hort. Berol, 1: 236 (1809); Boissier in Fl. Orient.: 284 (1879); E. & P. Pf.: 22 (1895); P.O.A. C: 352 (1895); Dammer in E.J. 38: 178 (1907)

?*S. incertum* Dunal in Hist. *Solanum*: 155 (1813); Dunal in DC Pror., 13(1): 57 (1852). Type:
Rheede, Hort. Malab.: 10, t. 73 (1686), iconotype!

S. nigrum L. subsp. *miniatum* (Willd.) Hartm. in Svensk & Norsk Exc.-Fl.: 34 (1846)

S. plebejum A. Rich. in Tent. Fl. Abyss. 2: 100 (1851) as *S. plebeium*; Engler in Hochgebirgsfl.
Trop. Afr.: 372 (1892) & P.O.A. C: 352 (1895); Dammer in E.J. 38: 179 (1907);
Chiovenda in N. Giorn. Bot. Ital. 26: 159 (1919); E.P.A.: 875 (1963). Type: Ethiopia,
Chiré, *Quartin-Dillon & Petit* s.n. (P!, lecto. designated by Lester in J.L.S. 125: 275
(1997); P!, isolecto.)

S. nigrum L. subsp. *puniceum* Kirschleger in Fl. Alsace 1: 532 (1852), pro *S. puniceum*
C.C.Gmelin quod est *nom. illegit.*

S. plebejum A. Rich. var. *subtile* Bitter in E.J. 49: 564 (1913). Type: Ethiopia, Lake Alemaia
[Hararmaja], *Ellenbeck* 462 (B†, holo.)

S. plebejum A. Rich. var. *brachysepalum* Bitter in E.J. 49: 564 (1913). Type: Ethiopia, Ego,
Ellenbeck 358 (B†, holo.)

S. patens Lowe in Man. Fl. Madeira 2(1): 74 (1872). Type: Portugal, Madeira, Rib. de Sta
Luzia, *Lowe* 547 (BM!, lecto. designated here; BM! sheet 2 isolecto.)

S. villosum Lam. var. *laevigata* Lowe in Man. Fl. Madeira 2: 78 (1872). Type: higher or
moister elevations in Madeira and in Lisbon, *Lowe* s.n. (BM!, lecto., designated here)

S. hildebrandtii A. Br. & Bouché in Ind. Sem. Hort. Bot. Berol., 8: 18 (1874); Dammer in E.J.
38: 177 (1906); E.P.A.: 868 (1963). Type: cultivated Berlin Botanic Garden, origin Yafir,
Somalia, *Hildebrandt* 865 (B†, holo.; BM!, lecto. designated here; L!, isolecto.)

S. nodiflorum sensu Hiern, Cat. Afr. Pl. Welw. 3: 745 (1898) pro parte

S. luteum Mill. subsp. *alatum* (Moench) Dostál, Květana ČSR: 1270 (1950); Hawkes &
Edmonds in Fl. Europ. 3: 197 (1972)

S. luteum sensu E.P.A.: 870 (1963), *non* Mill.

S. nigrum L. subsp. *humile* (Willd.) Wu & Huang in Acta Phytotaxonomica Sinicae 16(2):
72 (1978)

S. villosum Mill. subsp. *puniceum* (Kirschl.) Edmonds in J.L.S. 78: 215 (1979)

NOTE. A full list of synonyms and citations for the two subspecies of the tetraploid *S. villosum*
are given in Edmonds (1979 & 1984), together with a lengthy discussion; many of these are
taxa described from Europe. Of these two subspecies the eglandular-haired subsp. *miniatum*
predominates throughout the whole of Africa, with the glandular-haired subspecies *villosum*
occurring predominantly northwards from Sinai. Typical *S. villosum* has condensed few-
flowered (3–5) umbellate racemose inflorescences in which the fruiting pedicels often
exceed the fruiting peduncles in length. Across the Flora area, and throughout much of
Europe, lax cymose inflorescences with 3–8 flowers sometimes occur, and indeed two
Tanzanian specimens had up to 11-flowered inflorescences on extended rachides. However,
the latter on *Greenway* 6997 (**T** 2) varied from 0–1.6 cm, with the berries being described as
orange-yellow, while the inflorescences on the orange-berried *Hepper* & *Field* 5500 (**T** 7)
varied from 4–11-flowered though the fruiting rachides only varied from 0–5 mm.

This plant is widely used as a vegetable in East Africa; the leaves and sometimes the whole
plants are used as a spinach in **K** 1–4, 6 & 7, in **T** 1, 2 & 7 and in **U** 2, while the berries are eaten
raw as a fruit (with varying reports of induced-illness) in **K** 2 & 4 and in **T** 1 & 7. It is said to be
less bitter than other section *Solanum* species. Cattle, game and goats reportedly graze the
plants in Kenya, with medicinal uses including the use of berry juice for sore eyes and poultices
made from ground soaked leaves for swellings in **T** 1. Occasional specimens have been
encountered which might belong to the glandular-haired subsp. *villosum* or to one of the other
African glandular-haired species of the section *Solanum*. A specimen collected from **K** 1 (*Hepper*
& *Jaeger* 6694) is densely villous but with eglandular hairs and unusually small flowers for this
species (anthers 1–1.3 mm); the berry colour was described as orange. The leaf morphology is
typical of another tetraploid species – the blackish-fruited *S. memphiticum* (see below) raising
the possibility that it may be a hybrid between these two tetraploid species.

The correct placement of *S. humile* is difficult; many European authors have treated it as a
synonym of, or described it as, an infra-specific taxon of either *S. nigrum* or of *S. villosum*.
Willdenow described the berries of his species as greenish but smaller than those in *S.
nigrum*; this size reference is suggestive of *S. villosum*, and the berries may have been
immature when he described them since those on the type specimens look blackish and
could have been reddish when fresh. Morphologically his specimens suggest conspecificity
with *S. villosum* subsp. *miniatum*. Dunal (1852) in his later treatment of this species cited a
number of specimens (all seen by the author), some of which are synonymous with the latter
and some with *S. nigrum* subsp. *nigrum*. Similarly, the infraspecific taxa described by various
authors and based on Willdenow's species are synonymous with either *S. nigrum* or *S. villosum*.

The very brief description of *S. luteo-virescens* described the berry colour as yellowish-green and "not yellow", and the habitat as Karlsruhe in Germany. These are suggestive of it being synonymous with *S. nigrum* subsp. *nigrum*. However, Gmelin mentioned the possibility of his taxon being a hybrid between *S. nigrum* and *S. luteum* (=*S. villosum*). Since it was based on *S. humile* Bernh., the type of which is clearly conspecific with *S. villosum* subsp. *miniatum* this taxon has been tentatively synonymised with the latter here.

Though it has been suggested that the species described as *S. incertum* Dunal (Hist. *Solanum*:155 (1813)) might be the correct name for *S. sinaicum*, a species occurring from Sinai through Saudi Arabia and Yemen to Ethiopia, it is now thought to be conspecific with *S. villosum* subsp. *miniatum*. The Malabaricus plate (t. 73) is somewhat stylised, while a Yemeni specimen (*Dunal* 30, MPU!) cited later by Dunal (1852) has much smaller flowers than those typical for *S. sinaicum*. Similarly an Indian specimen from Calcutta (*Dunal* 400) in P, which has *S. incertum* written on the label in Dunal's hand and which has long been labelled as a type specimen, also has small flowers. The berries of the Dunalian species were described as pale orange and they look longitudinally ovoid in the Malabaricus plate; this species is therefore considered to be a synonym of *S. villosum* subsp. *miniatum*, a taxon common in India.

The similarity of *S. plebejum* A. Rich. to both *S. villosum* subsp. *miniatum* and *S. sinaicum* is mentioned below. The small flowers seen on the type material, however, suggest that this species is synonymous with the former. When Bitter (1913) redescribed this species, he cited *Schimper* 509 (W); and a duplicate of this specimen at K is identical to var. *miniatum*.

Though the holotypes of Bitter's two varieties of "*S. plebejum*" were destroyed, there is little doubt from the protologues that both var. *subtile* Bitter and var. *brachysepalum* Bitter are conspecific with *S. plebejum*, and that they are also synonymous with *S. villosum* subsp. *miniatum*.

Lowe did not give any specimen details or numbers for the types of his new Madeiran taxa. There are two sheets labelled *Lowe* 547 at the BM; one is composed of two specimens both labelled as having been collected from the type locality Rib. de Sta Luzia. Both specimens are mophologically identical to *S. villosum* subsp. *miniatum*, and this sheet has been selected as the lectotype of *S. patens*. The second sheet is composed of three specimens. All have differing collection details which roughly correspond to those given in Lowe's description of habitats in which this species was found. Lowe described the lobes as having distinct purple veins, a feature which can occur in several Section *Solanum* species, but which particularly characterises two other tetraploid African species – the northern *S. sinaicum* Dunal and the southern *S. retroflexum* Dunal. The latter always has dull purple berries whereas *S. sinaicum* is characterised by yellowish berries which turn black on drying, and are not translucent; Lowe described the berries of his new species as being dull reddish and showing the seeds inside which is characteristic of *S. villosum*. He also described the habit of *S. patens* as suffrutescent with stiffly stout almost woody branches – features which could also be applicable to *S. sinacium* – but the berry colour of the latter does not match Lowe's description. The overall morphology of all three specimens is again indicative of synonymy with *S. villosum* subsp. *miniatum*, and the morphological variation exhibited by the different sheets is that expected in this taxon.

Lowe's type specimens of his two varieties of *S. villosum* are mounted on the same sheet. The lower of these is labelled var. *laevigata*; it has *S. miniatum* and Lisbon on the label, and is conspecific with S. *villosum* subsp. *miniatum*. Indeed in his protologue, Lowe gave *S. miniatum* as a synonym of his new variety. The upper specimen, *Lowe* 722, is labelled var. *velutina* and is conspecific with the var. *villosum*.

The species *S. hildebrandtii* was described from seeds collected in Somalia by Hildebrandt and presumably cultivated in the Berlin Botanic Garden. Bitter (1913) later redescribed the species, citing *Hildebrandt* 865 (B†) as the holotype, whilst also mentioning some other specimens. There are duplicates of *Hildebrandt* 865 at the BM and L, while a plant from Berlin, now at HBG and annotated as *S. hildebrandtii* is probably part of the original live collection from the Berlin Botanical Garden. These specimens are all conspecific with *S. villosum* subsp. *miniatum*, with which *S. hildebrandtii* has therefore been synonymised.

13. **Solanum americanum** *Mill.*, Gard. Dict. ed. 8: *Solanum* No. 5 (1768); Edmonds in J. Arnold. Arbor. 52: 634 (1971); Hawkes & Edmonds in Fl. Europ. 3: 197 (1972); D'Arcy in Ann. Missouri Bot. Gard. 60(3): 735 (1973); Symon in Journ. Adelaide Bot. Gard. 4: 37 (1981) & 8: 25 (1985); Purdie *et al.*, Fl. Austral. 29: 95 (1982); Jaeger, Syst. studies *Solanum* in Africa: 293 (1985, ined.); Hepper in Fl. Ceylon 6: 368 (1987); Bukenya & Hall in Bothalia 18(1): 82 (1988); D'Arcy & Rakotozafy, Fl. Madag.: 62 (1994); Bukenya & Carasco in Bothalia 25(1): 47 (1995); Edmonds & Chweya, Black

Nightshades: 22 (1997); Mansfeld, Encycl. Ag. & Hort. Crops 4: 1806 (2001); Schippers in Afr. Indig. Veget.: 187 (2000) & Légum. Afr. Indig.: 377 (2004); Olet, Taxonomy *Solanum* section *Solanum* in Uganda: papers i-vi (2004); Gonçalves in F.Z. 8(4): 81 (2005); Edmonds in Fl. Eth. 5: 119 (2006) & in Fl. Somalia 3: 207 (2006); Manoko, Syst. Study African *Solanum* sect. *Solanum*: chapters 1, 2, 6–8 (2007); Manoko *et al.* in Plant Syst. & Evolution 267: 1 (2007). Type: cultivated Chelsea Physic Garden, introduced from Virginia, *Miller* s.n. (BM!, holo)

Annual herb, 0.3–1.5 m high, occasionally suffrutescent at base, erect, procumbent or scrambling, spreading up to 2 m; stems green to densely purple, angular or smooth with edentate or slightly dentate ridges, all parts pilose with appressed eglandular hairs, often glabrescent. Leaves often membranaceous, light to dark green, sometimes densely purple, lanceolate, ovate-lanceolate or ovate, (1.9–)3–8(–15) × 1.1–4.8(–10.8) cm, sometimes smaller on short-shoots, bases cuneate and decurrent, margins entire, sinuate or sinuate-dentate with deep usually obtuse lobes, apices acute to acuminate; surfaces pilose to glabrescent; petioles 0.3–4(–6) cm. Inflorescences simple condensed umbellate cymes, rachides absent or < 2 mm, usually extra-axillary, 3–7(–10)-flowered; peduncles erect and 0.4–2(–3.5) cm long in flower, 1.2–2.8 cm long in fruit; pedicels erect to reflexed and 3–9 mm long in flower, typically erect and splayed and 7–14 mm in fruit but often reflexed and nodding in the Flora area; axes pilose. Calyx campanulate, 1.1–2(–2.5) mm long, pilose externally; lobes broadly ovate, 0.5–1.1 × 0.5–1 mm, acute to obtuse, often unequal; enlarging to 0.7–1(–1.5) × 1–1.5 mm in fruit when usually reflexed away from berry bases. Corolla white, sometimes with purple tinge and occasionally purple median veins to the lobes, with translucent to yellowish-green basal star, deeply stellate, 6–10(–12) mm diameter, tube 0.5–1(–1.3) mm long; lobes spreading to reflexed, ovate with acute apices, 2.5–4(–5) × 1.5–2.5 mm, shortly pilose externally with papillate margins. Filaments free for 0.5–1.4 mm, villous internally; anthers yellow, 0.8–1.5(–2) × 0.5–0.7 mm, connivent. Ovary brownish, 0.5–1 × 0.5–0.9 mm; style 2–3(–4.2) mm long, lower half shortly pilose, exserted up to 1 mm before anthesis after which absent, straight or geniculate; stigma capitate, 0.2–0.3 mm diameter. Berries typically borne erect but fruiting pedicels sometimes reflexed in the Floral area, black to purplish-black (occasionally dark green outside Africa) with shiny opaque cuticle, globose, rarely broadly ovoid, 4–7(–9) mm diameter, surrounded basally by reflexed calyx lobes; deciduous from calyces. Seeds (3–)15–80(–101) per berry, yellow, light or dark brown, obovoid, discoidal to orbicular, 0.8–1.4 (1.8) × 0.8–1.3 mm, reticulate; sclerotic granules present (when up to 4 per berry) or absent. Fig. 17/12–14, p. 126.

UGANDA. Lango District: Lira Central, Erute county, Ereda, 3 km SE Lira town, 15 Jan. 2001, *Olet* 066!; NW slope Mount Elgon, ridge between Kajeri and Sisi rivers, 13 Mar. 1993, *Niaga* 515!; Mengo District: East Mengo, Mukono, Kyage County, Naminya N of Njeru town council, 22 Aug. 2000, *Olet* 001!
KENYA. Fort Hall District: Kimakia Forest Reserve, 28 July 1958, *Kerfoot* 638!; Kwale District: near Inepanga, 22 Mar. 1902, *Kassner* 434!; Kilifi District: Magarini, 1 Nov. 1982, *Robertson* 3454!
TANZANIA. Tanga District: Pongwe–Machui, 29 Nov. 1967, *Faulkner* 4068!; Pangani District: Hale Estate, 3 Oct. 1967, *Faulkner* 4036!; Kilosa District: Mikumi Game Lodge, 24 Oct. 1970, *Batty* 1100!; Zanzibar, Massazine, 30 Oct.1962, *Faulkner* 3120!
DISTR. U 1, 2, 4; K 1, 3, 4, ?6, 7; T 2, 3, 6, 7; Z; Cape Verde, Senegal, Gambia, Sierra Leone, Liberia, Guinea, Ivory Coast, Ghana, Togo, Nigeria, Cameroon, São Tóme, Bioko, Congo-Kinshasa, Burundi, Sudan, Ethiopia, Somalia, Angola, Zambia, Malawi, Mozambique, Zimbabwe, South Africa; Madagascar, W Indian Ocean Islands, Europe, the Mediterranean, Asia, Malesia, N, C and S America, the West Indies, New Zealand, New Guinea and Australia. Controversy surrounds its place of origin; generally considered to be an adventive in much of the Old World
HAB. Common weed of cultivation (maize and cassava plots, banana plantations, gardens), old cultivation areas, waste land and disturbed places, grassland, open bush and forest clearings, shrubland, stony cleared land, and upland rain forest; sea-level to 3200 m

SYN. *S. nigrum* L. var. *patulum* L., Sp. Pl. 1: 186 (1753). Type: seeds originally from Barbados, based on *Solanum procerius patulum vulgaris fructu*, Dill. Hort. Elth.: 367, t. 275, f. 355 (1732), *Dillenius* 441 (OXF!, lectotype designated here)

S. nodiflorum Jacq., Ic. Pl. Rar. 2: 11, t. 326 (1789); Dunal in Synopsis: 12 (1816); G. Don, Gen. Hist. Dichlam. Pl. 4: 411 (1837); Dunal in DC., Prodr. 13(1): 46 (1852); C.H. Wright in F.T.A. 4, 2: 218 (1906); Dammer in E.J. 38: 177 (1906); F.P.U.: 129 (1962); Henderson in Contrib. Queensland Herb. 16: 28 (1974); Durand & Durand, Syll. Fl. Cong.: 394 (1909); Moore in J.L.S. 40: 151 (1911) pro parte; T.T.C.L.: 577 (1949); E.P.A.: 874 (1963); Ross, Fl. Natal: 308 (1972). Type: Mauritius, Jacquin, Icon. Pl. Rar. 2: t. 326 (1789)!

?*S. triangulare* Lam., Tab. Encycl. Meth. 2: 18 (1793); Dunal, Hist. *Solanum*: 155 (1813); Don, Gen. Hist. Dichlam. Pl. 4: 413 (1837); Dunal in DC., Prodr. 13(1): 53 (1852). Type: East Indies, no collector given (P-LAM!, holo.)

S. nigrum L. subsp. *patulum* (L.) Pers., Syn. Pl., 1: 224 (1805)

S. dillenii Schult., Oestr. Fl. 1, ed. 1: 393 (1814); G. Don, Gen. Hist. Dichlam. Pl. 4: 411 (1837); Dunal in DC., Prodr. 13(1): 47 (1852). See Note

S. nigrum L. subsp. *dillenii* (Schult.) Nyman, Conspect. Fl. Europ, 3: 526 (1881)

S. depilatum Bitter in E.J. 49: 566 (1913) & in F.R. 8: 88 (1913). Type: Madagascar, Fort Dauphin, *Paroisse* 10 (P!, holo.)

S. imerinense Bitter in E.J. 49: 566 (1913). Type: Imerina, Madagascar, *Hildebrandt* 3796 (B†, holo.; M!, lectotype, designated here; BREM!, G!, P!, W!, isolecto.)

S. sancti-thomae Bitter in E.J. 49: 560 (1913). Type: São Tóme, *Quintas & Moller* 47 (B†, holo.; COI, iso.)

S. dillenianum Polgár, in Acta Hort. Gotoburg, 13: 281 (1939). Based on, and with type as, *S. nigrum* var. *patulum* L.

S. nodiflorum Jacq. subsp. *nutans* Henderson in Contrib. Queensland Herb. 16: 30 (1974). Type: Australia, *Henderson* 518 (BRI, holo.; K!, NSW, MEL, iso.)

NOTE. This is probably the most widespread and morphologically variable species in the section *Solanum* and is diploid (2n=2x=24). A more complete list of synonyms of this cosmopolitan species is given in Edmonds (1972) and Mansfeld (2001). The species is typically characterised by umbellate cymes of very small flowers, which are succeeded by small globose berries typically on erect, spreading or splayed pedicels. In Africa the fruiting pedicels are often reflexed and indeed Henderson (1974) recognised a similar variant as the subsp. *nutans* in Australia. However, extensive growth of many duplicates of *S. americanum* collected from throughout world indicated that many of the features characterising the species can be extremely variable including fruiting pedicels which can often be reflexed in field-grown plants, but erect in glasshouse- grown ones. Moreover, in some plants, the lower pedicel of the inflorescence can arise up to 2 mm below the condensed umbel. Other characters showing great variation include plant habit; pigmentation; leaf size, shape, and margin shape; flower colour and size; berry shape, colour and sclerotic granule presence or absence (cf. Jardine & Edmonds in New Phytol., 73: 1259 (1974) & Edmonds in J.L.S., 76: 27 (1978)). A number of infraspecific variants of this species have been formally described. Edmonds (1971, 1977) recognised two ecogeographical varieties of this species in South America with var. *americanum* being pilose and var. *patulum* (L.) Edmonds being glabrescent. This distinction is not so apparent elsewhere in the world including Africa, where both types occur and are best identified as *S. americanum* sensu lato. Though sclerotic granules are generally absent from the berries of South American plants of this species, they are often found in the African representatives.

Olet (2004) recognised two morphological forms of *S. americanum* in Uganda, which she proposed could be given infraspecific recognition, though her molecular work suggested that there might be three groups, one of which could have been composed of hybrid plants. The synonymy of *S. nodiflorum* with *S. americanum* was based on examination on the type material of both species (Edmonds, 1971). Using material of this taxon from throughout the world, Manoko (2007) & Manoko *et al.* (2007) concluded from their molecular and crossability results that these two species were distinct. However, it is thought that the material which they identified as *S. americanum*, derived from the USA, is the distinct species *S. ptycanthum* Dunal. The clustering behaviour of all their other accessions, with the exception of a few from Brazil, confirmed their conspecificity, and in my view support their identification as *S. americanum* sensu stricto.

There are reports of the use of this species as a vegetable in Uganda (U 1), Kenya (K 3) and Tanzania (T 2, 3 & 6) including its sale in local markets. It is also frequently used as a vegetable in Madagascar and for its fruits, though there are reports of the toxic alkaloids in the raw fruits causing nausea, diarrhoea, constipation, apathy, circulatory or respiratory problems (cf. D'Arcy & Rakotozafy, 1994). There are a number of specimens in SW Ethiopia and East Africa which appear florally intermediate between this taxon and the next species *S. tarderemotum*, which usually has slightly larger flowers and lax cymose many-flowered inflorescences in which the fruiting pedicels are reflexed.

A new species *S. umalilaense* Manoko (Manoko et al. in Manoko, Syst. Study of African *Solanum* L. Sect. *Solanum* (Solanaceae), Chap. 4: 69–73 of Doctoral Thesis presented to Radboud University, Njimegen (2007) was recently described for tetraploid plants possibly endemic to the Umalila Forest Reserve region of the southern highlands of Tanzania. This was previously identified as *affin. S. americanum*, with which it is superficially similar. However, among the characters by which Manoko differentiated it are terminal inflorescences, shortly accrescent fruiting calyces, and small yellowish translucent berries which remain attached to the plants at maturity; its distinction from other African section *Solanum* species was supported by his AFLP analyses. It is *species A* in Schippers (2004), and the type material cited was *Gereau et al.* 5084 (DSM, holo.; K!, MO, NHT, iso.). This taxon is a popular vegetable which is extensively grown by the Umalila people. It is thickly sown in large plots, and the leaves harvested three to four weeks after planting until flowering commences when they become too bitter. It is widely sold in local markets. Morphologically the plants are somewhat intermediate between the diploid *S. americanum* and tetraploid *S. tarderemotum*. It could be an autopolyploid form of the former, but recent (2010) field work suggests that it is a good species which is widespread in cultivation, though no wild populations were found (van der Weerden, pers. comm.). Further experimental work on these collections is currently underway before the new species is formally published.

It is probable that *S. triangulare* is synonymous with *S. americanum*. The protologue is brief but the possible holotype has a similar inflorescence structure with erect pedicels and bears tiny flowers with the corolla being described as pale violet and the anthers ± 1mm long. The specimen later cited by Dunal (1852) for this species – *Gaudichaud s.n.* (G-DC) is not conspecific and seems to belong to *S. villosum* subsp. *miniatum*.

It is generally accepted that Schultes' *S. dillenii*, though based on Dillenius' *Solanum procerius patulum vulgaris fructu* was described from a mixture of specimens – representing some quite different species (see Thellung in Botan. Exch. Club Reports, 8: 186 (1926–1928); Polgár in Botan. Közlemenyek 23: 3 (1926) & in Acta Hort. Got. 13: 281 (1939)). To clarify the confusion surrounding this species, Polgár later described the Swedish adventive which he thought identical to the Dillenian plant as *S. dillenianum* (see above).

The protologue and morphology of the holotype of *S. depilatum* clearly demonstrates its conspecificity with *S. americanum*, though sclerotic granules are absent from the berries in this case.

Several duplicates of the type collection of *S. imerinense* have been traced and although the specimens have sinuate-dentate leaf margins and some reflexed fruiting pedicels, the remaining inflor- and infruct-escence characters are identical to those found in *S. americanum*.

Although a type specimen of *S. sancti-thomae* has not yet been seen, the plant habit and tiny flowers described in the protologue suggest conspecificity with *S. americanum*.

14. **Solanum tarderemotum** *Bitter* in F.R. 10: 547 (1912); Jaeger, Syst. studies *Solanum* in Africa: 304 (1985, ined.); Bukenya & Carasco in Bothalia 25(1): 48 (1995); Schippers in Légum. Afr. Indig.: 381 (2004); Olet, Taxonomy *Solanum* section *Solanum* in Uganda: papers i-vi (2004); Gonçalves in F.Z. 8(4): 83 (2005); Edmonds in Fl. Eth. 5: 121 (2006); Manoko, Syst. Study African *Solanum* sect. *Solanum*: chapters 1–8 (2007). Type: Tanzania, Kilimanjaro, Marangu, *Winkler* 3856 (WRSL, holo.)

Erect, procumbent or scrambling herb 0.3–1(–1.5) m high, often much-branched, occasionally with procumbent branches spreading to 3 m, rarely shrubby and woody basally; stems smooth to dentate; all parts pilose often glabrescent, with simple appressed and spreading usually long eglandular hairs. Leaves membranaceous, pale to dark green, lanceolate, 4.5–10(–14) × 2.5–5(–7) cm, bases cuneate and decurrent, margins entire, sinuate, or sinuate-dentate, apices acute to acuminate; surfaces pilose to glabrescent as stems, with hairs denser on lower surfaces, especially on veins and midribs; petioles 1–3.5(–5) cm, often winged above from decurrent lamina. Inflorescences simple cymes, often umbellate in flower becoming lax and extended with rachides 0.4–1.4 cm long in fruit, occasionally also forked, 5–14-flowered; peduncles usually erect, 1–1.5 cm long in flower, 1.5–3.7 cm long in fruit; pedicels erect and 4–6 mm long in flower, recurved to spreading and 6–12 mm long in fruit; axes often densely pilose. Calyx campanulate, 1–2.5 mm long, pilose externally; lobes broadly ovate to ligulate, obtuse, 0.5–1.3 × 0.4–1 mm, becoming broadly ovate or triangular and 1–2 × 0.8–2.3 mm in fruit when adherent becoming reflexed. Corolla white or sometimes pale purple, with yellow or translucent basal star, sometimes with purple median vein to lobes, stellate, 6–12(–16) mm diameter, tube ± 1 mm long; lobes spreading to reflexed at anthesis, ovate to triangular, 2–4(–6) × 0.7–1.5(–2) mm, pilose externally with shortly pilose (to papillate) margins, sometimes with band of long hairs basally around tube mouth. Filaments 0.5–1 mm long, glabrous to villous internally; anthers yellow to yellow-orange, 1.2–2(–2.5) × 0.5–1 mm. Ovary brown, 0.6–1.5 × 0.5–1.2 mm, glabrous; style often sigmoid or geniculate, 2.5–4(–5) × 0.2–0.3 mm, pilose or villous in lower half to two thirds, exserted up to 2 mm; stigma capitate, 0.1–0.3 mm diameter. Berries yellowish-green to dark purple or black, globose, 4–6(–7) mm diameter; usually shed still attached to pedicels leaving scars on rachides. Seeds 13–56 per berry, light to dark brown or yellowish, obovoid to discoidal, 1.3–1.9 × 1–1.5 mm, reticulate; 1–3 ovoid sclerotic granules usually present, 0.2–0.7 × 0.2–0.6 mm, occasionally absent. Fig. 17/1–9, p. 126.

UGANDA. Toro District: Fort Portal, Toro, 4 Apr. 1932, *Hazel* 219!; Mbale District: Budadiri, Bugisha, Jan. 1932, *Chandler* 458!; Mengo District: km 19 Kampala to Entebbe Road, May 1932, *Eggeling* 415!
KENYA. Nakuru District: near division forest office, Londiani, 15 Nov. 1967, *Perdue & Kibuwa* 9065!; Embu District: below Castle Forest Station, 4 Apr. 1970, *Gillett & Mathew* 19099!; Fort Hall District: Kimakia Forest Reserve, E Aberdares, 28 July 1958, *Kerfoot* 637!
TANZANIA. Tanga District: W slope of E Usambaras between Ngua and Magunga estates, 17 July 1953, *Drummond & Hemsley* 3347!; Ufipa District: Nsanga Forest, Ufipa, 8 Aug. 1960, *Richards* 13004!; Iringa District: Ruaha N Park, Mbagi, 14 Feb. 1966, *Richards* 21308!
DISTR. U 1–4; K 1, 3–7; T 1–4, 6, 7; Equatorial Guinea, Cameroon, Congo-Brazzaville, Congo-Kinshasa, Rwanda, Burundi, Sudan, Ethiopia, Angola, Zambia, Malawi, Zimbabwe, Botswana, Namibia and South Africa
HAB. In disturbed vegetation, moist forest margins, -clearings and -paths, gallery forest, streamsides, bushland, rough grassland, on wasteland, on rocky hills, weed of cultivation, often in wet areas; 550–2950 m

SYN. *S. pentagonocalyx* Bitter in F.R. 10: 544 (1912). Type: Tanzania, Lushoto District: Usambara, Kwa Mshusa [Kwa Mstuza], *Holst* 9021 (B†, holo.; M!, lecto. designated here; HBG!, W!, isolecto.)
?*S. tetrachondrum* Bitter in E.J. 49: 565 (1913). Type: Tanzania, Kilimanjaro, near Marangu, *Volkens* 623 (?B†, holo.)
?*S. tetrachondrum* Bitter var. *subintegrum* Bitter in E.J. 49: 566 (1913). Type: Tanzania, Marangu, Kilimanjaro, *Volkens* 622 (?B†, holo.)
S. viridimaculatum Gilli in Ann. Naturhist. Mus. Wien 77: 43 (1973). Type: Tanzania, Njombe District: Madunda, *Gilli* 499 (W!, holo.)
S. eldorettii of Schippers in Légum. Afr. Indig.: 381 (2004); Manoko, Syst. Study African *Solanum* sect. *Solanum*: chapter 3 (2007), *nom. nud.*

NOTE. This fairly well defined taxon has been frequently recorded from Ethiopia to Tanzania and Malawi, where it has invariably been identified as *S. nigrum*. It conforms to the protologue of *S. tarderemotum* although isotype material of this species has not yet been located. Vegetatively it is extremely variable, with the pubescence varying from glabrescent to conspicuously pilose, and the leaf margins from entire to sinuate dentate. As with *S. nigrum* sensu stricto and with the following species *S. florulentum*, the berry colour can vary from yellowish green to purple or blackish. Olet (2004) recognised three morphological forms/variants in Uganda, though she did not give them formal taxonomic recognition. Manoko (2007) later concluded that some of her plants represented introgressed hybrids.

The species is used as a vegetable or pot-herb throughout its range (cf. Schippers, 2002) but especially in Kenya, Tanzania and parts of Ethiopia. Decoctions of the leaves are used to treat malaria; raw roots are eaten as a treatment for stomach ache in T 4 and fresh leaf juice used to treat conjunctivitis in T 6. However, there are reports that specimens of this species are poisonous to stock in K 1. It is close to the small-flowered diploid species *S. americanum* (see above), but is tetraploid (cf. Olet, 2004; Manoko, 2007). The mature infructescences illustrated in Schippers' (2004) are very similar to those of the South American diploid *S. chenopodioides* Lam. which, as far as is known, has only been identified in South Africa (cf. Edmonds & Chweya 1997), possibly indicating a common genome. Like this species, the fruiting peducles of *S. tarderemotum* can be deflexed, the fruiting pedicels splayed and the berries fall still attached to the pedicels.

Jaeger's (1985) suggestion that this species is synonymous with *S. sarrachoides* is erroneous. This glandular-haired South American species is only found sporadically in South Africa, and its fruiting sepal dimensions, for example, are far greater than those cited. Two Tanzanian specimens (*Richards* 9735 and 9775) collected from 2340 m, are probably conspecific with *S. tarderemotum*. They have larger flowers than those usually associated with this species, with anthers ranging from 2.2–2.5 mm long. However, they also have long flowering peduncles (–3.5cm), with slightly extended cymose inflorescences to 8-flowered. Their flower size might be due to the shady habitat at high elevation from which they were collected; overall they show more affinity with *S. tarderemotum* than any of the other section *Solanum* species found in East Africa.

The inflorescences on the duplicate type material of *S. pentagonocalyx* are umbellate to extended racemose cymes with up to 9 flowers, whose floral dimensions and stylar exsertion indicate the conspecificity of Bitter's species with *S. tarderemotum*.

Bitter did not cite the locality of his holotypes for either *S. tetrachondrum* or its variety *subintegrum* and no duplicate specimens of these two taxa have yet been found. The long 5–10- flowered peduncles and all the floral and berry dimensions given in the protologue of *S. tetrachondrum* generally match those found in *S. tarderemotum*, and all three taxa were collected from Marangu near Kilimanjaro. The dense pubescence and the dentate leaf margins cited are within the margins of variability exhibited by this species – especially in the habitat cited. However, the pedicels are described as being congested on the top of the peduncle. This species has therefore only been provisionally synonymised.

Bitter differentiated his var. *subintegrum* from *S. tetrachondrum* largely on the basis of its entire rather than dentate leaf margins. Leaf margins in typical *S. tarderemotum* are entire to sinuate, and this variety has also been provisionally synonymised with the latter.

The holotype of *S. viridimaculatum* has sinuate-dentate leaves, dentate stems and slightly larger flowers than those typical of *S. tarderemotum*. However, the anthers are slightly smaller on the holotype (2.4–2.5 mm long) than those cited in the protologue (2.5–3mm), and all floral dimensions are within the overall morphological variability displayed by this species throughout its geographical range; they are considered to be conspecific.

According to Schippers (2004) the name *eldorettii* or *eldoretii* was introduced in 1987 by "Mtotomwema" for green-fruited plants frequently cultivated in Eldoret, Kenya, without any published description. He later correctly identified these plants as *S. tarderemotum*, and this was confirmed by the AFLP clustering patterns derived by Manoko (2007) during his multidisciplinary analyses of this group.

15. **Solanum florulentum** *Bitter* in F.R. 10: 544 (1912); Jaeger, Syst. studies *Solanum* in Africa: 303 (1985, ined.); Bukenya & Carasco in Bothalia 25(1): 48 (1995); Schippers in Légum. Afr. Indig.: 378 (2004); Olet, Taxonomy *Solanum* section *Solanum* in Uganda: papers i-vi (2004); Manoko, Syst. Study African *Solanum* sect. *Solanum*: chapters 1–8 (2007). Type: Tanzania, Lushoto District: Kwai, *Albers* 189 (B†, holo.)

Erect or scrambling herb to 1.5 m high, occasionally sub-shrubby or scandent; stems sometimes woody basally, but usually succulent and angulate to winged, the ridges or wings usually dentate, pilose with simple long eglandular hairs and especially young parts sometimes appearing ochraceous, older parts glabrescent. Leaves fleshy to membranaceous, pale to dark green, occasionally ochraceous, lanceolate to ovate-lanceolate, 5–10(–16) × (1.5–)3–8 cm, bases cuneate and decurrent often to stem, margins entire to sinuate, rarely sinuate-dentate with 3–8 lobes, apices acute; pubescent as stems with hairs denser on veins and midribs especially below; petioles 1–4(–6) cm. Inflorescences extra-axillary, usually forked, often multiply so, extended cymes, sometimes also simple, 10–32-flowered; peduncles usually erect, 9–26 mm long in flower, 1–3(–5) cm long in fruit; pedicels erect becoming reflexed, 3–6 mm long in flower, 7–10 mm long in fruit; axes pilose. Calyx cupulate, 1–2(–2.5) mm long, pilose externally; lobes shallowly triangular to obovate, often unequal, 0.5–1 mm, obtuse; in fruit with broadly triangular obtuse persistent calyx lobes enlarging to 0.8–2 × 1.3–3 mm. Corolla white with yellow basal star, often with median purple or green vein to lobes, stellate, 6–12 mm diameter, tube 1–1.3 mm long; lobes spreading or reflexed after anthesis, ovate to narrowly triangular, 2.2–4 × 1.3–2.5 mm, pubescent externally with shortly pilose margins, glabrous internally. Filaments 0.3–0.8 mm long, pilose/villous internally; anthers yellow to yellow/orange, 1.6–2(–2.2) × 0.5–0.9 mm. Ovary brown, 1–1.6 × 0.7–1.7 mm; style geniculate to straight, whitish, 2.8–5 mm, lower half pilose, exserted up to 2.5 mm; stigma capitate, 0.1–0.3 mm diameter. Berries purplish black or greenish yellow, globose, 4–6 mm diameter, basal adherent calyx appearing pentagonal; usually falling with pedicels attached. Seeds up to 46 per berry, yellow to brown, obovoid to discoidal, 1.4–1.5(–1.8) × 0.9–1.3 mm, reticulate; with up to 6 small (0.3–0.7 mm diameter) sclerotic granules per berry. Fig. 17/10 & 11, p. 126.

UGANDA. Kigezi District: Kigezi, DFI, 28 Aug. 1972, *Goode* G2/72!; Mengo District: Kampala, Aug. 1931, *Lab. Staff* (Dept. Agriculture) 2153! & Kituza, 56 km SE Kampala Agric. Dept. Coffee Res. Station, June 1957, *Griffiths* 47!
KENYA. Northern Frontier District: Kichich, 23 Dec. 1959, *Newbould* 3542!; Naivasha District: Lake Naivasha, 31 Oct. 1965, *E. Polhill* 145!; Kericho District: SW Mau Forest Reserve, camp 7, 1 Aug. 1949, *Maas Geesteranus* 5603!
TANZANIA. Moshi District: Weru-Weru Gorge, just above ford Lyamungu to Machame, 22 Feb. 1955, *Huxley* 116!; Rungwe District: Rungwe, N of Lake Nyasa, 19 Sep. 1932, *Geilinger* 2482!; Songea District: Matagoro Hills just S of Songea, 22 Feb. 1956, *Milne-Redhead & Taylor* 8869!
DISTR. U 2, 4; K 1, 3–5; T 2, 3, 7, 8; Congo-Kinshasa and northern Namibia
HAB. Grassland and clearings, especially in damp habitats such as stream- or river-banks and lake-shores, and on steep or rocky hillsides, wasteland, roadsides, cultivated ground, scrubland; also a weed of cultivation; 1100–2450 m

NOTE. This species seems to be indigenous to parts of East Africa – particularly to Kenya where it is widely cultivated, though the leaves from plants occurring spontaneously are eaten as spinach in K 3, U 2 and U 4. It is morphologically close to *S. tarderemotum* but the inflorescences are generally complexly branched lax cymes, though simple inflorescences can occur on the same plant. Indeed Olet (2004) described inflorescences with up to 5 branches in Ugandan plants of this species, where she also recognised three morphological variants without giving them formal taxonomic recognition. Manoko (2007) recorded up to 48 flowers on inflorescences of this species. Jaeger (1985) suggested that although *S. tarderemotum* and *S. florulentum* were recognisable entities, they might be subspecies of the same taxon. However, Olet (2004) found that these two species were well differentiated from each other in her numerical analyses, though not so in molecular studies. Later Manoko (2007) proposed that his systematic results indicated that these two taxa are distinct species. Both *S. florulentum* and *S. tarderemotum* are tetraploid (cf. Olet, 2004; Manoko, 2007) and it is likely from their morphology that they possess a common genome from the widespread diploid *S. americanum*.

Bitter described *S. dasytrichum* Bitter (in E.J. 49: 568 (1913); type: *Eick* 227, B†, holo.) from Kwai, the same Tanzanian type locality as that cited for *S. florulentum*. From its protologue, the description of dense pubescence on juvenile parts which later become glabrescent, the floral dimensions, peduncle lengths and berry characteristics are all similar to those found in both *S. florulentum* and *S. tarderemotum*, though the description of a pentagonal fruiting calyx is more typical of *S. florulentum*. However, no mention was made of the inflorescence forking. Jaeger (1985) thought that *S. dasytrichum* showed an affinity with *S. americanum*, presumably since the protologue described densely congested pedicels at the apex of the peduncle. The floral and fruit dimensions are larger than those usually associated with the latter, and its true status awaits future collection of material from the type locality.

16. **Solanum pseudospinosum** *C.H. Wright* in F.T.A. 4, 2: 220 (1906); Heine in F.W.T.A. 2nd ed., 2: 335 (1963); Bitter in F.R. 10: 546 (1912); Jaeger, Syst. studies *Solanum* in Africa: 303 (1985, ined.). Type: Cameroon, Mt Cameroon, *Mann* 1938 (K!, holo.)

Erect, decumbent, semi-prostrate or scrambling herb 0.4–1.9 m high, occasionally shrubby; stems woody basally, with dentate ridges often surmounted by prominent hairs, often angular, sometimes dark purple, densely pilose/villous with simple long eglandular hairs mixed with glands, glabrescent. Leaves coarse to membranaceous, light to dark green, lanceolate to ovate-lanceolate, usually small, 2.2–5.5(–11) × 1.4–2.8(–5) cm, bases cuneate and decurrent for whole length of petiole, margins entire to sinuate, but often sinuate-dentate in FTEA area with shallow obtuse lobes, apices acute; surfaces prominently strigose/pilose, denser on margins, veins, midribs and lower surfaces, hairs as on stems; petioles 0.5–2(–4) cm. Inflorescences simple, leaf-opposed to extra-axillary umbellate becoming extended cymes, 3–5(–9)-flowered; peduncles erect and 0.3–1.1(–1.8) cm long in flower, erect to deflexed and 0.8–2.3 cm long in fruit; pedicels erect and 4–10 mm long in flower, spreading to reflexed and 8–15 mm in fruit; axes densely pilose/villous. Calyx campanulate, 1.3–2.8 mm long, pilose externally with long spreading eglandular hairs; lobes often unequal, obovate, 0.5–1.5 × 0.5–1.5(–3) mm, obtuse; persistent and enlarging to 0.8–4 × 1–3 mm when in fruit. Corolla white, sometimes with median purple vein to lobes, stellate, 8–18 mm diameter, tube 0.6–1.3 mm long; lobes spreading or reflexed after anthesis, ovate to narrowly triangular, 3–7 × 1.5–3 mm basally. Stamens equal; filaments 0.5–1 mm long, densely villous; anthers lemon-yellow, 1.5–2(–2.5) × 0.5–0.8 mm, connivent. Ovary brown, 0.8–1.3 × 0.6–1.3 mm, glabrous; style geniculate becoming straight, 3–6 × 0.2–0.3 mm, lower half shortly pilose, exserted up to 2.5 mm; stigma capitate, 0.1–0.4 mm diameter. Berries usually borne erect, purple to black, globose, 4–5(–6) mm diameter, smooth, surrounded basally by adherent acutely triangular calyx lobes; deciduous from calyces. Seeds 4–25 per berry, light to dark brown, obovoid, discoidal to orbicular, 1.5–2 × 1.3–2 mm, reticulate; with 2–9 ovoid to spherical sclerotic granules 0.2–0.8 × 0.1–0.7 mm, often visible through cuticle. Fig. 17/21–25, p. 126.

UGANDA. Toro District: Ruwenzori, Bujuku valley below Nyabitaba Hut, slope down to bridge, 16 Jan. 1967, *G.H.S. Wood* 815!; & above Kichuchu, Mitaku Valley, 15 July 1962, *Ross* 576! & River Nyamagasani, western ridge, 23 Aug. 1952, *Osmaston* 2328!
KENYA. Mt Elgon, Jan. 1931, *Lugard & Lugard* 492! & SW Elgon, 12 Aug. 1958, *Symes* 392!: Kiambu District: Limuru, 5 June 1918, *Snowden* 564!
TANZANIA. Masai District: Ngorongoro Conservation area, Empakaai Crater outside eastern slope, 10 Aug. 1972, *Frame* 24!; Morogoro District: Uluguru Mountains, Lukwangulu Plateau, 19 Sep. 1970, *Thulin & Mhoro* 1048!
DISTR. U 2–3; K 3–5; T 2, 6, ?7; Ghana, Cameroon, Equatorial Guinea, Congo-Kinshasa, Rwanda, Malawi. Possible specimens have also been seen from Ethiopia and Sierra Leone
HAB. Montane grassland and bushland, *Hagenia* woodland, bamboo thickets; 2100–3600 m

SYN. *S. nigrum* L. *forma* Hook. in J.L.S. 7: 209 (1864). Type: none indicated, un-named form
?*S. pachyarthrotrichum* Bitter in F.R. 10: 542 (1912). Type: Cameroon, *Deistel* 631 (B†, holo.)
?*S. hypopsilum* Bitter in F.R. 10: 543 (1912). Type: Cameroon, Buea, *Lehmbach* 175 (B†, holo.)
S. molliusculum Bitter in F.R. 10: 546 (1912). Type: Cameroon, Buea, *Preuss* 740a (B†, holo.)

NOTE. This species was considered to be endemic to Cameroon when Wright described it in F.T.A. (1906), and the Cameroonian specimens are all morphologically very similar. However, it seems to occur sporadically at high altitudes across central Africa from the Cameroon and Equatorial Guinea in the west through to Congo-Kinshasa and Malawi in the east, though these specimens are more variable morphologically. Such variability is evident in the inflorescence and infructescence structures, number of flowers and the indumentum density. From the numbers of berries setting, the inflorescences are typically few (3–5)-flowered, but scarring on the fruiting rachides suggests that they can be up to 9-flowered. Several specimens collected from 1900 m in the Mount Loma area of Sierra Leone also seem to be conspecific with this taxon, together with possible collections from high altitudes in Ethiopia. Its chromosome number is not yet known.

Bitter (1912) described three new species from Cameroon, with two coming from Buea which is also the type locality. Though no isotype material of these species has yet been located, from their protologues they are all considered to be probable synonyms of *S. pseudospinosum* (see above).

From its protologue, *S. pachyarthrotrichum* is considered to be synonymous with *S. pseudospinosum*, though Bitter did not give any locality details for the holotype. However, his description of thick 'jointed' hairs together with the floral and berry dimensions, the markedly enlarged fruiting calyces, the occurrence of relatively few seeds and 10 sclerotic granules in the fruits, all indicate conspecificity. Moreover, he also noted that the two lower pedicels were separated from the remaining three to four, which is suggestive of the peculiar lax cymes which characterise *S. pseudospinosum*.

Many of the characters described in Bitter's protologue of *S. hypopsilum* are very similar to those found in *S. pseudospinosum*. However, Bitter described its indumentum as glabrous to sparsely pilose in his new species whereas *S. pseudospinosum* is characterised by a distinct pilose-strigose pubescence on the stems and leaves, though the stems can become glabrescent as they mature. It is possible that the *Lehmbach* specimen was collected from an old plant or from secondary growth but until type material has been located this species is only provisionally synonymised with *S. pseudospinosum*.

Bitter (1912) mentioned that although he had not seen Mann's type specimen of *S. pseudospinosum*, he considered that the minor differences he noted between Wright's protologue and the *Preuss* specimen warranted the recognition of the *S. molliusculum* as a distinct species. However, the morphological features mentioned in the protologue of this species are similar to those found in *S. pseudospinosum*.

17. **Solanum memphiticum** *Gmel.*, Syst. Nat., 2 (1): 385 (1791), *non* Sendtn.; E.P.A.: 872 (1963); Jaeger, Syst. studies *Solanum* in Africa: 303 (1985, ined.); Edmonds in K.B. 62: 657 (2007) & in Fl. Eth. 5: 122 (2006) & in Fl. Somalia 3: 208 (2006). Type: herb. *Forsskål* Sheet 421 (C!, lectotype designated by Edmonds in K.B. 62: 665 (2007))

Annual or perennial herb to 1 m high, erect, ascending, spreading or decumbent, with dense branches often arising from woody base or rootstock; stems often light green, densely viscid with all parts covered with mixture of pale long spreading glandular-headed and shorter eglandular-headed hairs; sand and soil particles adhering to these often make the indumentum appear brownish. Leaves usually light green, ovate, rhomboidal or lanceolate, 2.5–6(–10.5) × 1.5–3.6(–7) cm, bases truncate and decurrent to stems, margins sinuate-dentate to incised with many small obtuse to acute lobes though upper leaves sometimes entire to sinuate, apices acute to obtuse; surfaces villous, hairs as on stems; petioles (1–)2–3.5(–5) cm long. Inflorescences extra-axillary, simple, umbellate in flower, sometimes becoming lax erect cymes in fruit, 3–5(–6)-flowered; peduncles erect, 0.7–1.9 cm in flower and 1.1–2.2(–3.5) cm in fruit; pedicels reflexed, 5–9 mm in flower and 9–13 mm in fruit; axes densely villous. Calyx campanulate, 2–3.5(–4) mm long, densely villous externally especially on margins of lobes; lobes spatulate to obovate, 0.8–2 × 0.3–0.9 mm, enlarging to triangular and 2.2–4(–5.5) × 1.5–3 mm in fruit, adherent becoming reflexed from berry bases. Corolla white with translucent, yellow or yellowish-green basal star, occasionally pale purple, sometimes with purple vein to lobes, stellate, 8–15 mm diameter, tube 1–1.3 mm long; lobes ovate to triangular, 2.5–4(–5) × 1.2–2.8(–3.5) mm, villous on outside surfaces, spreading after anthesis. Filaments light green, 0.5–1(–1.5) mm long, glabrous; anthers yellow, 1–2(–2.5) × 0.5–0.8 mm. Ovary green, 0.8–1.5 × 0.8–1.4 mm, bilocular; style geniculate becoming straight, green, (2–)2.5–4.5 mm long, lower half pilose or villous, exserted to 1 mm; stigma green, capitate, 0.1–0.4 mm diameter. Berries purple to black or yellowish-green when cuticles transulucent, globose to broadly ovoid, 5–9 mm diameter, bases surrounded by adherent acutely triangular calyx lobes; eventually falling with pedicels, often splitting open whilst still on plant. Seeds up to 63 per berry, yellow to yellowish-brown, obovoid to discoid, 1.4–2 × 1.1–1.6 mm; 1–5 spherical sclerotic granules often present, 0.5–0.7 mm diameter. Fig. 17/26–30, p. 126.

UGANDA. Kigezi District: Kachwekano Farm, Sep. 1949, *Purseglove* 3121! & Kigezi D.F.T., Orushwiga, 28 Aug. 1972, *Goode* G5/72!; Mbale District: Budadiri, Bugushi, Jan. 1932, *Chandler* 405!

KENYA. Northern Frontier District: Mathews Range, Ol Doinyo Lengio, 20 Dec. 1958, *Newbould* 3289!; Naivasha District: W shore of Lake Naivasha, 14 Feb. 1971, *Gillett* 19300!; Nairobi District: Mathare Valley, Mathare River, between Mathare Police Station and Eastleigh Section One, at eastern edge of Upper Mathare Village, 11 Sep. 1971, *Mwangangi & Kasyoki* 1797!

TANZANIA. Ngara District: Murugwanza, Bugufi, 20 Jan. 1961, *Tanner* 5617!; Mbulu District: Lake Manyara National Park, Marera River, 9 Mar. 1964, *Greenway & Kanuri* 11332!; Kondoa District: Great North Road, Kondoa, 11 Jan. 1962, *Polhill & Paulo* 1130!

DISTR. U 1–3; K 1–6; T 1, 2, 5; from Egypt and Arabia to Sudan, through Somalia, Ethiopia and Eritrea to tropical East Africa; also in Cameroon, Congo-Kinshasa, Malawi and South Africa

HAB. Forest, bushland, scrub, grassland, lake- and river-banks, lava/rocky and stony areas, disturbed land and roadsides, weed of cultivation, gardens and pasture; 950–2450 m

SYN. *S. aegyptiacum* Forssk. var. b) fructu nigro; foliis integris, villosissimus, Forssk. Fl. Aegypt.-Arab.: 46 (1775). Type: as for *S. memphiticum*
S. nigrum L. var. *hirsutum* Vahl, Symb. 2: 40 (1791); Willd. Sp. Pl. 1: 1036 (1798). Based on, and with same type as, *S. aegyptiacum* Forsk. var. b
S. nigrum L. subsp. *hirsutum* (Vahl) Pers., Syn. Pl.: 224 (1805)
S. hirsutum (Vahl) Dunal, Hist. Solan.: 158 (1813); Don, Gen. Hist. Dichlam. Pl. 4: 413 (1837); Dunal in DC., Prodr. 13(1): 58 (1852); Olet, Taxonomy *Solanum* section *Solanum* in Uganda: papers i, ii, v & vi (2004)
S. grossidentatum A. Rich., Tent. Fl. Abyss. 2: 101 (1851); Engler in Hochgebirgsfl. Trop. Afr.: 372 (1892); Jaeger, Syst. studies *Solanum* in Africa: 301 (1985, ined.); Bukenya & Carasco in Bothalia 25(1): 48 (1995); Gonçalves in F.Z. 8(4): 82 (2005); Manoko, Syst. Study African *Solanum* sect. *Solanum*: chapters 1, 3, 6–8 (2007). Type: Ethiopia, Tchélikote, *Quartin-Dillon & Petit* s.n. (P, lecto., P, isolecto. designated by Lester in J.L.S. 125: 275 (1997); G! GOET! P! Z! isolecto.)
S. hirsutum Dunal var. *abyssinicum* Dunal in DC. Prodr., 13(1): 58 (1852); Dammer in E.J. 38: 179 (1905–7). Type: Ethiopia, Adua [Adoam], *Schimper* 46 (G-DC!, lecto. designated here; herb. Moric., G!, BM!, CGE!, K!, MPU!, P!, iso.)
S. subuniflorum Bitter in F.R. 10: 546 (1912). Type: Tanzania, Marangu near Kilimanjaro, *Volkens* 2108 (B†, holo.; BR!, lecto., designated here)
S. plebejum A. Rich. var. *grossidentatum* (A. Rich.) Chiov. in N. Giorn. Bot. Ital. 26: 159 (1919)
S. memphiticum Gmel. var. *abyssinicum* (Dunal) Cufodontis, E.P.A.: 872 (1963)
S. sarrachoides sensu Schippers in Légum. Afr. Indig.: 379 (2004), *non* Sendtner

NOTE. Ethiopian specimens exhibiting a dense viscid pubescence and prominently stellate fruiting calyx lobes have often been identified as *S. grossidentatum*, though the species named earlier as *S. hirsutum* by Dunal (1813), and based on *S. nigrum* L. var. *hirsutum* Vahl (1791) is conspecific. Both this variety and *S. memphiticum* Gmel. (1791) - the earliest relevant epithet at specific rank - were based on Forsskål's *Solanum aegyptiacum* b) *Fructo nigro* (1775). Of the three specimens in Forsskål type folder of this species, specimens 421 and 422 are similar to one another and to the types of *S. hirsutum* and *S. grossidentatum*, while the third, no. 405 is totally different and seems to be conspecific with *S. nigrum* sensu stricto. The latter has therefore been rejected as a syntype; specimen 421 has been selected as the lectotype of *S. memphiticum* and this typification is discussed in Edmonds (2006). The species is tetraploid (cf. Olet, 2004; Manoko, 2007). Despite being characterised by a dense hispid glandular pubescence and having densely pubescent styles, *S. memphiticum* is unique among African section *Solanum* species in having glabrous filaments.

There are three specimens collected from Kenya which superficially belong to this taxon, but for which the berries are described as red (*Glover & Samuel* 3110 – K 6; and *Meyerhoff* 21 – K 2) and as reddish-orange (*Brodhurst Hill* 197 – K 3). Either these citations are erroneous – perhaps based on the fruits after they had dried, or the specimens belong to the glandular-haired subspecies of *S. villosum*. However, no other specimens of this latter subspecies have been identified from East Africa and their hispid pubescence, characteristic leaf shape and stellate fruiting calyces are more suggestive of *S. memphiticum*.

Lester (in J.L.S. 125: 285 (1997)) stated that although the synonymous *S. grossidentatum* was originally spelled with an "e" by Richard, it should be corrected to *S. grossidentatum*. Manoko (2007), thought that his AFLP clustering analysis demonstrated that *S. grossidentatum* was a distinct species and not a synonym of *S. memphiticum*; he particularly mentioned differences in berry colour and calyx accrescence. There is considerable variation in the berry colour from blackish-purple to greenish-yellow (when the cuticles are translucent) in the taxon identified as *S. memphiticum*, which might be geographically related. Analagous berry colour variation is found in *S nigrum* sensu stricto and in *S.*

tarderemotum and *S. florulentum*. However, future field work might indicate that the taxon treated as *S. memphiticum* here is indeed composed of two entities which might warrant formal taxonomic recognition.

The earlier suggestion in Edmonds & Chweya (1997) that this synonymous taxon might be conspecific with *S. retroflexum* Dunal (DC, Prodr., 13(1): 50 (1852)) is now considered erroneous. This latter purple-berried tetraploid relative originally described from South Africa is now thought to have a much more restricted distribution, probably not occurring further north than the Flora Zambesiaca region. Superficially, and especially in some herbarium material, specimens of *S. memphiticum* resemble those of the South American diploid *S. physalifolium* Rusby var. *nitidibaccatum* (Bitter) Edmonds. The latter has been sparingly introduced into southern Africa and their similarity might indicate a common genome.

The leaves of *S. memphiticum* are used as a vegetable in Kenya (**K** 4–6), and ripe berries are reportedly eaten in **U** 2 and in Ethiopia (Shashamene). Tweedie (1586 collected in 1958 from **K** 3/5) thought that this taxon was indigenous to that area, since it was very common in her original garden site which was then virgin bush.

Though the altitudinal collection details differ slightly on the *S. subuniflorum* BR duplicate of Volkens' type collection from those cited (1550 m instead of 1580 m), this specimen has been selected as the lectotype of *S. subuniflorum*. This specimen exhibits all the morphological features generally associated with *S. memphiticum*, including the dense glandular-haired pubescence, and is clearly conspecific with it.

18. **Solanum pseudocapsicum** *L.*, Sp. Pl. 1: 184 (1753); Dunal, Synopsis: 11 (1816); G. Don, Gen. Hist. Dichlam. Pl.: 411 (1837); Dunal in DC., Prodr. 13(1): 152 (1852); Wright in Fl. Cap.: 90 (1904); Bitter in E.J. 54: 497 (1917); Lawrence in Baileya 8: 34 (1960); Polhill, *Solanum* in E and NE Africa: 12 (ined., 1961); E.P.A.: 874 (1963); Bailey, Man. Cult. Plants: 868 (1966); Verdcourt & Trump, Common Poisonous Pl. E. Afr.; 168 (1969); D'Arcy in Ann. Missouri Bot. Gard. 60: 714 (1973); Symon in Journ. Adelaide Bot. Gard. 4: 97 (1981); Hawkes & Edmonds in Fl. Europ. 3: 198 (1972); Schönbeck-Temesy in Fl. Iranica 100: 6 (1972); Jaeger, Syst. studies *Solanum* in Africa: 331 (1985, ined.); Purdie *et al.*, Fl. Australia 29: 116 (1982); Huxley *et al.*, New RHS Dict. Gard. 4: 319 (1992); Hepper in Fl. Egypt 159, Solanaceae: 43 (1998); Gonçalves in F.Z. 8 (4): 70 (2005); Friis in Fl. Eth.: 124 (2006). Type: cultivated in Uppsala, "habitat in Madeira", Herb. Linn. 284.4 (LINN!, lecto.), designated by Schönbeck-Temesy in Fl. Iranica 100: 6 (1972) though usually attributed to D'Arcy in Ann. Missouri Bot. Gard. 60: 714 (1973) [See Knapp & Jarvis in J.L.S. 104: 351 (1990) for full discussion of typification]

Herb or erect shrub to 2 m high, stems green when young becoming woody and brown with smoothish striated bark and numerous leaf scars; all parts glabrous or sparsely pilose with scattered dendritic hairs mixed with simple few-celled multicellular hairs, small stalked glands usually present on stems and leaves. Leaves alternate or opposite with one of the pair often much smaller, usually membranaceous, dark green with a prominent lower midrib, linear- or narrowly elliptic to linear-lanceolate, 3.5–10.5 × 0.6–2 cm, bases cuneate and sometimes decurrent, margins entire to sinuate and often inrolled especially towards the petiole, apices acute to acuminate; surfaces glabrous to sparsely pubescent often with scattered stalked glands on both surfaces; petioles 0.6–1.8 cm long. Inflorescences leaf-opposed to extra-axillary, simple to extended cymes, (1–)2–3-flowered; peduncles erect, vestigial or to 8 mm; pedicels erect or sometimes recurved, 5–7 mm long in flower, thickened to woody and 8–13 mm long in fruit especially beneath calyx; axes glabrous to sparsely pubescent. Calyx campanulate, (3.5–)4–7 mm long, glabrous to sparsely pubescent externally; lobes obovate, linear-lanceolate to narrowly triangular, 2.3–5 × 1–1.8 mm, acute; persistent and enlarging to 4–7 × 1.1–3.2 mm in fruit. Corolla white, with median vein to lobes, stellate, 1.2–1.6 cm diameter, tube 1–1.5 mm long; lobes spreading or reflexed after anthesis, broadly ovate to lanceolate, 5–6 × 2–3.5 mm. Stamens equal;

filaments 0.5–0.7 mm long, glabrous; anthers yellow to orange-yellow, 2.4–3.4 × 0.8–1 mm, connivent. Ovary ± 1.6 × 1.5 mm, glabrous; style straight, green, 3.5–5.5 × 0.3 mm, glabrous, exserted to 2 mm; stigma capitate, ± 0.3 mm diameter. Fruit always borne erect, smooth, red to orange globose and often glossy berries 8–18 mm diameter, surrounded by accrescent narrowly triangular calyx lobes which become basally adherent at maturity, probably deciduous from calyces. Seeds many (67–74) per berry, yellow to yellowish-brown, obovoid, discoidal to reniform, 3–3.8 × 2.5–3.2 mm, flattened with raised margin, foveolate; sclerotic granules absent.

KENYA. Kiambu District: Lari, Uplands Forest Station, cult., 14 Sept. 1954, *Verdcourt* 1146!; Nairobi (cult.), Apr. 1963, *Dale* H128!; Kericho District: Kericho Township, 5 Jan. 1967, *Magogo* 17!
TANZANIA. Morogoro District: Kibuko Old Mica Camp, Mar. 1955, *Mgaza* 24!
DISTR. **K** 3, 4; **T** 6; cultivated throughout Africa including South Africa and Madagascar, possibly naturalised in Eritrea and Ethiopia; also in Madeira, the Oceanic Islands, Panama, Egypt, Iran, Pakistan, Australia and Europe
HAB. Usually found as an escape from cultivation; recorded in *Thuja-Araucaria* plantation; 1600–2150 m

SYN. *S. diphyllum* sensu Forssk. XIII No. 134 (1775), *non* L., *nom. nud.* [see Hepper & Friis, Plants of Forsskål's Flora Aegyptiaca-Arabica: 233 (1994)]
  S. microcarpum Vahl, Symb. Bot., 2: 40 (1791). Based on *"Solanum diphyllum* Forssk. Catal. Pl. Aegypt.: 63, n. 134 (1775)". Type: Egypt, *Forsskål* 1751 ex Herb. Hofman Bang, with field label *"Solanum diphyllum* Ro.8, type of *S. microcarpum* (C! - Microf. 101: III, 7, 8, see Hepper & Friis, Plants of Forsskål's Flora Aegyptiaca-Arabica: 233 (1994))
  S. pseudocapsicum L. var. *microcarpum* (Vahl) Pers., Syn. Pl . 1: 224 (1805)
  ?*S. diflorum* Vell., Fl. Flumin.: 84 (1829) & Icon. 2: t. 98 (1831). Type: Velloso, Fl. Flumin. Icones 2: fig. 102 (1831) lecto., designated by Morton, Revision Argentine species of Solanum, 1976)
  ?*S. capsicastrum* Schauer in Allegem. Gartenz. 1: 228 (1883). Type: Uruguay, Montevideo, *Sellow* s.n. (location not cited)
  S. pseudocapsicum L. var. *diflorum* (Vell.) Bitter in E.J. 54: 498 (1917).

NOTE. Though Linnaeus gave the type locality as Madera (Madeira), it is unlikely that the plant was native there. Knapp & Jarvis (1990) considered that historical evidence pointed to Linnaeus deriving his specific epithet from a much earlier work based on a Spanish name which indicated that it is native to the New World, probably to the drier parts of Mexico, SE Brazil and northern S America, from where it may have been introduced to Europe by Spanish and Portugese traders. It is an attractive ornamental which has been introduced throughout the world and is now widely cultivated in warm regions, where it has often become a troublesome weed. The species is commonly known as the Jerusalem-, Winter- or Madeira- Cherry, and is often a popular decorative shrub. Verdcourt & Trump (1969) reported that the fruits contain solanine, solanidine and solanocapsine and that whilst it is often eaten, it should be classed as mildly poisonous since it causes a slowing action on the heart. Many subspecies and varieties of *S. pseudocapsicum* have been formally described, mostly from countries in S. America. They have not been included here as none of these names have been encountered in African literature.
  S. capsicastrum is another *Solanum* described from temperate eastern S America widely cultivated as an ornamental, which also occurs as a casual. It is almost certainly a synonym of the *S. diflorum* Vell described earlier from Brazil. Hawkes & Edmonds (1972) considered that *S. capsicastrum* was a separate species distiguishable from *S. pseudocapsicum* by its dense dendritic pubescence, its shorter stems, and larger berries. However, many other authors (e.g. Gonçalves, 2005) have since synonymised these two species.
  Bitter differentiated his var. *diflorum* from *S. pseudocapsicum* on the basis of the dense indumentum of branched hairs found on all vegetative parts. This character has often been used to differentiate the latter species from *S. capsicastrum*. Most African specimens of *S. pseudocapsicum* seem to be glabrous or only very sparsely pubescent becoming glabrescent. It is possible those specimens found in the Americas more typically exhibit the dense pubescence described for this variety and for *S. capsicastrum*.

19. **Solanum mauritianum** *Scop.*, Delic. Fl. & Faun. Insubr.: 3: 16, t. 8 (1788); Bitter in E.J. 54: 486 (1917); Lawrence in Baileya 8: 34 (1960); Polhill, *Solanum* in E and N Africa: 12 (ined., 1961); Heine in F.W.T.A. 2ⁿᵈ. ed.: 332 (1963); Hawkes & Edmonds in Fl. Europ. 3: 197 (1972); Roe in Brittonia 24: 239 (1972); Agnew in U.K.W.F.: 526 (1974); Heine in Fl. Nouv.-Caled., 7: 137 (1976); Symon in J. Adelaide Bot. Gard. 4: 95 (1981); Troupin in Fl. Pl. Lign. Rwanda, Ser. 8, No. 2: 657 (1982) & Fl. Rwanda 3: 380 (1985); Hepper in Fl. Ceylon 4: 369 (1987); Jaeger, Syst. studies *Solanum* in Africa: 328 (1985, ined.); Huxley *et al.*, New RHS Dict. Gard. 4: 319 (1992); K.T.S.L.: 582 (1994); D'Arcy & Rakotozafy, Fl. Madagascar, Solanaceae: 110 (1994); Purdie *et al.*, in Fl. Austral. 29: 115 (1982); Hepper in Fl. Ceylon 6: 370 (1987); K.T.S.L.: 582 (1994); U.K.W.F. ed. 2: 243 (1994); Bukenya & Carasco in Bothalia 25(1): 50 (1995); Mansfeld, Encycl. Agric. & Hort. Crops: 1818 (2001); Gonçalves in F.Z. 8(4): 86, t. 17 (2005). Type: cultivated in Hortus Botanicus Ticinensis, Pavia from seed probably originating in Mauritius; type specimen unknown, t. 8, Scopoli (1788) designated as a lectotype by Roe (1972)

Shrub to 3 m or small tree to 7 m, evergreen, foetid; branches erect, yellowish-green and velvety through dense tomentose/flocculose pubescence with mostly stellate hairs, either sessile or long, multiseriate-stalked (sometimes with a long central ray), mixed with some uniseriate simple hairs and small four-celled glands. Leaves solitary to alternate, coriaceous, dark green above, whitish- to yellowish-green below, usually lanceolate, 19–34+ × 6–9 cm, bases cuneate and decurrent, margins entire, apices acute to acuminate; pubescent with predominantly sessile stellate hairs above, tomentose/flocculose below with predominently stalked stellate hairs, especially dense on lower midribs and veins; petioles 2.5–4.5 cm long, subtended by two small sessile ovate auricular pseudo-stipules, 1–3.5 × 0.9–1.8 cm. Inflorescences terminal becoming lateral and axillary, dichotomously forked, many-flowered erect compound corymbs 14–25 × 7–14 cm, usually > 100-flowered; peduncles 10–18 cm long, thick; pedicels erect, 3–4 mm long in flower, 4–5 mm long in fruit, articulated; axes densely tomentose/flocculose. Calyx cupulate, 4–6.5 mm long, yellowish-green and densely stellate-tomentose, glabrescent internally; lobes broadly ovate to broadly triangular, sometimes unequal, 2–4 × 1.5–2.5 mm, obtuse to acute, adherent in fruit and 2.5–5 × 1.5–3.5 mm. Corolla flowers purple to lilac with pale basal star, stellate to pentagonal, 1.4–2.3 cm diameter; tube ± 1.5 mm long; lobes broadly ovate to - triangular, 3.5–6.5 × 2.5–6 mm, acute, stellate-tomentose externally but less so than calyx, glabrous internally, margins sometimes inrolled, laterally reflexed exposing anthers after anthesis. Stamens equal to subequal; filaments free for 1–1.2 mm, glabrous; anthers yellow to yellow-orange, 2–3 × 0.9–1.2 mm, free. Ovary bilocular, 1–2 × 1.7–2 mm, densely pubescent; style straight to geniculate, exserted to 4 mm, 5.5–7 × 0.3–0.5 mm, stellate-pubescent at least in lower part; stigma clavate, 0.6–1.5 × 0.4–0.9 mm. Berries yellow to yellowish-green, globose, with coriaceous pericarp softening with maturity, to 10 mm diameter, stellate-scabrid and glabrescent, probably shed with pedicels leaving scarring on inflorescence branches. Seeds usually 100–150 or more per berry, yellow to yellowish-orange, orbicular, ovoid, spherical or discoid, 1.8–2 × 1.5–1.8 mm, foveolate; up to two globular sclerotic granules 0.8 mm diameter present.

UGANDA. Mengo District: Entebbe, June 1937, *Chandler* 1749! & Mar. 1923, *Maitland* 597! & Mabira Forest near Jinja, Feb. 1965, *Tweedie* 2996!
KENYA. Nairobi District: Nairobi, Dec. 1931, *Napier* 1615! & Nairobi beyond Fisheries Dept. off Ngara Estate near branch of Nairobi River, 17 Aug. 1971, *Mwangangi & Kanuri* 1634!; Kiambu/Nairobi District: Karura Forest just beyond city limits of Nairobi, 21 Nov. 1966, *Perdue & Kibuwa* 8042!
DISTR. U 4; K 3, 4; native to southeastern Brazil, Uruguay and NE Argentina, now widespread and often naturalised throughout the tropics: Sierra Leone, ?Nigeria, Cameroon, Rwanda, Angola, Zambia, Mozambique, Zimbabwe and South Africa; Mauritius, Madagascar, Mascarenes; adventive in India, Nepal and Sri Lanka; naturalised in the Atlantic Islands including Madeira and the Azores; a troublesome weed in many parts of Australia, and often cultivated in Europe

HAB. Escape from cultivated land, especially along forest paths and -margins, river banks and
in dry river beds; 1150–2800 m

SYN. *S. auriculatum* Aiton, Hort. Kew., 1: 246 (1789); Willd., Enum. Pl. Hort. Berol: 233 (1809);
Dunal in Hist. *Solanum*: 166 (1813) & in Synopsis: 17 (1816); Nees von Esenbeck in
Trans. Linn. Soc., 17: 46 (1834); Sendtn. in Fl. Bras. 10: 40 (1846); Don, Gen. Hist. Pl.,
4: 415 (1837); Baker, Fl. Mauritius: 215 (1877); Dunal in DC., Prodr. 13(1): 115 (1852);
C.H. Wright in Fl. Cap. 4: 94 (1904); Dammer in E.J. 38: 188 (1906): Bullock in F.W.T.A.
2: 206 (1931); Marloth, F.S.A. 3: 121 (1932); Watt & Breyer-Brandtwijk, Med. & Pois. Pl.
S & E Afr., ed. 2: 990 (1962). Type: *L'Heritier* s.n. (G-DC holo? fide Roe in Brittonia 24:
253 (1972)
S. verbascifolium sensu C.H. Wright in F.T.A. 4, 2: 221 (1906) pro parte, *non* L.; Heine in
F.W.T.A. 2nd ed., 2: 332 (1963)

NOTE. This species is a highly successful pantropical weed now enjoying widespread
distribution thoughout the tropics, largely due to the attraction of its succulent berries to
birds causing rapid dispersal. Roe (1972) suggested that the dispersal of this and related
species from their native South America to Africa coincided either accidentally or purposely
with old Portugese and Spanish trade routes begun during the early- to mid- sixteenth
century. Roe also noted that *S. mauritianum* is a successful coloniser of open and disturbed
ground in a variety of habitats, and that it can also reproduce by adventitous shoots arising
from shallow roots to form large colonies of plants. Jaeger & Hepper (Review of the genus
Solanum in Africa, in D'Arcy (ed.), Solanaceae – Biology and Systematics: 48 (1986)) reported
that *S. mauritianum* is the most notable of all the non-native Solanums introduced into Africa,
now growing in western, central and eastern parts, and that is particularly widespread and
troublesome in South Africa where it is thought to be the host plant for the larval stage of
the Natal fruit-fly (*Pterandrus rosa*). It is often known as the Wild Tobacco- or Bug- tree and is
reportedly poisonous to pigs and cattle in Australia. In South Africa, Watt & Breyer-Brantwijk
(1962) cited early reports of fruits of the synonymous *S. auriculatum* causing fatalities in
humans in the Natal area as well as cattle poisoning, though birds eat the seeds without
apparent ill effect. These authors also reported allergic skin reactions when the plants were
handled together with various medicinal uses of this synonymous species. These included the
treatment of Manioc poisoning by leaf juice application; leaf decoctions as a lotion for
haemaerrhoid treatment, and the use of seeds mixed with coconut oil as an ointment to
alleviate rheumatism. It is cultivated in New Zealand as a rootstock for the tree tomato
(*Solanum betaceum* – see page 104)

Sendtner (1846) placed *S. mauritianum* in synonymy with *S. auriculatum* although the latter
had been described one year later.

20. **Solanum wendlandii** *Hook. f.* in Bot. Mag. 113: t. 6914 (1887); Polhill, *Solanum*
in E & NE Africa: 11 (ined., 1961); Seithe in E.J. 81: 292, 330 (1962); D'Arcy in Ann.
Miss. Bot. Gard. 60: 685 (1973). Heine, Fl. Nouvelle-Calédonie et Dependences 7:
144 (1976); Symon in J. Adelaide Bot. Gard. 4: 69 (1981); Whalen in Gentes Herb.
12: 208 (1984); Jaeger, Syst. stud. *Solanum* in Africa: 336 (1985, ined.); Troupin, Fl.
Rwanda 3: 382 (1985); Bukenya & Carasco in Bothalia 25: 51 (1995); Hepper in Flora
Egypt 6, Family 159, Solanaceae: 52 (1995); Mansfeld, Encycl. Ag. & Hort. Crops 4:
1811 (2001); Gonçalves in F.Z. 8, 4: 71 (2005). Type: cultivated in England, Kew, from
plants received from Herrenhausen Garden Hannover, *Wendland* s.n., originally
collected in Costa Rica, *Wendland* s.n. (K!, holo.; K!, iso.)

Vigorous climber or liana or more rarely a creeper to 7 m high; stems glabrous or
with scattered long simple or branched minute hairs when young, with trunk to 8 cm
diameter and pale brown bark; scattered stalked club-glands often present
throughout, with or without small stout sharp recurved prickles. Leaves alternate,
bright green, membranaceous; variable from simple and ovate, 9–15 × 5–12 cm, to
ovate-lanceolate when three-lobed with equal or unequal leaflets to 4 × 2.4 cm above,
pinnate on lower branches with 1–6 pairs of ovate acute leaflets and with the lowest
pair sometimes petiolulate, bases cordate to cuneate, often unequal (oblique),
margins entire, apices obtuse, acute or acuminate; both surfaces typically glabrous
but occasionally with sparse long spreading hairs; lower midribs sometimes with

small prickles; petioles 3–7 cm long, sometimes with small retrorse prickles to 1.5 mm long. Inflorescences terminal, pendulous, multiforked, > 50-flowered lax scorpioid or paniculate cymes to 17 cm diameter; flowers ?andromonoecious; pedicels erect and 0.4–1.5(–2) cm long in flower; axes usually glabrous, occasionally sparsely pilose with long eglandular hairs, and scattered stalked glands. Calyx pale green, cupulate/campanulate, 4–6 mm long, occasionally with scattered hairs basally, otherwise glabrous; lobes ovate, usually equal, 2.2–4 × 2–3.5 mm, acute to apiculate, margins thickened; fruiting posture and dimensions not known. Corolla pale blue, lilac or purple with a distinct contrasting star reaching to the lobe apices, pentagonal-rotate, 2.8–7 cm diameter, shallowly lobed with lateral posture exposing androecium; tube 1.5–2.2 mm long, glabrous internally and externally; lobes broadly triangular, 2–4 × 6–10 mm, acute, glabrous apart from apices with small erect eglandular hairs. Stamens usually unequal within same flower; filaments white to cream, free for 0.5–3(–5) mm, glabrous; anthers poricidal, yellow, orange or blue, 7–10.5 × 1.3–1.8(–2.4) mm, usually connivent. Ovary brownish, 0.4–0.8(–2) × 0.8–1.3(–2) mm, glabrous, bilocular; style vestigial or very short within base of the staminal tube, light brown, 0.3–1.5 mm, glabrous; stigma light brown, capitate, sometimes bilobed, 0.2–0.4 mm diameter. Fruit in clusters at the ends of elongated branches, dark bottle green becoming yellow to red, globose, 1–1.5 cm diameter, glabrous. Seeds many, greenish-white, reniform, 1–1.2 × 1 mm, flattened; sclerotic granules absent.

UGANDA. (fide Bukenya & Carasco, 1995)
KENYA. SE slopes of Mt Elgon, 4 May 1953, *Padwa* 42!; Nairobi District: Nairobi, Hort. Hemming, July 1961, *Polhill* in EA 12424! & road to Baptist church off Ngong Road, 17 Apr. 1977, *Kahurananga* 837!;
DISTR. U 4; K 3, 4; native to Costa Rica, this species is now widely cultivated as an ornamental from tropical to temperate regions of the world for its showy flowers, including Egypt, Cameroon, Rwanda, Zambia, Malawi, Mozambique, Zimbabwe; Europe through India and Sri Lanka to New Caledonia and Australia, Central America and Jamaica
HAB. Cultivated in parks and gardens, found occasionally naturalised on roadsides and in montane forests; 1650–1800 m

NOTE. This is commonly known as the Potato-vine, the Giant Potato Creeper, Paradise Flower or Costa Rican Nightshade. In the discussion of his *Solanum wendlandii* group Whalen (1984) described the flowers as dimorphic and andromonoecious; insufficient material was available to verify these observations – though in the flowers examined, filament length varied from stamen to stamen within the same flower, with one filament generally being longer that the others. Moreover, the styles of all flowers dissected were either vestigial or extremely small and embedded at the base of the corollas, though the stigmas seemed well-developed. Symon (1981) commented that in cultivation plants only seem to develop male flowers and that fruit descriptions were rare and contradictory, with reported sizes ranging from 4 cm diameter to that of an apple. However, he ascertained from a Costan Rican source that seed germination of *S. wendlandii* is poor and that the plants are mostly propagated by cuttings. The fruit and seed characters given above are taken from Symon (1981).

21. **Solanum giganteum** *Jacq.*, Collect., 4: 125 (1791) & Ic. Pl. Rar., 2: t. 328 (1793); Nees von Esenbeck in Trans. Linn. Soc., 17: 47 (1832); Don, Gen. Hist. Dichlam. Pl. 4: 430 (1837); Dunal, Hist. *Solanum*: 202 (1813) & in Syn. Solan. Hist.: 36 (1816) & in DC., Prodr. 13(1): 258 (1852); Sims in Bot. Mag. 44: t. 1921 (1817); Clarke in Fl. Brit. Ind. 4(1): 233 (1885); Engler, Hochgebirgsfl. Trop. Afr.: 373 (1892); Kuntze in Rev. Gen. Pl. 3, 2: 226 (1893); Wright in Fl. Cap. 4: 94 (1904); C.H. Wright in F.T.A. 4, 2: 229 (1906); Dammer in E.J. 38: 191 (1906); Pax in E.J. 39: 648 (1907); Bitter in E.J. 57: 257 (1921); Chiovenda in Racc. Bot. Miss. Consol. Kenya: 88 (1935); I.T.U.: 233 (1940); T.T.C.L.: 578 (1949); I.T.U.: 414 (1952); Lawrence in Baileya 8: 30 (1960); K.T.S.: 538 (1961); Polhill, *Solanum* in E & NE Africa: 13 (ined., 1961); Seithe in E.J. 81: 316 (1962); Watt & Breyer-Brandwijk, Med. & Pois. Pl. S. & E. Africa, ed. 2: 993 (1962); Heine in F.W.T.A. 2nd ed., 2: 332 (1963); E.P.A., 2: 866 (1963); Symon in Journ. Adelaide Bot. Gard. 4: 117 (1981); Troupin, Fl. Pl. Ligneuses Rwanda: 657 (1982) & Fl.

Rwanda: 378 (1985); Purdie *et al.*, Fl. Australia 29: 75, 87 (1982); Whalen in Gentes Herb. 12: 212 (1984); Jaeger, Syst. stud. *Solanum* in Africa: 347 (1985, ined); Blundell, Wild Fl. E. Afr.: 190 (1992); U.K.W.F., 2ⁿᵈ ed.: 243 (1994); Bukenya & Carasco in Bothalia 25: 56 (1995); K.T.S.L.: 580 (1994); Mansfeld, Encycl. Agric. & Hort. Crops, 4: 1818 (2001); Gonçalves in F.Z. 8(4): 88 (2005); Friis in Fl. Eth. 5: 125 (2006); Welman in Bothalia 38(1): 40 (2008). Type: cultivated in Hortus Vienna, originally from South Africa (? Cape of Good Hope fide Bot. Mag. (1817)), fide Cufodontis (1963); Hort. Schonbr., Herb. Jacquin s.n. (W, syn., photos !) fide Lester (unpublished 2005); t. 328 in Jacquin, Icones plantarum rariorum, 2: (1790) fide Friis (2006)

Shrub to 4 m high, sometimes straggling, occasionally tree to 8 m tall (trunk reported "as thick as an arm" by Wright, 1904); much branched, branches ascending; with dense white floccose indumentum of sessile stellate hairs with ± 12 equal eglandular rays 0.1–0.3 mm long together with small simple hairs and stalked glands, glabrescent exposing brown bark; main stems and branches with stout conical, straight or recurved prickles up to 5.5 × 5.5 mm basally tapering to ± 0.2 mm apically, laterally compressed, tomentose/floccose basally, glabrescent and finally yellowish-brown, often leaving scars on stems and branches. Leaves alternate, coriaceous, dark green above, white to yellowish-white below, ovate, obovate or elliptic, sometimes narrowly lanceolate, (7.5–)12–36 × (4–)5–13 cm, bases cuneate often decurrent for half of petiole, margins usually entire, occasionally repand, apices usually acute sometimes acuminate; lower surfaces covered with dense floccose pubescence of stellate hairs as on stems, rarely with a few small prickles on lower part of midrib, upper surfaces glabrous but for the often densely floccose midribs where stalked glands sometimes also visible; petioles 1.5–6 cm long, sometimes with 1–2 leaf-like elliptic to ovate pseudostipules basally, sometimes with small prickles. Inflorescences subterminal becoming lateral, forked (usually multiply so), 100–200-flowered, lax, scorpioid cymes borne in compact pyramidal corymbs 7–14 cm wide; flowers usually nodding; peduncles 1.2–6 cm long in flower, 3.5–5 cm long and woody in fruit, often with small curved prickles; pedicels erect to recurved and 0.8–1.5 × 1.1–2.1 cm in flower, erect in fruit and spreading, often becoming woody and thickened beneath calyx; axes all densely floccose with stellate hairs as above. Calyx usually cupulate, sometimes campanulate, (3–)4–6 mm long, densely floccose externally, glabrescent internally where prominent venation often visible, lobes often unequal, triangular, 1–2.5(–3) × 1–2.5(–3) mm, acute, occasionally apiculate; adherent basally in fruit, later becoming reflexed, (2.5–)3–6(–6.5) × 1.5–4 mm, glabrescent. Corolla violet, purple, pale lilac to whitish, sometimes with yellow or greenish central star, stellate, 1.2–1.8(–2) cm diameter; tube 1–1.6 mm long, glabrous externally; lobes lanceolate, (3–)4.5–8 × 1.3–3 mm, densely stellate-floccose externally, sparsely stellate-pubescent or glabrescent on median veins internally; lobes reflexed and splayed between the calyx lobes after anthesis. Stamens equal; filaments free for 0.3–0.8 mm, glabrous; anthers with small apical pores, yellow to brownish, 2.3–3.5 × 0.8–1.1 mm, connivent. Ovary brownish, 0.8–1.7 × 0.8–1.6 mm, glabrous, bilocular; style straight sometimes curved apically, pale green, 4.5–7 × 0.3–0.4 mm, glabrous, exserted up to 3 mm; stigma pale green, capitate, 0.3–0.4(–0.7) mm diameter. Berries smooth, red or orange, rarely yellow, globose to ovoid, glabrous, glossy, (5–)6–9(–10) mm diameter. Seeds 23–37 per berry, yellow to yellow-orange, obovoid, elliptic or reniform, 2.3–2.8(–3.2) × 1.5–2.5 mm, foveolate to reticulate; sclerotic granules absent.

UGANDA. Bunyoro District: Bundongo Forest, 24 Nov. 1938, *Loveridge* 95!; Mengo District: Kajansi Forest, Entebbe Road, May 1935, *Chandler* 1215! & 2 km E Bujuko on Mubende Road, 26 Feb. 1969, *Lye* 1953!
KENYA. Northern Frontier District: Mt Marsabit, 28 Feb. 1963, *Bally* 12557!; Meru District: Nyambeni Hills, base of Kirima, 8 Oct. 1960, *Polhill & Verdcourt* 267!; North Kavirondo District, Kakamega Forest Station, 16 Sept. 1949, *Maas Geesteranus* 6247!
TANZANIA. Lushoto District: Kungului Forest Reserve, 17 Nov. 1971, *Magogo* 193!; Morogoro District: Ulugurus, Kitundu, 22 Nov. 1934, *E.M. Bruce* 188!; Lushoto/Tanga District: E Usambaras, Ubili [Uberi], 25 Nov. 1935, *Greenway* 4183

DISTR. U 2–4; **K** 1, 3–5, 7; **T** 1–4, 6–8; probably native to the Cape region of South Africa, but now found from Nigeria to Cameroon, Gabon, Congo-Kinshasa, Rwanda, Sudan, Ethiopia, Malawi, Mozambique, Zimbabwe, Swaziland and South Africa; India and Sri Lanka, Australia and S America

HAB. Grassland, secondary bushland, thickets, moist and riverine forest and -margins, bushland, shambas, evergreen thickets, bamboo zone; 800–2450 m

SYN. *S. niveum* [Vahl, Symb. Bot., 2: 41(1791) ex] Thunb., Prodr. Pl. Cap. 1: 36 (1794) & Fl. Cap. 2: 59 (1818). Type: South Africa, Cape Province, *Thunberg* s.n. (UPS-THUNB, holo.; microfiche 1036, no. 5209!)

 S. farinosum Wall. in Roxb. Fl. Ind., ed. Carey & Wall., 2: 255 (1824). Type: India, "specimens are present in Dr Heyne's Herbarium, labelled as "*S. argenteum*; Babobad a name which has already been applied to another species" fide Wallich (1824) (K!, holo.)

 S. argenteum Wall. Cat. n. 2610, page 80 (1831), *nom. nud.* based on *Heyne* s.n., Babobad, K-WALL 2160!

 S. seretii De Wild., Miss. E. Laurent.,1: 439 (1907) & plate 122 (CXXII); Bitter in Engler in E.J. 57: 260 (1921) & in F.R. 16: 311 (1923). Type: Congo-Kinshasa, Bima to Bambili, *Seret* 166 (BR!, holo.; BR!, iso.)

 S. muansense Dammer in E.J. 48: 243 (1912); Bitter in E.J. 57: 261 (1921); T.T.C.L.: 579 (1949). Type: Tanzania, Mwanza [Muansa], *Stuhlmann* 4504 (?B†, ?holo.)

 ?*S. bequaertii* de Wild. in F.R. 13: 141 (1914). Type: Congo-Kinshasa, Lubumbashi [Elisabethville], *Bequaert* 219 (BR!, holo.; BR!, iso.)

 S. sordidescens Bitter in E.J. 57: 260 (1921); T.T.C.L.: 579 (1949). Type: Tanzania, Kilwa District: Matumbi Mts, Chumo [Tschumo], *Busse* 3097 (?B†, ?holo.; BR!, EA!, iso.)

NOTE. This species is commonly known as the Giant Nightshade, Red Bitter Apple or Red Bitter Berry, and as African Holly in Australia. It is often cultivated as an ornamental shrub or small tree in both the northern and southern hemispheres, or as a bedding or hedge plant (Welman 2008). It is characterised by a dense snowy white floccose indumentum on the lower leaf surfaces which strikingly contrasts with the glabrescent leathery upper surfaces, though Ugandan specimens do not seem to be so densely floccose. While scattered stellate hairs can often be seen on the mature berries, they are loose and will have become dislodged from the floccose floral and vegetative parts; the ovaries are always glabrous. Lawrence (1960) noted that the species was sparingly cultivated in southern California for its showy red berries in the 1930's. It is used as a shade tree in Kerala, India, with the fruits and leaves being used in traditional African medicine to treat sores (cf. Mansfeld, 2001). Medicinal reports include the use of its leaves to dress ulcers in southern and eastern Africa with the woolly undersurfaces being used to clean the lesions and the upper smooth surfaces to heal the wounds, together with the leaves and fruits for ulcer treatment and the leaves for insomnia in Uganda (cf. Welman, 2008). Plants of *S. giganteum* with a strong unpleasant smell have been reported from **K** 4, whilst berries have been reported as sweetly aromatic in **T** 1. The berries are eaten by larger birds such as bulbuls, doves and go-away-birds particularly in late winter when food is in short supply in southern Africa, where the species has also been listed as an occasional ruderal or silvicultural weed (cf. Welman 2008). The species may be of use as a source of raw materials for the steroid industry; various alkaloids have been isolated from the leaves including solasodine in both the fruits and leaves (see Welman 2008).

 Bitter (1921) placed this species together with *S. sordidescens, S. muansense, S. schumannianum* and *S. ulugurense* into a new series *Giganteiformia* Bitter. These species were all characterised by prickly stems, a dense stellate farinose/tomentose pubescence, large leaves, multiflowered forked corymbose inflorescences and smallish flowers. They are all considered to be synonyms of *S. giganteum* here. Dunal (page 258 (1852)) recognised two varieties of *S. giganteum* namely var. *tenuifolium* Dunal and var. *longifolium* Dunal – both from India and both probably synonymous with this species.

 Jaeger (1985) considered *S. muansense* and *S. sordidescens* to be part of the "*S. kagehense* Dammer group". However, he recognised this group for convenience due to the uncertainty of which species a number of specimens belonged to, the definitive taxonomy of which pended examination of their types. Lester (ined.) determined many Kew specimens of *S. giganteum* from across Africa variously as *S. seretii, S. sordidescens, "S. bequaertii* (= *S. sordidescens)*" and *S. muansense* in 1998. Apart from less dense pubescence on the under sufaces of the leaves of some of these specimens there is little to indicate the basis on which he differentiated them from *S. giganteum* sensu stricto. The sepal size and shape seems to be particularly variable in this taxon; the calyces are usually cupulate with broadly triangular calyx lobes. On some specimens (e.g. *Juniper et al.* 1937 from **T** 4) the calyces are more campanulate with narrowly triangular calyx lobes twice as long. This may have been among the factors on which Lester based his conclusions. However, both calyx and sepal size can also very considerably on same plant in this species with sepal size sometimes also varying in the same flower; these variables do not seem to be correlated with any other distinguishng features.

The species occurs in a variety of habitats throughout Africa; Bukenya & Carasco (1995) reported it to be fairly common in Uganda with a disjunct montane distribution southwards from Ethiopia to South Africa and west to Cameroon. Blundell (1992) reported that it is an uncommon plant in wet montane forests at higher altitudes in Kenya and in lake areas in both Kenya and Tanzania and all Ugandan regions with the exception of northern parts.

The protologues of both *S. niveum* Vahl and *S. farinosum* Wall. (see below) are quite informative, and include the most important distinguishing characters of *S. giganteum* with which they have been synonymised. Moreover *S. niveum* was treated as a synonym of this species by many earlier authors and a *Drège* (s.n., K!) specimen typical of *S. giganteum* was annotated by N.E. Brown in 1883 who stated that "this sheet matches Thunberg's type specimen" and that "Thunberg collected it in the woods at Essenbosch in Lange Kloop, Humansdorp Div. W Cape".

There are two *Heyne* specimens which might be duplicates of the same collection. One in K-WALL (No. 2160) is without a date, though the catalogue number date is 1831, and has two labels – one says "*S. argenteum* Babobad" and the other – which is the original Wallich label- says "*S. farinosum* Wall. ex Herb. Heyne". The specimen in the general Kew Herbarium was collected by Heyne and is labelled Babobad September 1816. This is the holotype of *S. farinosum*. Both specimens clearly belong to *S. giganteum*; *S. argenteum* Heyne ex Wall. is a nomen nudum.

The type material of *S. seretii*, together with de Wildeman's informative protologue and excellent plate leave little doubt that this species is synonymous with *S. giganteum*. Bitter (1921, 1923) actually thought that it was similar to and probably a variety of the latter species.

Bitter differentiated *S. sordidescens* and *S. muansense* from *S. giganteum* largely on the basis of their leaf sizes and shapes, pubescence colour, prickle coarseness and berry sizes, all of which are within parameters of morphological variability exhibited by *S. giganteum*. These two species have therefore been synonymised with the latter from the morphological characteristics given in their extensive protologues and from the isotypes of *S. sordidescens* (see below).

The type material of *S. bequaertii* indicates that this species is either very close to *S. giganteum* or a synonym of it. Bitter's protologue of *S. bequaertii* cited anthers 4.5 mm long, linear corolla lobes 6–7 mm long, and large (12 mm) red berries (becoming black). These features are all smaller on the two *Bequaert* specimens examined with the anthers 3.5–4 mm, the lobes narrowly triangular and ± 5.5 mm long, and the berries only 5–6 mm diameter – all measurements within the variability acceptable in *S. giganteum*. The only obvious disparity concerns the lower leaf surfaces being mealy rather than floccose, and the occurrence of small prickles on the upper midribs. However, many of the leaf petioles are associated with the pseudostipules characteristic of *S. giganteum*. Bitter (1914) suggested that *S. bequaertii* was related to *S. torvum* Sw. and final clarification of its taxonomic placement requires further work on related species.

22. **Solanum goetzei** *Dammer* in E.J., 28: 473 (1900); C.H. Wright in F.T.A. 4, 2: 218 (1906); Bitter in E.J. 57: 269 (1921); T.T.C.L.: 579 (1949); Polhill, *Solanum* in E & NE Africa: 16 (ined., 1961); Whalen in Gentes Herb. 12: 215 (1984) [as *S. goetzii*]; Jaeger, Syst. stud. *Solanum* in Africa: 352 (1985, ined.); K.T.S.L.: 580 (1994); Gonçalves in F.Z. 8(4): 91 (2005); Welman in Bothalia 38(1): 42 (2008). Type: Tanzania, Bagamoyo District: Kikoka, *Stuhlmann* 127 (?B†, syn.); Morogoro District: Ukutu [Khutu-steppe], *Goetze* 112 (K!, syn.); Handeni District: Kiwanda, *Fischer* 409 (?B†, syn.)

Perennial herb or shrub to 3 m high, erect, basal stems becoming woody; prickles absent throughout; stems initially ochraceous and covered with dense indumentum of interlocking sessile stellate hairs to 0.2 mm diameter with ± 8 equal eglandular rays and erect median ray, intermixed with short stalked glands, occasionally young stems purplish due to hair pigmentation. Leaves opposite or alternate, often **membranaceous, sometimes coriaceous, dull to dark green above, paler below,** ovate, obovate or lanceolate, 5–9(–15) × 2–5(–7.5) cm, bases cuneate and often decurrent for half of petiole, margins usually entire, occasionally repand, apices

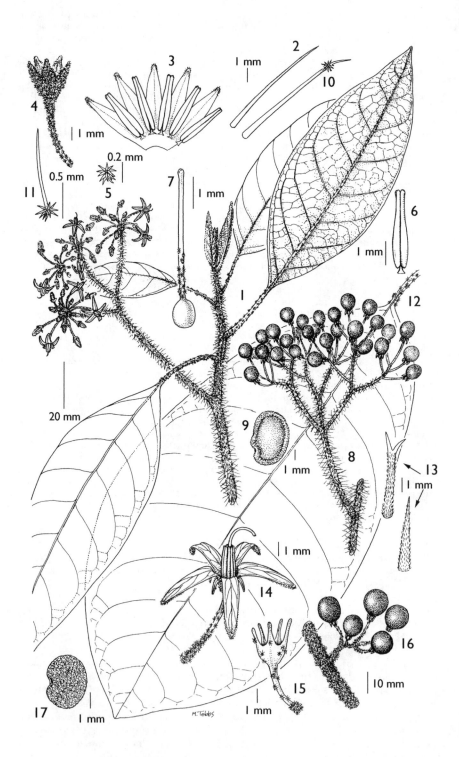

M. Tebbs

acute or sometimes acuminate; both surfaces stellate-pubescent, usually glabrescent, with glands mixed with stellate hairs to 0.4 mm diameter with up to 14 lateral rays to 0.2 mm, and erect prominent median rays, all denser on midribs and veins especially on the lower surfaces, venation usually prominent; petioles 1–3.8 cm long. Inflorescences subterminal to lateral, leaf-opposed or extra-axillary, simple or simple and forked, 5–20-flowered lax scorpioid cymes 3.5–5 cm broad; flowers 5-merous, usually nodding; peduncles 2–12 mm long in flower, 2–14 mm and usually woody in fruit; pedicels erect to recurved and 4.5–9 mm long in flower, erect and spreading and 6–15 mm in fruit, often becoming woody and thickened beneath calyx; axes initially densely pubescent, stellate hairs as above. Calyx campanulate, 2.2–4(–4.5) mm long, with scattered hairs becoming denser towards the pedicels, glabrous internally; lobes often unequal, narrowly triangular becoming subulate, 1.5–2.5 × 0.3–1.5 mm, acute to apiculate; adherent basally in fruit, later becoming strongly reflexed, subulate, 3–5(–8) × 1–3 mm, glabrous. Corolla violet, purple, pale lilac or white, sometimes with purplish yellow central star, stellate, deeply lobed, 1–2(–2.8) cm diameter; tube ± 1 mm long, glabrous externally; lobes lanceolate, 5–9 × 1.5–3.5 mm, acute, with prominent median veins, densely stellate-pubescent to glabrescent externally, sparsely stellate-pubescent on veins internally, denser toward the apices, lobes eventually strongly reflexed. Stamens usually equal; filaments free for 0.5–1.8 mm, glabrous; anthers poricidal, yellow often drying brown, 3–4 × 0.6–0.8(–1) mm. Ovary brownish, 1.1–2 × 1.2–1.3 mm, glabrous but often with small stalked glands apically, bilocular; style straight below, often curving downwards apically, white, 5–9 × 0.2–0.5 mm, ± glabrous, exserted up to 3 mm; stigma green, capitate or bilobed, 0.3–0.5 mm diameter. Berries smooth, bright red, globose, glossy, 6–10(–12) mm diameter, usually glabrous, occasionally with odd scattered stellate hair, often with translucent cuticle through which seeds visible. Seeds (8–)17–26 per berry, yellow to brown, obovoid, elliptic or reniform, 3.5–4.2(–4.8) × 3–3.8(–4.2) mm, laterally compressed, deeply reticulate to deeply reticulate-foveolate; sclerotic granules absent. Fig. 18/14–17, p. 152.

KENYA. Kwale District: Mrima Hill, 21 Sep. 1982, *Polhill* 4818!; Kilifi District: Sabaki, 6 km N of Malindi, 12 Nov. 1961, *Polhill & Paulo* 750! & ± 2 km N of Mariakani along Mariakani–Bamba road, 16 Dec. 1973, *B.R. Adams* 76!
TANZANIA. Mwanza District: Deda, 23 Feb. 1933, *Rounce* 250!; Lushoto District: Segora Forest, 30 July 1966, *Faulkner* 3834!; Tanga District: Rugwe–Maweni, 29 Nov. 1967, *Faulkner* 4067!
DISTR. **K** 7; **T** 1, 3, 4, 6, 8; Zambia, Malawi and Mozambique
HAB. Dry to moist forest and riverine forest, including clearings and -margins, coastal thicket, woodland, wooded grassland, *Acacia-Commiphora* bushland, thicket, cleared ground, roadsides, in regenerating cultivation; sea-level to 750(–1250) m

SYN. *S. muha* Dammer in E.J. 38: 186 (1906). Type: Tanzania, Lushoto District: Usambara, Amboni Hill, *Holst* 2731 (?B†, syn.); Morogoro District: Ukami, E of Kitondwe, *Stuhlmann* 8267 (?B†, syn.); Bagamoyo District: Ukwere, Kissemo, *Stuhlmann* 8408 (?B†, syn.)
S. pauperum sens auctt. e.g. C.H. Wright in F.T.A. 4, 2: 217 (1906) saltem pro parte quad specimen ex Malawi, *Kirk* s.n. (which = *goetzei*), *non* C.H. Wright
?*S. bagamojense* Bitter & Dammer in F.R. 13: 92 (1914). Type: Tanzania, Bagamoyo District: Kingani Delta, *Hildebrandt* 990b (BREM, W, iso. (photo!))
?*S. goetzei* Dammer var. *bagamojense* (Bitter & Dammer) Bitter in E.J. 57: 270 (1921); T.T.C.L.: 579 (1949)

FIG. 18. *SOLANUM SCHUMANNIANUM* var. *SCHUMMANIANUM* — **1**, flowering habit; **2**, stem bristle; **3**, dissected flower; **4**, calyx and pedicel; **5**, equal-rayed stellate hair; **6**, anther and filament; **7**, gynoecium; **8**, infructescence; **9**, seed. Var. *SUBULATUM* — **10**, stem bristle; **11**, stellate hair with long median ray. *S. SCHLIEBENII* — **12**, leaf; **13**, simple and forked bristles. *S. GOETZEI* — **14**, complete flower; **15**, calyx and pedicel; **16**, infructescence; **17**, seed. 1 from *Verdcourt & Polhill* 2734; 3–7 from *Drummond & Hemsley* 4309; 2, 8–9 from *Gilbert & Rankin* 4819; 10–11 from *Thulin & Mhoro* 27416; 12–13 from *Paulo* 39; 14–15 from *Faulkner* 1724; 16–17 from *Archbold* 2722. Drawn by Margaret Tebbs.

NOTE. This is essentially a lowland species usually found in wooded and forest areas from the coast to elevations of around 700 m. Jaeger (1985) considered that *S. goetzei* showed affinities with the West African species *S. anomalum* Thonn., though the stems and branches of the latter usually bear prickles and the flowers are tetramerous. He also considered it close to the predominantly Angolan *S. pauperum* C.H. Wright (K.B. 1894: 127 (1894); C.H. Wright in F.T.A. 4, 2: 217 (1906). Type: Angola, *Welwitsch* 6054 & 6075 (K!, syn.)). The syntypes cited by Wright (1894, 1906) are indeed similar to *S. goetzei* florally, but the indumentum is composed of small densely- and multi-rayed stellate hairs and the inflorescence structure is very different with long slender pedicels arising in subterminal umbels which are virtually pedunculate. Such morphology is found in many Angolan specimens, and are sufficiently divergent to warrant the maintainance of *S. pauperum* as a distinct species which does not occur in East Africa. However, the Malawian specimen *Kirk* s.n. cited by Wright (1906) is synonymous with *S. goetzei*.

Gonçalves (2006) listed large seeds, lack of prickles and sparse pubescence as useful characteristics of this species. Whilst the latter two characters may be useful, there is considerable overlap in seed size between with this species and *S. schummannianum*, which is closely related to *S. goetzei*.

Plants of *S. goetzei* are grazed by cattle in **K** 7; its leaves are used as a vegetable in **T** 3 and medicinally in **K** 7, where hot poultices are used to reduce swellings and draw out abscesses and to treat whitlows on fingers. It is also listed in PROTA (11: 308 (2002)) as a medicinal plant in tropical Africa.

Bitter (1921) gave *S. muha* as a synonym of *S. goetzei*, and would presumably have seen the *Stuhlmann* specimens on which it was based in Berlin. No duplicate type material has yet been located, but the protologue suggests that Bitter was right and that this species is indeed a synonym of *S. goetzei*.

The correct placement of *S. bagamojense* is difficult. Bitter (1914) considered it to belong to the section *Dulcamara* (Dunal) Bitter but may later have changed his mind when he formally recognised it as a variety of Dammer's species *S. goetzei*. The characters given in the protologues of these two taxa imply synonymy with *S. goetzei*, apart from the citation of 40-flowered inflorescences - clearly apparent on Hildebrandt's syntype (W, photo!). It was presumably on this basis that Bitter recognised it varietally. However, in view of the complicity of all other morphological features described in the protologues and the shortness of the peduncle on the syntype, these two taxa have provisionally been synonymised with *S. goetzei*.

23. **Solanum schumannianum** *Dammer* in P.O.A. C: 352 (1895) & in E.J. 38: 190 (1906); C.H. Wright in F.T.A. 4, 2: 234 (1906); Bitter in E.J. 57: 263 (1921); T.T.C.L.: 578 (1949); Polhill, *Solanum* in E & NE Africa: 14 (ined., 1961); Seithe in E.J. 81: 326 (1962); Jaeger, Syst. stud. *Solanum* in Africa: 350 (1985, ined.); Blundell, Wild Flowers E. Africa: 191, t. 842 (1992); U.K.W.F., 2ⁿᵈ ed.: 243 (1994); K.T.S.L.: 582 (1994); Gonçalves in F.Z. 8(4): 90 (2005). Type: Tanzania, Lushoto District: Usambara, Magamba, *Holst* 3841 (?B†, holo.)

Shrub or subshrub 1–4 m high, sometimes straggling, occasionally small tree, rarely herbaceous; stems with raised lenticels, usually covered with indumentum ***either*** of clustered sessile or shortly stalked stellate hairs 0.1–0.4 mm diameter, with multiple short equal and eglandular rays, appearing mealy-tomentose when young, ***or*** of larger stellate hairs with long median rays (–1.5 mm), appearing hirsute; both types often interspersed with small stalked brown glands, or plants virtually glabrous; main stems and flowering branches also with soft, dense to sparse, glabrous bristles up to 11.5 × 1 mm, purple becoming brown, spreading and straight or curved apically, either tapering to an apical point or with many tapering to a stellate cluster of short rays with an extended median transparent ray; occasionally bristles absent. Leaves alternate, dark green, usually membranaceous, occasionally coriaceous, ovate, oblanceolate to lanceolate, (6.5–)8.5–19.5(–25) × 2.5–8.5(–10) cm, bases cuneate often decurrent, sometimes asymmetrical, margins usually entire, occasionally sinuate, apices acute to acuminate; lower surfaces often ochraceous and stellate-pubescent, usually with a small bristles and small stalked glands on lower part of midrib and larger veins, sometimes glabrous; upper surfaces glabrous or stellate-pubescent, with or without scattered bristles especially on midribs; petioles 1.5–5(–7) cm long, with or without bristles. Inflorescences subterminal becoming lateral, multiply forked, 15–100+-flowered, lax, umbellate cymes borne in pyramidal corymbs 5.5–12 cm broad; flowers usually 5-merous, occasionally 4-merous; axes with bristles interspersed with stellate-haired pubescence, occasionally glabrous, often becoming woody in fruit; peduncles 1.5–5 cm long in flower, 2.5–5 cm long in fruit; pedicels always without bristles, erect to recurved and 4–12 mm in flower, erect and spreading and 10–15 mm in fruit, thickened beneath calyx. Calyx cupulate to campanulate, 1.5–3.5(–5) mm long, usually stellate-pubescent externally, glabrescent internally; lobes triangular to ovate, 0.6–1.8(–2.5) × 0.5–1.5(–2) mm, acute (apiculate); adherent basally in fruit, becoming strongly reflexed, 1.5–3(–4.5) × 1.5–3 mm, glabrescent. Corolla white, bluish, pale purple or mauve, usually nodding, stellate, 5–10 mm diameter; tube 0.6–1.5 mm long, glabrous; lobes narrowly lanceolate or narrowly triangular, 3.5–7(–8.5) × 1.3–2.4 mm, stellate-pubescent especially towards apices, glabrescent, strongly reflexed. Stamens equal; filaments whitish, free for 0.2–0.5 mm, glabrous; anthers poricidal, yellow, orange or brownish 3–4.5(–5.5) × 0.6–1(–1.2) mm, sometimes connivent. Ovary 1–1.6 × 0.8–1.5 mm, ± glabrous, bilocular; style usually straight, 5–7(–9) × 0.2–0.4 mm, hairy in lower part, exserted up to 2 mm; stigma yellow to greenish, capitate, 0.2–0.3 mm diameter. Berries smooth, usually red, sometimes purplish or blackish (?on drying) globose, 6–10(–12) mm diameter, glabrous or with scattered sessile stellate hairs, pericarp often leathery. Seeds 7–19 per berry, light brown to yellowish-orange, obovoid, elliptic or reniform, 3.5–5 (?–6) × 3–4 (?–5) mm, flattened, minutely vesiculate; sclerotic granules absent.

NOTE. Whilst this high altitude species is often easily recognisable by its soft straight bristle-like prickles, these vary in their density, length and spread, not only on the vegetative and floral stems, but also on the petioles, leaf midribs and veins. Moreover, they may be completely absent from plants otherwise morphologically identical. Specimens in which bristles are absent are quite common in the Mbeya (T 7) region of Tanzania (cf. *Richards* 14122) as well as in neighbouring T 6 (the Ulugurus) and in Malawi. Moreover, the actual indumentum hairs can vary greatly. Plants can appear densely mealy – especially on young stems and lower leaf laminas, when the indumentum is composed of dense to sparse clusters of small stellate equal-rayed hairs. Other plants can appear to be hirsute with larger stellate hairs having a very long median ray making the young inflorescences, buds and calyces in particular appear silky and to glisten. At the other extreme entire plants can be virtually glabrous. Several authors and collectors have speculated on the variation found in this species suggesting that it should be recognised infraspecifically. This indumentum variation seems to be correlated eco-geographically and broadly separable into a north-south divide in East Africa. Two varieties of *S. schummannianum* have been recognised in this account, though some other authors have recognised one very variable species; future field work may justify the recognition of the variation apparent in this species at subspecific level. Unfortunately in this already taxonomically complex genus, this has necessitated a new combination since none of the varieties already published are applicable to the hirsute variant.

a. var. **schummanianum**

Plants either glabrescent or covered with a mealy pubescence of clusters of small equal-rayed stellate hairs, particularly dense on young stems and leaves, on peduncles, forks and calyces; basal parts of styles covered with scattered small equal-rayed stellate hairs; all bristles tapering to a sharp point. Fig. 18/1–9, p. 152.

KENYA. Meru District: NW Mount Kenya, Marimba Forest, 14 Oct. 1960, *Polhill & Verdcourt* 314!; Kiambu District: Karirana Estates between Kiambu and Limuru, 3 Sept. 1960, *Polhill* 258!; Teita District: Taita Hills, Yale Peak, 13 Sept. 1953, *Drummond & Hemsley* 4309!
TANZANIA. Moshi District: Kilimanjaro Timbers area, 15 June 1992, *Grimshaw* 92/76B!; Lushoto District: Baga–Bumbuli road, 2 km NE Sakarani, W Usambaras, 6 May 1953, *Drummond & Hemsley* 2415! & Shume Forest Reserve, Nov. 1957, *Semsei* 2720!
DISTR. **K** 4–5, 7; **T** 2, 3, 6, 7; not known elsewhere

Hab. Semideciduous and evergreen forest (*Ocotea, Podocarpus, Albizia*) including clearings and -margins and secondary forest, grassland, bushland, disturbed places, roadsides, margins of cultivation; 1350–3000 m

Syn. *S. ulugurense* Dammer in E.J. 38: 191 (1906); T.T.C.L.: 578 (1949). Type: Tanzania, Morogoro District: Uluguru, Lukwangulo Station, *Stuhlmann* 9111 (?B†, holo.; K!, iso. fragment)
    S. schumannianum Dammer var. *stolzii* Dammer in E.J. 53: 330 (1915); Bitter in E.J. 57: 265 (1921); T.T.C.L.: 578 (1949). Type: Tanzania, Rungwe District: Kyimbila, Rungwe rainforest, *Stolz* s.n. (?B†, holo.; K! fragment, M photo!, W!, WAG photo! [all *Stolz* 1566], all ?iso.)
    S. schumannianum Dammer var. *austerum* Bitter in E.J. 57: 264 (1921). Type: Kenya, Kiambu District: Limuru [Lamuru], *Scheffler* 256 (?B†, holo.; K!, W, photo!, iso.)
    S. lignosum Werd. in Notiz. K. Bot. Berlin, 12: 93 (1934); T.T.C.L.: 579 (1949). Type: Tanzania, Morogoro District: Uluguru mountains, NE side, *Schlieben* 3150 (B!, holo.; BM! BR!, M photo!, P!, iso.), *non* Sloboda, 1852
    S. ulugurense Holub., Preslia, 46: 169 (1974), *nom. nov.* pro *S. lignosum* Werd., *nom. illeg. non* Dammer

Note. Specimens of this variety are closely clustered in neighbouring regions of the districts in which they occur in central and eastern Kenya and neighbouring N and E Tanzania. This variety is predominant in **T** 3, while numerous specimens have also been collected from **T** 2, with additional specimens occurring in **T** 6 and 7. Most of the specimens recorded from **K** 4 are glabrescent (e.g. *Polhill* 258). The fruits of this species are recorded to be bitter but are eaten by the Washambaa in the Usambaras (**T** 3). Ruffo *et al.* (Edible Wild Plants of Tanzania, 2002) also reported widespread use of *S. schumannianum* in Tanzania; in addition to the fruits being eaten, the plants are used for food, fodder, hedging and boundary marking and medicinally as a cure for constipation and intestinal worms.

The taxa synonymised with this variety have generally been considered to be conspecific by other authors. All were recognised on minor morphological features which are insignificant in this very variable species, with the protologues and type material, where available, all confirming that *S. ulugurense, S. schumannianum* Dammer var. *austerum* and *S. lignosum* should be regarded as synonyms of the mealy var. *schumannianum*.

Dammer (1915) differentiated his Tanzanian variety *stolzii* on the basis of its more robust habit, larger leaves, and more frequent bristles on the petiole and leaf laminas. However, the variation in these characters noted from herbarium material falls within the vegetative variability found within *S. schumannianum*. Moreover probable isotype specimens of this variety (*Stolz* 1566) also exhibit vegetative and floral characters typical of this species, with the pubescence being typical of the mealy small stellate-haired var. *schummanianum*. Bitter (1921) later elaborated on his variety adding that it displayed strikingly shorter floral parts than those of *S. schumannianum*. While the floral measurements given by Bitter together with those on the *Stolz* 1566 specimens are towards the lower ends of the ranges given above, they are still within those acceptable within this species. Dammer's variety has therefore been synonymised with the variety *schummanianum*.

Type material of Bitter's variety *austerum* is almost completely glabrous in line with most other specimens of the subspecies *schummanianum* collected from **K** 4 (see above).

The name *S. lignosum* published by Werderman (1934) had previously been used by Slobada (Rostlinictví: 358 (1852)) for a maritime race of *S. litorale* Raab in the Dulcamara group (cf. Jaeger, 1985). Holub (1974) recognised this as a homonym giving this taxon the new name *S. ulugurense* without realising that this combination had already been used by Dammer (1906). Both of these names are synonyms of *S. schumannianum*. Photographs of the holo- and the iso-type specimens of *S. lignosum* clearly demonstrate the absence of bristles in this conspecific species.

The isotype fragment of *S. ulugurense* bears a few berries with seeds slightly larger (5.5–6 × 4.5–5 mm) than those found in other specimens of *S. schummanianum*. However, Dammer's protologue leaves little doubt that his species is synonymous with the latter.

b. var. **subulatum** (*Wright*) *Edmonds* **comb. nov.** Type: Malawi, Masuka Plateau, *Whyte* 280 (K!, holo.; BM!, iso.) [The holotype specimen of *S. subulatum* exhibits stellate hairs with long median rays especially on the lower midribs, floral stems, flower buds, calyces and external lobe surfaces, though Wright's protologue merely mentioned the occurrence of scurfy stellate hairs on young stems and leaves. This Malawian species is clearly conspecific with the hirsute variety of *S. schummanianum*, necessitating a new combination for this taxon]

Plants hirsute and covered with stellate hairs with long median rays (up to 1.5 mm), particularly dense on young inflorescences, buds and calyces - these often appear silky and glistening; basal areas of styles often hirsute from adherent long median rays of stellate hairs; bristles usually taper to a stellate cluster of short stout rays with an extended median transparent ray which may slough off with maturity. Fig. 18/10 & 11, p. 152.

Tanzania. Kilosa District: Ukaguru Mountains, W slopes of Mnyera Mt, 31 May 1978, *Thulin & Mhoro* 2741b!; Mbeya District: Kikonso Camp, Poroto Mountains, 21 Jan. 1961, *Richards* 14122!; Iringa District: Malakala, 18 km from Johns Corner on the Mufindi Road, 13 Mar. 1962, *Polhill & Paulo* 1747!

Distr. **T** 4, 5, 6–8; Malawi

Hab. Upland evergreen forest (e.g. *Podocarpus*) and forest margins, *Hagenia* woodland, secondary scrub and scattered tree grassland; 1650–2600 m

Syn. *S. subulatum* C.H. Wright in F.T.A. 4, 2: 221 (1906); Bitter in E.J. 54: 496 (1917)
    S. hirsuticaule Werd. in N.B.G.B. 12: 91 (1934); T.T.C.L.: 579 (1949). Type: Tanzania, Songea District: Mbejera, Mapala, *Schlieben* 973 (?B†, holo., BR!, iso.)

Note. This variety does not occur in Kenya, but is the predominant variety of *S. schumannianum* found in **T** 7. It is also fairly common in neighbouring **T** 6 with occasional specimens recorded from **T** 4, 5 and 8. There are a few extremely ochraceous specimens on which the many of the stellate hairs are stalked (– 0.4 mm long) collected from **T** 6 (e.g. *Polhill & Wingfield* 4684 from the Ulugurus and *Rounce* 565 from Morogoro). These seem to be precursors of the bristles terminating in stellate clusters typical of the variety *subulatum*, which are also visible on these specimens. It is the only variety which occurs in Malawi, though Gonçalves (2005) did not mention these long-rayed hairs in his account of *S. schumannianum*, perhaps including them in his description of rays modified as bristles.

   Although Werderman's protologue of *S. hirsuticaule* gives little information on the actual stellate hair composition, an isotype clearly exhibits long median rays on the stellate hairs of the floral parts, together with short glabrous bristles on the stems on which occasional apical stellate clusters are visible. This species was described from Songea in **T** 8 which fits with geographical distribution of this variety.

24. **Solanum schliebenii** *Werd.* in N.B.G.B. 12: 92 (1934); T.T.C.L.: 578 (1949); Polhill, *Solanum* in E & NE Africa: 15 (ined., 1961); Seithe in E.J. 81: 326 (1962); Jaeger, Syst. stud. *Solanum* in Africa: 351 (1985, ined.). Type: Tanzania, Morogoro District: Uluguru Mts, NW side, *Schlieben* 3415 (?B†, holo.; BR!, BM!, M, P!, all iso. labelled *Schlieben* 3415a)

Shrub or small tree to 8 m high; stems sparsely branched, becoming thick and lenticellate; young stems and branches covered with dense small sessile stellate ± equal-rayed hairs ± 0.2 mm diameter, glabrescent, also with spreading soft bristles to 7 × 0.2–1 mm tapering to an acute point, usually simple, sometimes forked, purple becoming yellow to brown, densely tomentose with short glistening hairs. Leaves alternate, dark green, membranaceous, ovate to broadly ovate, (16.5–)20–40 × 8–14 cm, bases cuneate sometimes decurrent and unequal, margins occasionally sinuate, apices acute to acuminate; lower surfaces of young leaves mealy and densely tomentose-stellate, hairs as on stems, becoming more confined to veins and midribs with maturity; upper surfaces glabrescent, both surfaces with prominent 7–11 alternate pairs of veins, bristles absent; petioles 3.5–9 cm long, bristles absent. Inflorescences terminal to subterminal becoming lateral, multiply forked, many-flowered (usually 100+) umbellate cymes borne in pyramidal candelabra-like corymbs up to 12 cm broad apically; flowers 5-merous, erect or nodding; flowering axes densely stellate-tomentose, bristles absent; peduncles 0–6 mm long in flower and in fruit, becoming glabrous and woody; pedicels erect to recurved and 3–9 mm long in flower, erect and 6–9 cm long in fruit, glabrous, woody, thickened beneath calyx. Calyx campanulate to cupulate, 1.8–2.5 mm long, stellate-tomentose in bud sometimes glabrescent with small brown glands externally, glabrescent internally, lobes triangular or ovate, 0.5–1.4 × 0.8–1.5 mm, obtuse, stellate-tomentose externally;

adherent basally in fruit, later becoming reflexed, 1.5–2.5 × 1.5–2 mm, glabrous. Corolla pale purple to violet, stellate, 12–18 mm diameter; tube 0.7–1 mm long, stellate-tomentose externally, glabrous internally; lobes strongly reflexed, lanceolate, 4–6.5 × 0.9–1.5 mm, acute, stellate-tomentose externally, often with small stalked glands internally. Stamens equal; filaments free for ± 0.3 mm, glabrous; anthers bright yellow becoming brown, poricidal, 2.8–3.2 × 0.7–1.1 mm. Ovary brownish, 0.5–0.8 × 0.3–0.6 mm, with scattered stellate hairs apically, ?bilocular; style straight, stout, 4.2–5.2 × 0.4–0.5 mm, stellate-pilose for lower half, glabrous above, exserted up to 1.5 mm; stigma capitate, slightly bifid, 0.4–0.5 mm diameter. Berries smooth, yellow to orange-red, globose, glabrous, 6–10 mm diameter, often falling from pedicels when ripe, pericarp somewhat leathery. Seeds 4–16 per berry, light to dark brown, obovoid, orbicular, elliptic or reniform, 3–3.3 × 2.6–3 mm, flattened, minutely foveolate; sclerotic granules absent. Fig. 18/12 & 13, p. 152.

TANZANIA. Morogoro District: Uluguru North Catchment Forest Reserve on path Tegetero–Luthungo, above first river down path E of Camp 3 Jan. 2001, *Jannerup & Mhoro* 0201!; & Ulugurus, Tanana, 24 Jan. 1924, *E.M. Bruce* 618!; & Uluguru mountains, Bunduki Forest Reserve, Mar. 1953, *Paulo* 39!
DISTR. **T** 6; restricted to Uluguru Mts
HAB. Moist forest, often on riverine paths or river-banks; 1450–1850 m

NOTE. Though only represented by a few specimens, most authors working in this geographical area have recognised that *S. schliebenii* is a distinct species. It is closely related to *S. schumannianum*, from which it can be distinguished by its characteristic yellowish densely pubescent bristles (on which stellate hairs with a slightly longer median ray are sometimes discernible) which are confined to the stems and branches, by its large ovate eventually glabrescent dark green leaves with prominent venation, and by its smaller buds and seeds. From the material examined, including several isotype specimens, the inflorescences of *S. schliebenii* seem to be virtually epedunculate (with the peduncle either absent or < 6 mm long), and several long forked branches arising almost basally – these may have been what Werdermann erroneously noted as 4 cm long peduncles in his diagnosis.

Werdermann's protologue cited the holotype as *Schlieben* 3415 without any type locality. Schlieben's first set was deposited in Berlin and was destroyed in 1943. There are a number of isotypes of this species – but with the collection number *Schlieben* 3415a. Lester in 2005 selected the BR specimen as the lectotype of this species and the BM specimen as the isolecto., but these were not published, and there are a number of extant isotypes.

25. **Solanum tettense** *Klotzsch* in Peters, Reise Mossamb. Bot.: 237 (1861); Dammer in P.O.A. C: 355 (1895); C.H. Wright in F.T.A. 4, 2: 212 (1906); Bitter in E.J. 57: 276 (1921); E.P.A.: 879 (1963); Whalen in Gentes Herb. 12: 215 (1984); Jaeger, Syst. stud. *Solanum* in Africa: 358 (1985, ined.); Gonçalves in Kirkia 16: 86 (1996) & in F.Z. 8(4): 92 (2005); Friis in Fl. Eth. 5: 125 (2006) & in Fl. Somalia 3: 209 (2006); Welman in Bothalia 38(1): 43 (2008). Type: Mozambique, Tette [Tete], *Peters* s.n. (?B†, holo.; BM!, EA! iso.)

Shrub to 4 m high or small tree, sometimes straggling, rarely climbing; branches often angular, lenticellate, with flakey bark, younger stems ochraceous/mealy or tomentose and *either* covered with dense small sessile stellate hairs ± 0.3 mm diameter with up to 12 equal often purple eglandular rays *or* with mixture of long and short spreading simple multicellular glandular and eglandular hairs together with stellate hairs of up to 1 mm diameter in which the median rays are often long (–1.3 mm) and either eglandular or glandular; stalked brown glands also present throughout but especially visible on the leaves; main stems and branches with pyramidal prickles 1.6–3(–6) × 2–3(–4.5) mm, acute, often stellate-haired towards the base, occasionally absent (especially in **K** 4). Leaves alternate or opposite, often membranaceous, dark green above, yellowish below, ovate, obovate to lanceolate or elliptic, 3.4–10(–15) × 1.4–5(–7) cm, bases cuneate often decurrent, sometimes unequal, margins entire, occasionally sinuate, apices obtuse to acute; occasionally with a few small prickles on midribs; lower surfaces covered with stellate equal-rayed hairs to 0.8 mm diameter, sometimes mixed with long median-rayed and simple hairs as on stems, especially on margins and veins; upper surfaces with scattered equal-rayed stellate hairs to 0.3 mm diameter sometimes mixed with short simple hairs; petioles 0.6–2.5(–4) cm long, occasional prickles sometimes present. Inflorescences subterminal becoming lateral, simple, forked or multiply-branched, many-flowered, lax, scorpioid or umbellate cymes, borne in compact obpyramidal corymbs up to 9.5 cm wide, often extremely lax in fruit; flowers 4–5(–6)-merous; peduncles erect, 1.2–4 cm long in flower, 1.8–4 cm long and glabrous and woody in fruit; pedicels erect to recurved and 5–12 mm in flower, erect and spreading and 6–16 mm in fruit; flowering axes usually ochraceous/mealy stellate-hairy, simple glandular hairs sometimes also present. Calyx pale green to purple, cupulate to campanulate, (1.6–)2–3(–4.6) mm long, densely stellate-tomentose externally, glabrous internally; lobes equal, ovate to triangular, 0.5–2(–3) × (0.8–)1.3–2(–3) mm, obtuse to acute; adherent basally in fruit becoming spreading or reflexed often unequal, 1.2–4 × 1.5–3 mm. Corolla purple or pale lilac, occasionally white, often with a yellow basal star, stellate, 1.2–2.4 cm diameter; tube 1–1.5(–2) mm long, glabrous externally; lobes lanceolate, 4–10 × 1.4–3.2 mm, acute, stellate-tomentose externally, sometimes with glandular hairs basally, usually with stellate hairs on veins internally, lobes strongly reflexed after anthesis. Stamens usually equal, sometimes differing in length from flower to flower on the same plant; filaments whitish, free for 0.3–1(–1.6) mm, glabrous; anthers yellow, orange or brownish, poricidal, 3.2–5(–6.5) × 0.8–1.5 mm, usually connivent, occasionally spreading. Ovary brownish, 0.4–1.4 × 0.8–1.3 mm, ± glabrous, bilocular; style straight or curved, 5.5–9 × 0.3–0.5 mm, glabrous, exserted up to 3 mm; stigma green, capitate, sometimes bilobed, 0.3–0.6 mm diameter. Berries smooth, glossy red or occasionally orange sometimes turning black on drying, globose, cuticle often transparent, sometimes leathery, 5.5–8(–11) mm diameter, glabrous. Seeds 3–30 per berry, yellow, orange or brown, obovoid, orbicular or reniform, (2.9–)3.5–4.8 × (2.2–)3–4.2 mm, foveolate to reticulate; sclerotic granules absent.

NOTE. During his revision of this species for the Flora Zambesiaca area, Gonçalves (1996) concluded that the three species *S. tettense* Klotzsch, *S. renschii* Vatke and *S. kwebense* C.H. Wright were conspecific, forming a single polymorphic species which exhibits "great morphological and ecological diversity", with Klotzsch's epithet taking precedence as the legitimate name for the species. Within this, however, he recognised two varieties distinguishable on their stellate hairs having a glandular (var. *tettense*) or an eglandular (var. *renschii*) central ray. He did not consider that these varieties were separable on geographical or ecological terms.

Examination of innumerable specimens from eastern and southern Africa for the FTEA treatment has verified that this taxon does indeed exhibit considerable morphological variation, but that some is eco-geographically associated. I consider that the variation exhibited by these taxa can be recognised morpho-geographically and that the two species *S. kwebense* and '*S. renschii*' should be recognised specifically, and that within each two infraspecific variants should probably also be given formal taxonomic recognition. Unfortunately, the isotypes of *S. tettense* are fragmentary specimens, but short viscid glandular hairs are clearly present, and the flowers on the BM specimen are 5-merous. This together with other characters given both in the protologue and in Bitter's (1921) later elaboration of this species, particularly the description of its pubescence, which he presumably derived from the actual holotype, suggest that *S. tettense* is conspecific with the glandular-haired variant of '*S. renschii*' and must therefore be adopted as the correct name for this species.

As Gonçalves found, this species exhibits considerable vegetative and floral variability. Though characterised by prickly stems, the prickles not only vary in size but can also be sparse, dense, or even absent either from young stems and branches or from older woody stems; in some regions (e.g. **K** 4) they can be completely absent. The prickles are usually confined to stems, but can occur sparingly on petioles and on the upper and/or lower midribs as well as on some of the larger leaf veins. Occasionally they can also occur sparingly on the peduncles. Though the major differences in pubescence have been given formal recognition, intermediate types are found on some specimens. Floral morphology too can be variable, and although flowers on the two varieties *tettense* and *renschii* are predominantly 5- and 4-merous respectively, in some areas (e.g. Zambia) both 4- and 5-merous flowers can occur on the same plant. Indeed on the isotype of *S. tettense*, a dissected flower had 5 anthers but only 4 calyx lobes and lobes. Anther length too can vary both within a single flower or between flowers on the same specimen.

a. var. **tettense**; *A.E. Gonç.* in Kirkia 16: 87 (1997) & in F.Z. 8(4): 92 (2005); Friis in Fl. Eth. 5: 125 (2006); Welman in Bothalia 38(1): 43 (2008)

Plants viscid especially on young stems, lower leaf veins, midribs and margins and on petioles and calyces, with long and short spreading few-celled simple glandular and eglandular hairs mixed with stellate hairs up to 1 mm diameter in which the median rays can be elongated up to 1.3 mm and either glandular or eglandular. Inflorescences simple or simple and forked, condensed or lax cymes with rachides 1–1.8 cm long, usually < 30-flowered; flowers (4–)5(–6)-merous; lobes narrowly lanceolate; berries with 16–30 seeds. Fig. 19/1 & 2, p. 161.

UGANDA. Ankole District: Nyabushozi, 16 Sep. 1941, *A.S. Thomas* 3988! & Ruizi River, 16 Nov. 1950, *Jarrett* 208!; Teso District: Serere near Lobori, June/July 1926, *Maitland* 1349!
TANZANIA. Bukoba (? Karagwe) District: between Bugene and Nyabyonza, Feb. 1958, *Procter* 847!; Ngara District: Buseke, Keza, Bushubi, 20 May 1960, *Tanner* 4928!; Mwanza District: Nyegezi, Bunegeji, 15 Oct. 1952, *Tanner* 1061!
DISTR. **U** 1?, 2–4; **T** 1, 8; Congo-Kinshasa, Rwanda, Zambia, Malawi and Mozambique
HAB. Riverine habitats, grassland and on termite mounds; 800–1400 m

SYN. *S. kagehense* Dammer in E.J. 38:187 (1906); Bitter in E.J. 57: 274 (1921); T.T.C.L.: 578 (1949); Polhill, *Solanum* in E & NE Africa: 16 (ined., 1961); Whalen in Gentes Herb. 12: 215 (1984); Jaeger, Syst. stud. *Solanum* in Africa: 348 (1985, ined.). Type: Tanzania, Mwanza District: Kagehi, *Fischer* 78 (?B†, holo.)
   S. muansense Dammer subsp. *mildbraedii* Bitter in E.J. 57: 262 (1921). Type: Rwanda, Buganza, south of Lake Mohasi, *Mildbraed* 616 (?B†, holo.)
   S. wittei Robyns in B.J.B.B. 17: 82 (1943) & Fl. Sperm. Parc Nat. Alb., 2: 211 (1947); Whalen in Gentes Herb. 12: 215 (1984). Type: Congo-Kinshasa, Parc National Albert, Kabasha, *de Witte* 1142 (BR!, holo.)
   S. renschii Vatke var. 1 of Polhill, *Solanum* in E & NE Africa: 15 (ined., 1961)
   S. renschii sensu Troupin in Fl. Pl. Lign. Rwanda: 657 (1982) & Fl. Rwanda 3: 380 (1985), *non* Vatke

NOTE. It is interesting to note that Polhill (1961) separated two varieties of *S. renschii* on the basis of their median ray size, commenting that his var. 1 which he noted as occurring in **U** 2 and **U** 4 seemed to have been described as *S. wittei* – a synonym of var. *tettense*. In **T** 1, **U** 2–4, Rwanda and neighbouring areas of Congo-Kinshasa, this variety is represented by a uniform suite of specimens with 5-merous flowers and a silky tomentose pubescence of large stellate hairs with long median rays which can be glandular.
   The roots of this variety are used to treat colds in **U** 2.
   Dammer's protologue of *S. kagehense* is short and uninformative, citing subumbellate cymose inflorescences and 4-merous flowers. Bitter (1921) later redescribed it from the holotype, when he mentioned glandular spreading hairs, 20–30-flowered inflorescences, and 4(–5)-merous flowers. He also placed this species in his series *Anomalum* of section *Torvaria*, together with and preceding *S. renschii* and *S. tettense*. Jaeger (1986) placed this species into a group with *S. muansense* Dammer, *S. sordidescens* Bitter and *S. wittei* Robyns, which he considered closely related but requiring further study before their taxonomic status could be clarified. In the absence of extant type material, from the morphological elaboration given by Bitter and the citation of glandular hairs with some 5-merous flowers, *S. kagehense* has been synonymised with the glandular-haired var. *tettense*.

FIG. 19. *SOLANUM TETTENSE* var. *TETTENSE* — **1**, 5-merous flower on viscid pedicel; **2**, stem/leaf indument of simple glandular hairs and long-median-rayed stellate hairs. *S. TETTENSE* var. *RENSCHII* — **3**, flowering and fruiting habit; **4**, stem/prickle indument of stellate hairs; **5**, stellate haired leaf indument; **6**, 4-merous flower; **7**, stamen; **8**, gynoecium; **9**, ovary and stylar base; **10**, fruit on stellate-haired pedicel; **11**, seed. *S. RUNSORIENSE* subsp. *BENDIRIANUM* — **12**, habit in flower; **13**, stem indument of simple hairs. *S. RUNSORIENSE* subsp. *RUNSORIENSE* — **14**, stem indument of stalked dendritic hairs; **15**, pedicel indument of sessile dendritic hairs; **16**, dissected flower; **17**, opened calyx; **18**, stamen; **19**, gynoecium. 1–2 from *Tanner* 1061; 3, 5, 10–11 from *Simba* 12901; 4, 6–9 from *Richards* 26384; 12–13 from *Jex-Blake* 28; 14 from *Liebenberg* 1637; 15–19 from *Thomas* 650. Drawn by Margaret Tebbs.

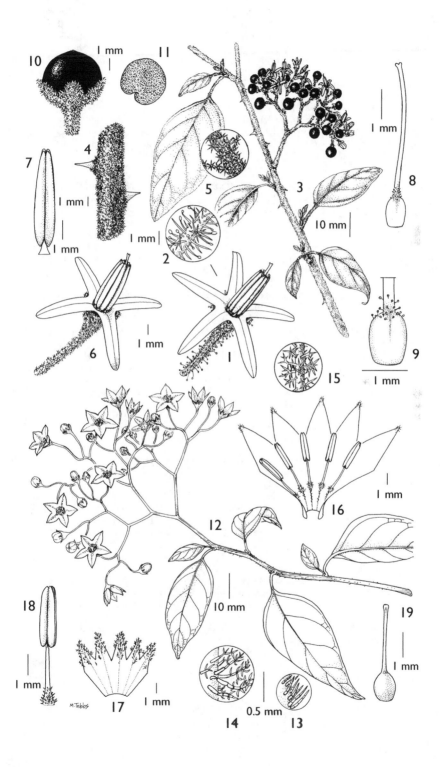

The protologue of *S. muansense* subsp. *mildbraedii* clearly mentions long slender rays to the stellate hairs, and apically glandular hairs on the calyces. Together with the general vegetative and inflorescence characters described by Bitter, the synonymy of this taxon with the glandular-haired variant of *S. tettense* is further confirmed by its collection in Rwanda. However, *S. muansense* Damm. itself is thought to be synonymous with *S. giganteum* (see page 148); indeed Dammer discusses the affinity of his new species with the latter. Clearly resolution of the correct taxonomic affinity of these two taxa is dependent on the location of duplicate type material.

There is little doubt that *S. wittei* is synonymous with this variety; the holotype is densely viscid with long and short simple glandular hairs together with long median-rayed stellate hairs. All vegetative, floral and fruiting characters are analogous to those found in this predominantly glandular-haired variety including the inflorescence being many-flowered and the flowers 5-merous.

b. var. **renschii** (*Vatke*) *A.E. Gonç.* in Kirkia 16: 89 (1997) & in F.Z. 8(4): 93 (2005); Friis in Fl. Eth. 5: 126 (2006); Welman in Bothalia 38(1): 44 (2008). Type: Kenya, Kitui District: Ukambama, *Hildebrandt* 2735 (K!, W, iso., photo!)

Plants non-viscid with a mealy ochraceous pubescence of equal-rayed eglandular stellate hairs 0.3–0.8 mm diameter, sometimes mixed with simple eglandular hairs. Inflorescences multiply-forked lax cymes with rachides up to 4 cm long, usually > 40-flowered; flowers 4(–5)-merous; lobes lanceolate. Berries with 3–12 seeds. Fig. 19/3–11, p. 161.

UGANDA. Karamoja District: Matheniko, Moroto River, May 1954, *Philip* 562!; & Moroto, 12 Sept. 1956, *Bally* 10788!; & foothills of Mount Moroto, 10 Oct. 1952, *Verdcourt* 809!
KENYA.Turkana District: Karapobot, Rift Valley, Nakuijit, Sept. 1970, *Tweedie* 3845!; Baringo District: 6 km N of Kampi ya Samaki, 13 June 1977, *Gilbert* 4742!; Kitui District: 3 km W of Mwingi, 30 Jan. 1972, *Bally & Smith* 15002!
TANZANIA. Mbulu District: track to Burungi Lake, 2 Mar. 1969, *Richards* 24384!; Iringa District: between Iringa and Dodoma, 26 Feb. 1961, *Verdcourt* 3073!; & Msembe–Kimiramatonge Circuit, km 6 from Msembe, 24 Feb. 1070, *Greenway & Kanuri* 13949!
DISTR. U 1; K 1–4, 4/6 & 7; T 2, 3 & 5–7; Ethiopia, Somalia, Angola, Zambia, Malawi, Mozambique, Zimbabwe, Namibia and South Africa
HAB. Grassland, wooded grassland, montane and/or rocky bushland, *Acacia-Commiphora-Eurphorbia* or *Acacia drepanolobium* bushland, thickets, degraded or riverine woodland or forest, river banks, waste-land, wadis, cultivation; 100–2100 m

SYN. *S. renschii* Vatke in Linnaea, 43: 328 (1882); Dammer in P.O.A., C: 352 (1895) & in E.J. 38: 190 (1906); C.H. Wright in F.T.A. 4, 2: 229 (1906) pro Nile Land; Bitter in E.J. 57: 275 (1921); T.T.C.L.: 579 (1949); Polhill, *Solanum* in E & NE Africa: 15 (ined., 1961); K.T.S.: 538 (1961); E.P.A.: 875 (1963); U.K.W.F., ed. 2: 243, t. 106 (1994); Whalen in Gentes Herb. 12: 215 (1984); Jaeger, Syst. stud. *Solanum* in Africa: 355 (1985, ined.); K.T.S.L.: 582 (1994); U.K.W.F., 2nd ed.: 243 (1994); Bukenya & Carasco in Bothalia 25: 56 (1995).
   S. diplocincinnum Dammer in E.J. 48: 240 (1912). Type: Kenya, Teita District: Voi, Buru Mountains (?Bura Bluff), *Engler* 1971 (?B†, holo.)
   S. koniortodes Dammer in E.J. 53: 332 (1915). Type: Kenya, Machakos District: Kibwezi, *Scheffler* 35 (?B†, holo.; K!, iso.)
   S. kibweziense Dammer in E.J. 53: 334 (1915); Whalen in Gentes Herb. 12: 215 (1987). Type: Kenya, Machakos District: Kibwezi, *Scheffler* 517 (B†, holo.; K!, iso.)
   S. baidoense Chiov., Result. Sc. Miss. Stefan.-Paoli:125 (1916). Type: Somalia, Baidoa, *Paoli* 1120 (FT, holo., photo!)
   S. munitum Bitter in E.J. 57: 284 (1921); E.P.A.: 873 (1963). Type: Ethiopia, Scheich-Huzein, *Ellenbeck* 1287 (B†, holo.)
   S. bifurcum Hochst. var. *baidoense* (Chiov.) Chiov., Fl. Somalia: 238 (1929)
   S. munitum Bitter var. *javellense* Lanza in Miss. Biol. Borana, Racc. Bot.: 200 (1939); E.P.A.: 873 (1963). Type: Ethiopia, Yavello at Acacia Spring, *Cufodontis* 379 (FT, holo., photo!; FT, iso., photo!)
   S. chiovendae Lanza in Miss. Biol. Borana, Racc. Bot.: 191 (1939); E.P.A.: 864 (1963). Type: Ethiopia, Moyale, *Cufodontis* 734 (FT, holo., photo!; FT, iso., photo!)
   ?*S. grewioides* Lanza in Miss. Biol. Borana, Racc. Bot.: 195 (1939); E.P.A.: 867 (1963). Type: Ethiopia, Yavello, Quota Littorio, *Cufodontis* 563 (FT, holo., photo!; W, iso., photo!)
   S. renschii Vatke var. *renschii*, Polhill, *Solanum* in E & NE Africa: 15 (ined., 1961)

NOTE. *Solanum tettense* var. *renschii* is a mealy eglandular stellate-haired taxon usually with 4-merous flowers, which is widely found throughout eastern and southern Africa. It is widespread throughout the FTEA area as well as in Ethiopia and Somalia, where it is the only variety that occurs. According to Bukenya & Carasco (1995) *S. renschii* is endemic to eastern Africa and is restricted to the arid regions of Karamoja in Uganda; the specimens cited in their account clearly belong to *S. tettense* var. *renschii*. However, *Dawkins* 643 at Kew, collected from U 1, is mophologically intermediate in many characters between the two *S. tettense* varieties, and might be a hybrid; the collection notes indicate that it was collected from a single individual in an exposed position.

Solanum tettense var. *renschii* is used medicinally in **K** 7; decoctions of soaked roots are used to treat skin rashes in **K** 2, boiled roots are used to cure stomach pains, menstrual cramps and vaginal discharge in **K** 3, while cut root sections are worn as charms by old men in **K** 4. The flowers are reported to be attractive to bees. Welman (2008) reported that *S. tettense* sensu lato, is eaten by eland, and that cooked roots are used to draw out pus from wounds, the roots are used for typhoid preventation and the fruits for the provision of arrow poison ingredients. The var. *renschii* is reportedly poisonous to cattle.

A number of species described from the FZ area are conspecific, and are closely allied to *S. renschii*. These centre around the Botswanan species *S. kwebense* C.H. Wright, whose flowers Wright (1906) described as 5-merous although those on the syntype specimens (*Lugard* 50 & 62) are clearly 4-merous. Other species considered synonymous with *S. kwebense* are *S. luederitzii* Schinz, *S. upingtoniae* Schinz, *S. chondrolobeum* Dammer, *S. kwebense* var. *acutius* Bitter and *S. kwebense* N.E. Br. var. *luederitzii* (Schinz) Bitter from Namibia, S. *tenuiramosum* Dammer from Botswana, and *S. kwebense* N.E. Br. var. *majorifrons* Bitter from Angola. All of these taxa together with the other specimens collected from the FZ area are characterised by small elliptic to obovate leaves (characteristically 2–4(–6) × 0.8–3 cm) with few-flowered (3–6(–10)) simple or shortly forked lax cymose inflorescences borne on peduncles 2–13 mm long. The prickles, when present, are small and narrow (acicular), and the flowers predominantly 4-merous but as in *S. renschii* occasionally 5-merous. Moreover, *S. kwebense* specimens display similar pubescence variation to that found in *S. renschii* with an eglandular variant occurring in Angola, Namibia, Botswana, Zimbabwe and northern South Africa, and a glandular viscid variant occurring in Botswana, Zimbabwe, Malawi, Zambia and occasionally in neighbouring Mozambique and South Africa, which could be recognised at an infraspecific level. From their protologues and available type material and photographs, all of the species cited above are thought to be eglandular-haired. If future work verifies that Gonçalves (2005) correctly treated *S. kwebense* as a synonym of *S. tettense* var. *renschii*, then these species should be included in the synonymy of that variety. Welman (2008) included *S. kwebense* and a number of the species allied to it as synonyms of var. *renschii*. Since none of these taxa have been described from the FTEA area, they are not considered further here.

Although Vatke described the corolla as "5-partite" in his protologue of *S. renschii*, those on the holotype (*Hildebrandt* 2735) are 4-merous.

Both Bitter (1921) and Jaeger (1985) considered *S. diplocincinnum* to be conspecific with *S. renschii*. Dammer described the pubescence as being composed of small stellate hairs and the flowers as 4-partite, features which suggest *S. diplocincinnum* is synonymous with the var. *renschii*.

The protologue and isotype specimen of *S. koniortodes* are completely identical in all vegetative and floral characters to the type of *S. renschii* and both Bitter (1921) and Jaeger (1985) considered it to be synonymous with this species. The small equal-rayed stellate-haired pubescence and the 4-merous flowers confirm this synonymy.

The protologue and isotype of *S. kibweziense* are completely identical in all vegetative and floral characters to the type of *S. renschii*, and again Bitter (1921) and Jaeger (1985) also considered it to be synonymous with this species. Though the isotype consists of woody small-leaved fragments on which prickles are absent, the pubescence and 4-merous flowers again clearly confirm this synonymy.

The citation of short, equal-rayed stellate hairs, multi-flowered inflorescences and 4-merous flowers in the protologue of *S. baidoense* all support the synonymy of this species with *S. renschii* var. *renschi*, though the anther length of 6 mm is at the higher end of the range exhibited by this species. Chiovenda (1916) mentioned the affinity of his new species *S. baidoense* to Dammer's *S. diplocincinnum*, while Friis (2006) concluded that *S. baidoense* was synonymous with *S. (tettense* var.) *renschii* after examing the holotype.

The synonymy of Chiovenda's new combination of *baidoense* as a variety of *S. bifurcum*, with var. *renschii* was correctly pointed out by Friis (2006) after he examined the holotype. *Solanum bifurcum* is a synonym of the diverse scandent or twining unarmed highland shrub *S. terminale* (see page 107).

With the exception of slightly longer anthers (6 mm), all vegetative and floral features given in the protologue of *S. munitum* are similar to those found in *S. tettense* var. *renschii*, with which it has been synonymised. In particular, Bitter's detailed indumentum description and his citation of 4-merous flowers justifies this synonymy. Moreover, Friis (2006) noted that this was the only variety of *S. tettense* commonly found in Ethiopia.

Lanza's brief protologue of var. *javellense* described minor morphological differences in prickle, leaf and peduncle sizes, and smaller flowers than those found in *S. munitum* itself; none of these are considered significant in this variable species. His illustration, the photograph of the holotype and the Ethiopian provenance leave little doubt that this variety is also synonymous with *S. tettense* var. *renschii*.

The synonymy of *S. chiovendae* with the var. *renschii* is based on its protologue, the accompanying illustration and photographs of type material. Lanza (1939) described the young branches of his new species as being densely stellate-tomentose becoming glabrescent, and with 4-merous flowers - features which confirm this synonymy. Cufodontis (1963) thought it to be near to S. *schimperianum*, but Friis (2006) also considered it to be a synonym of *S.* (*tettense* var.) *renschii*.

Lanza (1939) noted that *S. grewioides* was similar in most characters to *S. chiovendae* and Friis (2006) considered that it clearly belonged to var. *renschii* after examining the holotype. Moreover the type location of Javello would also suggest the synonymy of these two (cf. *S. munitum* var. *javellense* above). However, since Lanza cited the flowers as being 5-merous, and the inflorescences as 10–15-flowered, *S. grewioides* has only been provisionally synonymised pending examination of type material.

26. **Solanum aethiopicum** *L.*, Cent. Pl. 2: 10 (1756); C.H. Wright in F.T.A. 4, 2: 217 (1906); Heine in F.W.T.A. 2ⁿᵈ ed., 2: 332 (1963). Type: Ethiopia, *J. Burser vol. 9 no. 17* (UPS, lecto. designated by Hepper & Jaeger in K.B. 40: 391, 1985)

Annual or perennial herb or shrub, 0.3–1 m, erect, unarmed; young stems glabrous to ± stellate-pubescent with trichomes porrect, translucent or orange-translucent, sessile or shortly stalked, stalks up to 0.1 mm, rays ± 8(–12), 0.15–0.3 mm, midpoints shorter than rays. Leaf blades drying concolorous to discolorous, yellow-green to dark red-brown or almost black, 2–3 times longer than wide, ovate (elliptic), 5–18 × 2.5–10 cm, base cuneate to rounded, often unequal or oblique, margin subentire to weakly lobed, the lobes up to 3 on each side, up to 1.5 cm long, broadly rounded (acute) and extending up to $^1/_4$ of the distance to the midvein, apex acute; glabrous to stellate-pubescent; trichomes on abaxial surface porrect, sessile or shortly stalked, stalks to 0.1 mm, rays ± 8(–12), 0.2–0.35 mm, midpoints shorter than rays, adaxially with thick stalks and reduced rays and midpoints, often with minute simple hairs; primary veins 3–6 pairs; petiole 1–4 cm, $^1/_4$–$^1/_3$ of the leaf length. Inflorescences not branched, with 1–2(10) flowers, 1–1.8 cm long; peduncle 0–0.5 mm long; rachis 0–1 cm long; pedicels 0.5–2 cm long in flower, in fruit 0.7–2.5 cm long. Flowers perfect, 5–10-merous. Calyx 3.5–9 mm long, lobes broadly deltate, 1.5–5 mm long, acute. Corolla white, 0.8–1.8 cm in diameter, lobed for ± $^2/_3$ of its length, lobes deltate, 2.5–8 × 1.5–4.5 mm. Stamens equal; anthers 2.3–4(–6) mm. Ovary glabrous or stellate-pubescent in the upper $^1/_5$; style 4.5–9 mm long. Berries 1–4 per infructescence, evenly green or striped when young, orange at maturity, spherical or variously shaped, (10–)15–25(–50) mm in diameter. Seeds 2.1–3.5 × 2–2.5 mm.

UGANDA. Kigezi District: Ruhinda, Jan. 1951, *Purseglove* 3562!; Mengo District: Kipayo, May 1915, *Dummer* 2477!
KENYA. Kiambu, 21 Mar. 1982, *Grumbler* 16601!
TANZANIA. Pare District: Same, Jan. 1956, *Trant* H184/55!; Kilosa District: Ilonga, 8 Jan. 1970, *Greensword* H2!
DISTR. U 1–4; K 1–7; T 1–8; widely distributed in cultivation: Senegal, Mali, Burkina Faso, Ivory Coast, Niger, Ghana, Togo, Benin, Nigeria, Cameroon, Gabon, Central African Republic, Chad, Equatorial Guinea, Congo-Kinshasa, Sudan, Eritrea, Ethiopia, Somalia, Angola, Zambia, Malawi, Mozambique, Zimbabwe, Botswana; Madagascar, and throughout South America and in some parts of Europe and Asia

HAB. Cultivated in woodland and wooded grassland zones; does not naturalise

SYN. *S. hybridum* Jacq., Hort. Bot. Vindob. 2: 51, t. 113 (1772–1773). Type: Austria; grown at Vienna from seeds of unknown origin (LINN 248.40!, lecto., designated by D'Arcy, 1970; BM!, possible isolecto.)

S. integrifolium Poir. in Lam., Tabl. Encycl. 4: 301 (1797). Type: cultivated in Mauritius [Île-de-France] (no specimens found)

S. scabrum Jacq., Pl. Hort. Schoenbr. 3, t. 333 (1798), *non* Mill., 1768. Type: cultivated in Vienna from seeds of unknown origin, *Jacquin* s.n. (BM000942568!, possible type)

S. scabrum Zuccagni, Cent. Observ. Bot. 50: [21] (1806), *non* Mill., 1768. Type: cultivated in Florence (no specimens found; Zuccagni's herbarium was originally at FI but apparently has been destroyed)

S. zuccagnianum Dunal, Hist. Nat. *Solanum* 149, t. 11 (1813), *nom. nov.* for *Solanum scabrum* Zuccagni

S. obtusifolium Willd., Enum. Pl. Suppl. [Willd.] 11 (1814). Type: cultivated in Berlin (B-W!, holo. [IDC microfiche 271–315.299:I.8])

S. gilo Raddi in Mem. Soc. Ital. Moden. Fis. 18: 410 (1820); T.T.C.L.: 581 (1949); Heine in F.W.T.A. 2nd ed., 2: 332 (1963). Type: cultivated in Brazil (no specimens found)

S. geminifolium Thonn. in Schumach., Beskr. Guin. Pl.: 121 (1827). Type: Ghana, cultivated in Dyrkes, *Thonning* 143 (C!, holo. [IDC microfiche 102:II.7, "316"]; GOET!, iso., fragment)

S. zuccagnianum Dunal var. *allogonum* Dunal in DC., Prodr. 13(1): 351 (1852). Type: cultivated in Geneva, 1841, *anonymous* s.n. (G-DC [G00131557!], lecto. designated here [best material of that cited in the protologue]; AV, G!, isolecto.)

S. naumannii Engl. in E.J. 8: 64 (1886); C.H. Wright in F.T.A. 4, 2: 216 (1906). Type: Congo-Kinshasa, Central Province, Boma, 5 Sep. 1874, *Naumann* s.n. (B†, holo.)

S. pierreanum Pailleux & Bois in Bull. Soc. Natl. Acclim. France 37: 483 (1890). Type not known (no specimens found)

S. olivare Pailleux & Bois, Potag. Cur., ed. 2, 386 (1892). Type: Congo-Kinshasa (no specimens found)

S. monteiroi C.H. Wright in K.B. 1894: 127 (1894). Type: Angola, Bonia, Jan. 1873, *Monteiro* s.n. (K!, holo.)

S. paaschenianum H.J.P. Winkl. in E.J. 41: 285 (1908). Type: Cameroon, between Njong Pass above Dehane and Kukue village, *Winkler* 860 (WRSL, holo.)

S. aethiopicum L. var. *integrifolium* (Poir.) O.E. Schulz, in Urb., Symb. Antill. 6: 211 (1909)

S. poggei Dammer in E.J. 48: 242 (1912). Type: Congo-Kinshasa, Lufubu, *Pogge* 1156 (GOET!, lecto., designated by Vorontsova & Knapp 2010: 1599)

S. sparsespinosum De Wild. in B.J.B.B. 4: 399 (1914). Type: Congo-Kinshasa, Ktobola, *Flamigni* 453 (BR0000008994233!, lecto., designated here [best material]; BR!, isolecto.)

S. subsessile De Wild. in B.J.B.B. 4: 407 (1914). Type: Congo-Kinshasa, Dundusana, *De Giorgi* 1146 (BR0000008992956!, lecto., designated here [best material]; BR!, isolecto.)

S. giorgii De Wild. in B.J.B.B. 4: 401 (1914). Type: Congo-Kinshasa, Likimi, *De Giorgi* 1481 (BR0000008993274!, lecto., designated here [best material]; BR!, isolecto.)

S. elskensii De Wild. in B.J.B.B. 4: 403 (1914). Type: Congo-Kinshasa, Prov. Kisangani, Yangambi, 7 Sep. 1913, *Elskens* s.n. (lectotype, BR8237033!, lecto., designated here [best material]; BR!, isolecto.)

S. brieyi De Wild. in B.J.B.B. 4: 404 (1914). Type: Congo-Kinshasa, Ganda-Sundi, 1913, *de Briey* s.n. (BR0000008992918!, lecto., designated here [best material and most duplicates]; BR!, isolecto.)

S. sudanense Hammerst. in Beih. Tropenpflanzer 19: 95 (1919). Type not known

S. ovatifolium De Wild., Pl. Bequaert. 1: 432 (1922). Type: Congo-Kinshasa, Penghe, *Bequaert* 2370 (BR!, holo.; BR!, iso.)

S. indicum L. subsp. *ambifarium* Bitter in F.R. Beih. 16: 37 (1923). Type: Togo, Sokodé Farm, *Schroeder* 94 (E!, lecto., designated by Vorontsova & Knapp 2010: 1597)

S. indicum L. subsp. *pervilleanum* Bitter in F.R. Beih. 16: 38 (1923). Type: Madagascar, Nosy Be [Nossi-bé], *Pervillé* 510 (K!, lecto., designated here [best material]; P!, isolecto.)

S. aethiopicum L. var. *armatum* Bitter in F.R. Beih. 16: 46 (1923). Type: ?Angola, Mutété, *Pogge* 1003 (no specimens found)

S. aethiopicum L. var. *giorgii* (De Wild.) Bitter in F.R. Beih. 16: 46 (1923)

S. aethiopicum L. var. *modicelobatum* Bitter in F.R. Beih. 16: 46 (1923). Type: Cameroon, *Tessmann* 1102 (no specimens found)

S. aethiopicum L. var. *paaschenianum* (H.J.P. Winkl.) Bitter in F.R. Beih. 16: 47 (1923)

S. *gilo* Raddi var. *ellipsoideum* Bitter in F.R. Beih. 16: 53 (1923); T.T.C.L.: 581 (1949). Type: Tanzania, Bukoba District: W bank of Lake Victoria, *Stuhlmann* 3662 (GOET!, neo., designated by Vorontsova & Knapp in Taxon 59: 1596, 2010)

S. *gilo* Raddi var. *erectifructum* Bitter in F.R. Beih. 16: 54 (1923). Type: Cameroon, N'Goto, on the road from Nola to M'Baiki, *Tessmann* 2095*a* (GOET!, lecto., designated by Vorontsova & Knapp in Taxon 59: 1596, 2010)

S. *gilo* Raddi var. *pierreanum* (Pailleux & Bois) Bitter in F.R. Beih. 16: 46 (1923)

S. *gilo* Raddi subsp. *megalacanthum* Bitter in F.R. Beih. 16: 59 (1923). Type: Congo-Kinshasa, below Tondoa, Congo River, *Büttner* 341 (GOET!, lecto., designated by Vorontsova & Knapp 2010: 1596; GOET!, isolecto.)

S. *gilo* Raddi var. *sparseaculeatum* Bitter in F.R. Beih. 16: 59 (1923). Type: Africa. "*Solanum* 4ß, Bullomshore", *Afzelius* s.n. (UPS, holo.)

S. *gilo* Raddi subsp. *monteiroi* (C.H. Wright) Bitter in F.R. Beih. 16: 56 (1923); T.T.C.L.: 581 (1949)

NOTE. *Solanum aethiopicum* is a widespread cultivated vegetable crop, and includes all cultivated forms of *S. anguivi*. *Solanum aethiopicum* can be identified and distinguished from *S. anguivi* by a combination of the following characters: annual herbaceous habit, frequent lack of stellate indumentum, smaller and more entire leaves, 1–2(–10) flowers per inflorescence, short and thick pedicels of fruit crop varieties, more than 5 perianth lobes, ovary enlarged even at anthesis, fruit over 1 cm in diameter and with more than two locules, not detaching easily from the pedicels, immature fruit striped and/or not green when immature, and seeds over 2.5 mm long. Leaf vegetable varieties are easily recognizable by their thin membranous leaves and dark red-brown, almost black drying colour; these have up to 10 small flowers. Precise limits of this species are difficult to specify due to the wide variety of infrequently occurring characters. Prickly forms of *S. aethiopicum* have been recorded from European botanical gardens (Lester & Niakan, In: Solanaceae: Biology and Systematics, pp. 433–456. 1986), but are not known in Africa.

Solanum aethiopicum, *S. gilo*, *S. olivare*, and *S. integrifolium* are all interfertile cultivated races of *S. anguivi*. The cultivars have been separated into four groups: Gilo group (with stellate hairs and large fleshy variably shaped fruits that taste like carrots or green beans, common in Brazil), Shum group (leaf vegetable with no stellate hairs or prickles, small fruits ± 2 cm in diameter, common in Africa), Kumba group (edible leaves that are almost glabrous as well as sweet broad deeply-grooved fruits), and Aculeatum Group (with prickles, broad and deeply furrowed fruit, common in European botanical gardens but not known in Africa). *Solanum anguivi* was maintained as a distinct wild to semicultivated species due to its largely distinct morphology, lack of wild intermediates, and a probable selective pressure against new hybrids between *S. aethiopicum* and *S. anguivi*.

Synonyms listed here are only those in common use or based on African types. Complete synonymy for *S. aethiopicum* can be found on the Solanaceae Source website (http://www.solanaceaesource.org).

27. **Solanum agnewiorum** *Voronts.* in Phytotaxa 10: 32 (2010). Type: Kenya, Meru District, Igembe, Kangeta, Nyambeni Forest, *Vorontsova, Ficinski, Kirika, Muthoka & Muroki* 196 (EA!, holo.; BM!, MO!, K!, NY!, iso.)

Prostrate or climbing shrub to 4 m, heavily armed; young stems sparsely stellate-pubescent with trichomes porrect, orange-translucent, sessile, rays 6–8, 0.1–0.15 mm, midpoints usually 0.5–1 mm; prickles curved, 1.5–3 mm long, 1–2.5 mm wide at base, round or flattened. Leaf blades drying concolorous to weakly discolorous, yellow-green to green-brown, elliptic, 6–14 × 4.5–9 cm, ± 2 times longer than wide, base cuneate, often oblique, margin lobed, the lobes 2–5 on each side, deltate, 1–1.7 cm long, acuminate and extending $^1/_4$–$^1/_3$ of the distance to the midvein, apex acuminate; sparsely stellate-pubescent on both sides with trichomes on abaxial surface porrect, sessile, rays ± 8, 0.1–0.25 mm, midpoints 0.3–1 mm, trichomes of adaxial surface smaller with midpoints elongated to 1 mm; primary veins 3–5 pairs; petiole 1.2–6 cm, $^1/_4$–$^1/_2$ of the leaf length. Inflorescences not branched, 2–3 cm long, with 1–4 flowers; peduncle 0–5 mm long; rachis 0–0.8 cm long; peduncle and rachis unarmed or with up to 5 prickles; pedicels 0.9–1.3 cm long in flower, in fruit 2.5–3.5 cm long, with 2–10 prickles. Flowers heterostylous, only 1–2 long-styled, 5-

merous. Calyx 3–4 mm long, the lobes deltate, 1–2 mm long, acuminate, unarmed or with up to 10 prickles. Corolla white, 1.3–1.6 cm in diameter, lobed for $^2/_3$–$^3/_4$ of its length, lobes deltate, 5–6.5 × 2.5–3.5 mm. Stamens equal; anthers 4–4.5 mm. Ovary with a few stellate trichomes; style 5.5–7.5 mm long on long-styled flowers. Berries 1(–2) per infructescence, striped when young, orange at maturity, spherical, 20–25 mm in diameter; fruiting calyx unarmed or with up to 10 prickles. Seeds 5.5–6 × ± 4.5 mm.

KENYA. Kiambu District: near Limuru Girls High School, 20 Aug. 1961, *Polhill* 455!; Aberdare National Park, E side, 8 Apr 1975, *Hepper & Field* 4923!; NE Mt Kenya, Marimba Forest, 14 Oct. 1960, *Polhill & Verdcourt* 2993!
DISTR. **K** 4; not known elsewhere
HAB. Moist montane forest understorey, *Croton–Brachylaena–Calodendrum* and *Ocotea* forests; 1800–2500 m

NOTE. Specimens of *S. agnewiorum* look deceptively similar to the common and hypervariable *S. anguivi* Lam. as its distinguishing characters in the fruit are easily overlooked due to scarcity of fruiting material. *Solanum agnewiorum* differs from *S. anguivi* by its climbing habit, prickles always curved, leaf lobes acuminate, 1–4 flowers and 1–2 fruits per inflorescence, fruits 20–25 mm diameter, and seeds 5.5–6 mm diameter.

28. **Solanum anguivi** *Lam.*, Tabl. Encycl. 2: 23 (1794); Blundell, Wild Fl. E. Afr.: 190, t. 590 (1992); U.K.W.F., 2nd ed.: 243 (1994); Gonçalves in F.Z. 8(4): 93 (2005); Friis in Fl. Eth. 5: 129, fig. 158.11.10–13 (2006). Type: Mauritius [Île de France], garden of Monsieur Cossigny, 1769, *Commerson* s.n. (MPU, lecto., designated by Hepper 1978: 290; P-JU, P, isolecto.)

Shrub, 0.7–2 m, erect, armed or unarmed; young stems stellate-pubescent; trichomes porrect, orange-translucent, sessile or shortly stalked, stalks to 0.1 mm, rays 6–8, 0.1–0.25 mm, midpoints same length as rays or to 2 mm; prickles up to 5(–10) × 1–4 mm at base, straight to curved, rounded to flattened. Leaf blades drying discolorous, green-brown, ovate to elliptic, (5–)11–25 × 5–17 cm, 1.5–2.5 times longer than wide, base cuneate to rounded, usually unequal or oblique, margin usually lobed, sometimes subentire, the lobes up to 5 on each side, broadly deltate, to 2(–3) cm long and rounded to acute, extending up to $^1/_3$(–$^1/_2$) of the distance to the midvein, apex acute(acuminate); sparsely to densely stellate-pubescent on both sides with trichomes on abaxial surface porrect, sessile or shortly stalked, stalks to 0.1 mm, rays 6–8, 0.1–0.3 mm, midpoints same length as rays or up to 1.5 mm long, adaxially with reduced rays and elongated midpoints; primary veins 4–7 pairs; petiole 1–5 cm, $^1/_4$–$^1/_6$ of the leaf length. Inflorescences 2.5–6 cm long, not branched or branched once, with 5–22 flowers, peduncle 2–20 mm long; rachis 0.6–5 cm long; peduncle and rachis unarmed; pedicels 0.5–1.2 cm long in flower, in fruit 0.8–1.3 cm long, usually unarmed. Flowers perfect, 5-merous. Calyx 3–5 mm long, the lobes deltate to somewhat ovate, 1–2 mm long, acute, unarmed or with 1–4 prickles. Corolla white to mauve, 0.8–1.5 cm in diameter, lobed for ± $^2/_3$ of its length, the lobes deltate, 3–6 × 2–3 mm. Stamens equal; anthers 3.5–5 mm. Ovary with a few stellate trichomes towards the apex; style 6–8 mm long. Berries 6–22 per infructescence, evenly green when young, red at maturity, spherical, 6–9 mm in diameter; fruiting calyx unarmed. Seeds 1.8–2.5 × 1.8–2.3 mm. Fig. 20, p. 168.

UGANDA. Karamoja District: Mt Moroto, Jan. 1959, *J. Wilson* 660!; Ankole District: Butale, 27 Apr. 1946, *A.S. Thomas* 4471!; Mengo District: Kirerema, 10 Aug. 1913, *Dummer* 107!
KENYA. Northern Frontier District: Mt Kulal, 9 Oct. 1947, *Bally* 5561!; Kiambu District: Karura forest, 14 Oct. 1967, *Mwangangi & R. Abdalla* 229!; Masai District: Enesambulai Valley, 13 Feb. 1971, *Greenway & Kanuri* 14838!
TANZANIA. Arusha District: Ngurdoto crater forest, 5 May 1965, *Richards* 20346!; Njombe District: Livingstone Mountains, foot trail from Bumbigi, 5 Mar. 1991, *Gereau & Kayombo* 4228!; Songea District: Matengo Hills, 9 Jan. 1956, *Milne-Redhead & Taylor* 8084!

Fig. 20. *SOLANUM ANGUIVI* — habits showing variation: **1**, common fruiting habit; **2**, subentire leaves and few flowers; **3**, deeply lobed leaves and branched inflorescence; **4**, densely pubescent leaves and a branched infructescence; **5**, with large curved prickles; **6**, abaxial leaf surface trichome; **7**, abaxial leaf surface trichome. 1, 6 from *Semsei* 4149; 2, 7 from *Bally* 5097; 3 from *Richards* 8475; 4 from *Faden* 68/514; 5 from *Kerfoot* 4721. Drawn by Lucy T. Smith. Scale bar 1, 2, 3, 4, 5 = 4 cm; 6, 7 = 0.5 mm.

DISTR. U 1–4; **K** 1–7; **T** 1–8; Senegal, Gambia, Guinea Bissau, Sierra Leone, Liberia, Ivory Coast, Ghana, Togo, Nigeria, Cameroon, Equatorial Guinea, Central African Republic, Congo-Kinshasa, Rwanda, Burundi, South Sudan, Ethiopia, Angola, Zambia, Malawi, Mozambique, Zimbabwe, South Africa; Madagascar

HAB. Forest and forest edges, wooded grassland, grassland, bushland, disturbed areas, old cultivation, and roadsides; (40–)1000–2200(–3100) m

SYN. *S. lividum* Link, Enum. Hort. Berol. Alt. 1: 188 (1821). Type: cultivated in Berlin, seeds from Madagascar, *Willdenow* s.n. (B-W!, holo.)

S. distichum Schumach. & Thonn. in Schumach., Beskr. Guin. Pl. 122 [142] (1827). Type: Ghana, Aquapim, *Thonning* 199 (C!, lecto. [IDC microfiche 2203 102: II.1–2, "252"] designated here [best material]; C!, isolecto., LE!, possible isolecto.)

S. senegambicum Dunal in DC., Prodr. 13(1): 194 (1852). Type: Senegal, Rio Nunez, *Heudelot* 713 (P!, holo.; MPU, P!,. iso.)

S. scalare C.H. Wright in J.L.S. 30: 93 (1894). Type: Angola, Pungo Adongo, *Welwitsch* 6104 (K!, lecto., designated here [best material]; BM!, isolecto.)

S. rohrii C.H. Wright in K.B. 1894: 128 (1894); C.H. Wright in F.T.A. 4, 2: 231 (1906). Type: Ethiopia, Ankober [Aferbeine], *Roth* 445 (K!, lecto., designated here [only extant material found])

S. carvalhoi Dammer in P.O.A. C: 355 (1895). Type: Mozambique, Gorongosa, *Carvalho* s.n. (COI!, holo.)

S. stuhlmannii Dammer in P.O.A. C: 354 (1895). Syntypes: Tanzania, Bukoba, *Stuhlmann* 1577 (B†, syn); *Stuhlmann* 3873 (B†, syn); *Stuhlmann* 3994 (B†, syn)

S. wildemanii De Wild. & T. Durand in Ann. Mus. Congo Belge, Bot., sér. 3, 1: 291 (1901). Type: Congo-Kinshasa, Lukolela, *Dewèvre* 547 (BR, holo.)

S. buettneri Dammer in E.J. 38: 59 (1905). Type: Togo, near Bismarckburg, Katamara farm, *Buettner* 655 (B†, holo.)

S. halophilum Pax in E.J. 39: 648 (1907). Type: Ethiopia, Ambo Mieda, 2170 m, 4 Apr. 1905, *Rosen* s.n. (WRSL, holo.)

S. batangense Dammer in E.J. 48: 239 (1912). Type: Cameroon, Grand Batanga, *Dinklage* 906 (B†, holo.)

S. dinklagei Dammer in E.J. 48: 241 (1912). Type: Cameroon, Kribi, *Dinklage* 630 (B†, holo.)

S. sakarense Dammer in E.J. 48: 247 (1912). Type: Tanzania, Lushoto District: West Usambara, above Sakare, *Engler* 950 (B†, syn.) & Usumbura, Kafunamavi, *Keil* 61 (B†, syn.)

S. pseudogeminifolium Dammer in E.J. 48: 249 (1912). Type: Tanzania, Lushoto District, West Usambara, Nguelo [Ngwelo], *Engler* 681 (B†, holo.)

S. schroederi Dammer in E.J. 48: 250 (1912). Type: Nigeria, Djibuland, Ischagamo, *Schlechter* 13004 (BR!, lecto., designated here [best duplicate distribution; best material of this duplicate]; K!, P!, isolecto.)

S. rederi Dammer in E.J. 48: 251 (1912). Type: Cameroon, Buea, *Eder* 946 (B†, holo.)

S. newtonii Dammer in E.J. 48: 251 (1912). Type: Angola, Huilla, Humpata, *Newton* 200 (B†, holo.)

S. nguelense Dammer in E.J. 48: 252 (1912). Type: Tanzania, Lushoto District, Nguelo, *Zimmermann* 1470 (EA!, lecto. designated here [only extant material known])

S. kandtii Dammer in E.J. 48: 253 (1912). Type: "Rwanda", Nyanza, 1700 m, *Kandt* 80 (no specimens found)

S. jaegeri Dammer in E.J. 48: 256 (1912). Type: Tanzania, Mbulu District, Iraku, Mama Isara, *Jaeger* 208 (B†, holo.)

S. spathotrichum Dammer in E.J. 48: 238 (1912). Type: Tanzania, Iringa District: Udzungwa [Utschungwe] Mts, 1600 m, *Frau Hauptmann Prince* s.n. (B†, holo.)

S. albidum De Wild. in B.J.B.B. 4: 396 (1914). Type: Congo-Kinshasa, between Buta and Bima, *Seret* 128 (BR!, holo.; BR!, iso.)

S. flamignii De Wild. in B.J.B.B. 4: 399 (1914). Type: Congo-Kinshasa, Kitobola, *Flamigni* 388 (BR!, holo.; BR!, K!, iso.)

S. ueleense De Wild. in B.J.B.B. 4: 400 (1914). Type: Congo-Kinshasa, Kisangani, Uele River, 1904, *Delpierre* s.n. (BR!, holo.)

S. yangambiense De Wild. in B.J.B.B. 4: 402 (1914). Type: Congo-Kinshasa, Yangambi, 10 Sep. 1913, *Elskens* s.n. (BR!, holo.)

S. cultum De Wild. in B.J.B.B. 4: 406 (1914). Type: Congo-Kinshasa, Mobwasa, *Reygaert* 407 (BR0000008993076!, lecto., designated here [follows unpublished lectotypification by R. Lester]; BR!, isolecto.)

S. jespersenii De Wild. in B.J.B.B. 4: 407 (1914). Type: Congo-Kinshasa, Mondombe, *Jespersen* s.n. (BR!, holo.)

S. dichroanthum Dammer in E.J. 53: 340 (1915). Type: Zimbabwe, Chirinda [Shirinda], *Swynnerton* 388 (K! [K000414096], lecto., designated here [best material]; BM!, K!, US, isolecto.)

S. olivaceum Dammer in E.J. 53: 341 (1915). Type: Tanzania, Rungwe District: Kyimbila, *Stolz* 631 (U!, lecto., designated here [material demonstrates leaf shape variation]; C, JE, K!, MO, W, WAG, Z, isolecto.)

S. grotei Dammer in E.J. 53: 342 (1915). Type: Tanzania, Lushoto District: Usambaras, Amani, *Grote* 3426 (B†, holo.)

S. keniense Turrill in K.B. 1915: 77 (1915). Type: Kenya, E Mt Kenya forests, 1350 m, *Battiscombe* 853 (K!, holo.; EA! iso.)

S. aurantiacobaccatum De Wild., Pl. Bequaert. 1: 421 (1922). Type: Congo-Kinshasa, Ruwenzori, Butagu valley, *Bequaert* 3738 (BR!, holo.)

S. albiflorum De Wild., Pl. Bequaert 1: 422 (1922). Type: Congo-Kinshasa, Penghe, *Bequaert* 2369 (BR!, holo.; BR!, iso.)

S. ruwenzoriense De Wild., Pl. Bequaert. 1: 435 (1922). Type: Congo-Kinshasa, Ruwenzori, Butagu valley, *Bequaert* 3699 (BR!, holo.; BR!, iso.)

S. indicum L. var. *lividum* (Link) Bitter in F.R. Beih. 16: 10 (1923)

S. indicum L. var. *maroanum* Bitter in F.R. Beih. 16: 10 (1923). Type: Madagascar, Maroa, Antongil, *Mocquerys* 294 (no specimens found)

S. indicum L. subsp. *distichum* (Schumach. & Thonn.) Bitter in F.R. Beih. 16: 13 (1923); Heine in F.W.T.A. 2ⁿᵈ ed., 2: 333 (1963)

S. indicum L. var. *immunitum* Bitter in F.R. Beih. 16: 14 (1923), *nom. nov.* for *Solanum batangense* Dammer

S. indicum L. var. *brevipedicellatum* Bitter in F.R. Beih. 16: 15 (1923). Type: Angola, Loanda, Camilungo, *Welwitsch* 6089-*b* (B†, holo.; BM!, lecto., designated here [only extant material known])

S. indicum L. var. *depauperatum* Bitter in F.R. Beih. 16: 15 (1923). Type: Cameroon, Campo, *Ledermann* 382 (no specimens found)

S. indicum L. var. *modicearmatum* Bitter in F.R. Beih. 16: 16 (1923), *nom. nov.* for *Solanum buettneri* Dammer

S. indicum L. var. *grandemunitum* Bitter in F.R. Beih. 16: 17 (1923), *nom. nov.* for *Solanum rederi* Dammer

S. indicum L. var. *monbuttorum* Bitter in F.R. Beih. 16: 17 (1923). Type: Sudan, Southern Ghasal-sources and upper Uelle region, Kussumbo, Monbuttu territory, *Schweinfurth* 3175 (K!, lecto. designated here [only extant material found])

S. indicum L. var. *dichroanthum* (Dammer) Bitter in F.R. Beih. 16: 18 (1923)

S. indicum L. var. *halophilum* (Pax) Bitter in F.R. Beih. 16: 18 (1923)

S. indicum L. subsp. *newtonii* (Dammer) Bitter in F.R. Beih. 16: 19 (1923)

S. indicum L. var. *brevistellatum* Bitter in F.R. Beih. 16: 21 (1923); T.T.C.L.: 582 (1949). Type: Tanzania, Lushoto District: Gonja, *Engler* 3374 (EA!, lecto., designated here [only material known])

S. indicum L. var. *grotei* (Dammer) Bitter in F.R. Beih. 16: 22 (1923); T.T.C.L.: 582 (1949)

S. indicum L. var. *pseudogeminifolium* (Dammer) Bitter in F.R. Beih. 16: 22 (1923); T.T.C.L.: 582 (1949)

S. indicum L. subsp. *rohrii* (C.H. Wright) Bitter in F.R. Beih. 16: 23 (1923); T.T.C.L.: 582 (1949)

S. indicum L. var. *bukobense* Bitter in F.R. Beih. 16: 24 (1923); T.T.C.L.: 582 (1949). Type: Tanzania, Bukoba District: Magharibi, *Stuhlmann* 1458 (no specimens found)

S. indicum L. var. *spathotrichum* (Dammer) Bitter in F.R. Beih. 16: 24 (1923); T.T.C.L.: 583 (1949)

S. indicum L. var. *breviaculeatum* Bitter in F.R. Beih. 16: 25 (1923); T.T.C.L.: 582 (1949). Type: Tanzania, Lushoto District, E Usambara, Amani, *Braun* 717 in herb. Amani (no specimens found)

S. indicum L. var. *kiwuense* Bitter in F.R. Beih. 16: 25 (1923). Type: Rwanda, Mountains on S end of Lake Kivu, 1700 m, *R.E. Fries* 1529 (UPS, holo.; GOET!, five possible iso.)

S. indicum L. var. *nguelense* (Dammer) Bitter in F.R. Beih. 16: 25 (1923); T.T.C.L.: 583 (1949)

S. indicum L. var. *sakarense* (Dammer) Bitter in F.R. Beih. 16: 25 (1923); T.T.C.L.: 583 (1949)

S. indicum L. var. *kandtii* (Dammer) Bitter in F.R. Beih. 16: 26 (1923)

S. indicum L. var. *profundelobatum* Bitter in F.R. Beih. 16: 26 (1923). Type: Congo-Kinshasa, Upper Katanga, Kantu, *Kässner* 2403 (B†, holo.; K!, lecto, designated here [only material known])

S. indicum L. var. *carvalhoi* (Dammer) Bitter in F.R. Beih. 16: 27 (1923)

S. indicum L. var. *jaegeri* (Dammer) Bitter in F.R. Beih. 16: 27 (1923); T.T.C.L.: 582 (1949)

S. indicum L. var. *suprastrigulosum* Bitter in F.R. Beih. 16: 28 (1923). Type: South Africa, Pondoland, Egosa forest, betweeen Mission Station and Dorkin, *Bachmann* 1192 (no specimens found)

S. indicum L. subsp. *olivaceum* (Dammer) Bitter in F.R. Beih. 16: 28 (1923); T.T.C.L.: 582 (1949)

S. indicum L. subsp. *grandifrons* Bitter in F.R. Beih. 16: 29 (1923); T.T.C.L.: 582 (1949). Type: Tanzania, Kilimanjaro, Gonja, Bulwa [Bulua], *Holst* 4340 (B†, holo.)

S. indicum L. var. *subquercinum* Bitter in F.R. Beih. 16: 33 (1923). Type: Ethiopia, Agrima, *Schimper* 143 (P00344947!, lecto, designated here [best material]; BR!, P, isolecto.)

S. indicum L. var. *mesodolichum* Bitter in F.R. Beih. 16: 34 (1923). Type: Ethiopia, Begemder, Gerra Abuna Tekla Haimanot, *Schimper* 1129 (K!, lecto., designated here [best material]; BM!, US, isolecto.)

S. indicum L. var. *arussorum* Bitter in F.R. Beih. 16: 35 (1923). Type: Ethiopia, Gallaland, Arussi Galla, Abinas Mts, *Ellenbeck* 1336a (B†, holo.)

S. indicum L. var. *eldamae* Bitter in F.R. Beih. 16: 35 (1923). Type: Kenya, two days march from Eldama Ravine, *Whyte* s.n. (B†, holo.)

S. indicum L. subsp. *zechii* Bitter in F.R. Beih. 16: 36 (1923). Type: Togo, Kete Kratschi, *Zech* 351/352 (B†, holo.)

S. indicum L. var. *busogae* Bitter in F.R. Beih. 16: 37 (1923). Type: Uganda, Busoga District, Busoga gardens, *Whyte* s.n. (B†, holo.)

S. indicum L. subsp. *clinocarpum* Bitter in F.R. Beih. 16: 37 (1923); T.T.C.L.: 582 (1949). Types: Tanzania, Bukoba District: Magharibi, *Stuhlmann* 1057 (no specimens found) & same locality, *Stuhlmann* 3648 (no specimens found)

S. orthocarpum Pic.Serm., Miss. Stud. Lago Tana: 129 (1951). Type: Ethiopia, Basin of Lake Tana, W of Furie (Zeghie), *Pichi-Sermolli* 2575 (FT, holo.; FT, iso.)

S. mesodolichum (Bitter) Pic.Serm., Miss. Stud. Lago Tana: 133 (1951)

S. indicum sensu auctt. mult. e.g. T.T.C.L.: 581 (1949); K.T.S.L.: 581 (1994)

NOTE. *Solanum anguivi* is the second most common and variable species of spiny *Solanum* in Africa (after *S. campylacanthum*), and encompasses a great deal of morphological and genetic variation. Numerous instances of single morphologically anomalous specimens from different parts of Africa fall within *S. anguivi* sensu lato and it is possible these will be recognised as independent taxa following further research.

Cultivated forms of *S. anguivi* are widespread and placed here in *S. aethiopicum*. *Solanum anguivi* is maintained as a distinct wild to semicultivated species due to its largely distinct morphology, lack of wild intermediates, and a probable selective pressure against new hybrids between *S. aethiopicum* and *S. anguivi*.

For the last 150 years the name "*Solanum indicum* L." has been applied to both the African *S. anguivi* and the Asian *S. violaceum* Ortega, as well as other superficially similar prickly taxa. Due to historic confusion and widespread misapplication the epithet "indicum" was rejected; extensive evidence has demonstrated that *S. anguivi* is not directly related to the Asian *S. violaceum*.

Synonyms listed here are only those in common use or based on African types. Complete synonymy for *S. anguivi* can be found on the Solanaceae Source website (http://www.solanaceaesource.org).

29. **Solanum cordatum** *Forssk.*, Fl. Aegypt.-Arab.: 47 (1775); Friis in Fl. Eth. 5: 134, fig. 158.12.3–4 (2006). Type: Yemen, Naqil Khailan between Beui Harath & Nehm, NE of Sanaa, *G.H.S. Wood* 2327 (K!, neo. designated by Wood in K.B. 39: 136, 1984; BM!, isoneo.)

Erect or sometimes scandent subshrub, 0.3–1 m, unarmed or armed; young stems with scattered trichomes visible with the naked eye as white dots on the dark brown steam; trichomes white, multangulate, sessile, rays 12–18, up to 0.1(–0.15) mm, midpoints usually reduced to a globose gland; prickles straight or sometimes curved, up to 4 × 0.1–2(–6) mm at base. Leaf blades drying concolorous, yellowish green, orbicular (ovate), 0.7–1.5(–3) × 0.7–1.3(–3.5) cm, length usually same as the width (–1.5 times longer than wide), base cordate to attenuate (decurrent), equal, margin entire, apex rounded to acute; trichomes on abaxial surface porrect and multangulate, sessile (stalked), stalks to 0.1 mm, rays 8–16, 0.05–0.1(–0.15) mm, midpoints short or a globular gland, glabrescent; primary veins not visible or 2–4 pairs; petiole 0.2–1.3 cm, $^1/_3$–$^2/_3$(–1) of the leaf length. Inflorescences 1.5–4 cm long, with 1(–2) flowers; peduncle absent; pedicels 0.5–3 cm long in flower, in fruit 1.5–3.5 cm long, unarmed. Flowers perfect, 4- or 5-merous. Calyx 2–4 mm long, lobes broadly deltate to oblong, 0.5–2 mm long, obtuse to acuminate, unarmed. Corolla mauve to purple, 0.9–1.6 cm in diameter, lobed for $^2/_3$–$^4/_5$ of its length, lobes deltate to lanceolate, 3–6.5 × 1.2–2 mm. Stamens equal; anthers 3–5 mm. Ovary glabrous or with 1–2 trichomes; style 8–9 mm long. Berries 1(–2) per infructescence, red at maturity, spherical, 6–8 mm in diameter; fruiting calyx unarmed. Seeds 1.8–3 × 1.5–2.2 mm, almost black.

Kenya. Northern Frontie District: Dandu, 14 Apr. 1952, *Gillett* 12790! & Marsabit, Ilanto, 26 Oct. 1976, *Sato* 173!

Distr. **K** 1; Ethiopia, Somalia; also Arabia, Afghanistan, Pakistan and India

Hab. Grassland, bushland and open woodland on silty, sandy or stony soil; 700–1200 m

Syn. *S. gracilipes* Decne. in Jacquem., Voy. Bot.: 113, t. 119 (1844). Type: India, "India Borealis Occidentalis", *Jacquemont* 63 (P!, holo.)
   S. sabaeorum Deflers in Bull. Soc. Bot. France 43: 122 (1896). Type: Yemen, N side of Mount Nakhai, Bilad Fodhli, *Deflers* 488 (K!, syn.; P!, 4 syn.)
   S. darassumense Dammer in E.J. 38: 57 (1905). Type: Somalia, Arussi-Galla, Darassuma, *Ellenbeck* 2024 (GOET!, lecto., designated by Vorontsova & Knapp in Taxon 59: 1596, 2010)
   S. obbiadense Chiov. in Boll. Soc. Bot. Ital. 1925: 106 (1925). Type: Somalia, Sultanate of Obbia, Biomal, *Puccioni & Stefanini* 1061 [605] (FT!, holo.)
   S. nummulifolium Chiov. in Boll. Soc. Bot. Ital. 1925: 107 (1925). Type: Somalia, Sultanate of Obbia, Biomal, *Puccioni & Stefanini* 551 [605] (FT!, holo.)

Note. *Solanum cordatum* can usually be recognised by its long filiform pedicels protruding beyond the leaves, and membranous glaucous glabrescent leaves with long decurrent narrowly-winged petioles. Unfortunately neither of these characters is constant and it is important to look at the indumentum under the microscope: trichomes on the young stems of *S. cordatum* are compact and multangulate, with rays almost always under 0.1 mm long, and a swollen gland usually present on the centre of the trichome on stems, young leaves or both.
   Solanum cordatum is frequently confused with the variable and largely sympatric *S. forskalii*, and can be distinguished by its multangulate trichomes with 12–18 rays under 0.15 mm long on the young stems (versus porrect trichomes with 6–10 rays over 0.15 mm long on the young stems of *S. forskalii*), 1–2 flowers per inflorescence (versus 1–20 flowers per inflorescence in *S. forskalii*), and anthers 3–5 mm long (versus anthers 4.5–7 mm long in *S. forskalii*).
   This species is adapted to arid environments by shedding foliage during drought and opportunistic growth in wet periods.

30. **Solanum cyaneopurpureum** *De Wild.*, Pl. Bequaert. 1: 425 (1922). Type: Congo-Kinshasa, Kabare, *Bequaert* 5333 (BR!, holo.; BR!, iso.)

Scandent or climbing subshrub, 0.5–2 m, armed or unarmed; young stems sparsely to densely stellate-pubescent; trichomes porrect, translucent to yellow-orange, stalked, stalks ± 0.1 mm, rays 6–8, 0.1–0.2 mm, midpoints shorter than rays; prickles broad-based, slightly curved, 1.5–2(–3) mm long, 0.5–1 mm wide at base, flattened. Leaf blades drying usually concolorous, yellowish, ovate, 2.5–5.5 × 1.5–2.5 cm, 2–3 times longer than wide, base cuneate, often oblique, margin entire, rarely weakly lobed, apex acute (obtuse); densely stellate-pubescent to glabrescent on both sides, trichomes on abaxial surface porrect, stalked, stalks ± 0.1 mm, rays 6–8, 0.1–0.2(–0.25) mm, midpoints shorter than rays, adaxially with fewer rays; primary veins 4–5 pairs; petiole 0.3–1.2 cm, $^1/_4$–$^1/_5$ of the leaf length. Inflorescences 2.5–3.5 cm long, not branched, with 3–10 flowers; peduncle 0–9 mm long; rachis 0.1–1.5 cm long; peduncle and rachis usually unarmed; pedicels 0.6–0.8 cm long in flower, in fruit 1.2–2 cm long, usually unarmed. Flowers perfect, 5-merous. Calyx 2–4 mm long, lobes deltate, 1–2 mm long, acute, usually unarmed. Corolla mauve to purple,

1.2–1.5(–2) cm in diameter, lobed for ± ²⁄₃ of its length, lobes narrowly ovate to deltate, (5–)6–8(–9) × 2–3(–4) mm. Stamens equal; anthers (4–)5–6.5 mm. Ovary with occasional trichomes; style 10–12 mm long. Berries 1–4(–7) per infructescence, striped when young, red at maturity, sometimes elongate when young, 6–10 mm in diameter. Seeds 2.5–3.5 × 1.8–3.2 mm.

UGANDA. Ankole District: Ruizi river, 18 Nov. 1950, *Jarrett* 209!; Masaka District: Kabula, Sep. 1945, *Purseglove* 1803!; Mengo District: 1–2 km east of Kikoma, 19 Oct. 1969, *Lye* 4435!
TANZANIA. Mpwapwa District: Kongwe, 3 km from Dodoma–Morogoro road, 12 Apr. 1988, *Bidgood et al.* 1023!
DISTR. **U** 2, 4; **T** 1, 5; Congo-Kinshasa, Rwanda, Burundi
HAB. Wooded grassland, forest margins, thickets, often found on termite mounds; 800–1500 m

SYN. *S. kitivuense* Dammer subsp. *ukerewense* Bitter in F.R. Beih. 16: 114 (1923); T.T.C.L.: 583 (1949). Type: Tanzania, Mwanza District: "Unfroimi-Schachi" near Neuwied-Ukerewe, *Conrads* 380 (K!, lecto., designated here [best flowering material]; BM!, iso.)
   S. tanganikense Bitter in F.R. Beih. 16: 132 (1923). Type: Burundi or Tanzania, "N side of Lake Tanganyika", *Kässner* 3160 (P!, holo., designated by Vorontsova & Knapp in Taxon 59: 1600, 2010; BM!, E!, GOET!, K!, iso.)

NOTE. *Solanum cyaneopurpureum* is part of a complex series of mostly climbing, small-leaved species in *Solanum* section *Oliganthes* and is the only species with an inland distribution range in the Lake Victoria Basin regional mosaic and the Guineo-Congolan phytochorion. *Solanum cyaneopurpureum* and *S. zanzibarense* are difficult to distinguish and no one character separates all the specimens. *Solanum cyaneopurpureum* and *S. taitense* share small subentire leaves, untidy appearance and stalked trichomes.
   Solanum cyaneopurpureum is sometimes cultivated as a medicinal plant in maize fields in central Tanzania, growing closer to the coast than normal (**T** 5). These plants have broader leaves and more numerous flowers.

31. **Solanum forskalii** *Dunal*, Hist. Nat. *Solanum*: 237 (1813); Friis in Fl. Eth. 5: 134, fig. 158.12.7–9 (2006) & in Fl. Somalia 3: 215 (2006). Based on and type as *Solanum villosum* Forssk., *non* Mill., 1768, *nom. illeg.*, later homonym of *S. villosum* Mill., 1768. Type: Yemen, Wadi Surdud, *Forsskål* 414 (C, holo. [IDC microfiche 2200, 02:II.7,8])

Erect or scandent subshrub, 0.5–1 m, densely armed; young stems densely stellate-pubescent, the epidermis not visible between the trichomes, trichomes porrect, translucent, sessile and occasionally stalked, stalks up to 0.15 mm, rays 6–10, 0.15–0.3(–0.5) mm, midpoints same length as rays or up to 1.5 mm, sometimes shorter than the midpoint; prickles straight or sometimes curved, 3–10 mm long, 1–3 mm wide at base. Leaf blades drying concolorous to discolorous, grey-green, ovate, 1–4(–6) × 0.5–3(–4) cm, 1–2 times longer than wide, base cordate to rounded, equal or oblique, margin entire or shallowly lobed, the broadly deltate lobes 2(–3) on each side, 2–4 mm long and apically rounded to obtuse, extending up to ¹⁄₄ of the distance to the midvein, apex rounded to obtuse; moderately stellate-pubescent on both sides, trichomes on abaxial surface porrect, sessile (stalked), stalks to 0.2 mm, rays 6–10, 0.2–0.4(–1) mm, midpoints ± same length as rays (1.5 mm), adaxially with fewer rays; primary veins 2–3 pairs; petiole 0.2–1.6(–2.5) cm, ¹⁄₄–²⁄₃ of the leaf length. Inflorescences often branched once, 2–6.5 cm long, with (1–)2–20 flowers; peduncle 1–4(–15) mm long; rachis 0–2(–3) cm long; peduncle and rachis usually unarmed; pedicels 0.2–1 cm long in flower, in fruit 0.7–1.6 cm long, usually unarmed. Flowers perfect, 5-merous. Calyx 2–4.5 mm long, the lobes deltate, 0.5–2 mm long, acuminate, rarely with 1–5 prickles. Corolla mauve to purple, 1.3–2.4 cm in diameter, lobed for ± ¹⁄₅ of its length, lobes lanceolate, 6–11 × 1.5–2 mm. Stamens equal; anthers 4.5–7 mm. Ovary glabrous; style 9–12 mm long. Berries 1–10 per infructescence, red at maturity, spherical, 6–9 mm in diameter; fruiting calyx usually unarmed. Seeds 2.5–4 × 1.8–3 mm, almost black.

KENYA. Northern Frontier District: Dadaab–Wajir road, 5 km SE of Sabule Airstrip, 29 Nov. 1978, *Brenan, Gillett et al.* 14811! & S of Laisamis ± 100 km N of Isiolo, 8 Dec. 1978, *Hepper & Jaeger* 7273!

DISTR. **K** 1; Senegal, Chad, Mali, Niger, Sudan, Eritrea, Ethiopia, Somalia; Egypt, Arabia and India

HAB. Scrub on stony ground and rocky slopes, often on granite; 150–1300 m

SYN. *S. villosum* Forssk., Fl. Aegypt.-Arab.: 47 (1775). Type as for *S. forskalii*
  S. macilentum A. Rich., Tent. Fl. Abyss. 2: 105 (1850). Type: Ethiopia, Choho, *Quartin-Dillon & Petit* s.n. (P000343696!, lecto., designated by Lester 1997: 286; P!, two isolecto.)
  S. albicaule Dunal in DC., Prodr. 13(1): 204 (1852); Heine in F.W.T.A. 2nd ed., 2: 333 (1963). Type: Sudan, Nubia, Cordofan Chursi, *Kotschy* 309 (G-DC, lecto., designated by Lester 1997: 287; B†, E, ER, GOET!, K!, LZ, M, MO!, MPU!, NY, P!, STU, TCD, W, WAG!, isolecto.)
  S. heudelotii Dunal in DC., Prodr. 13(1): 205 (1852). Type: Senegal, near Gabor, *Heudelot* 417 (45) (P!, lecto., designated by Lester 1997: 287; K!, MPU, P!, UPS, isolecto.)
  S. hadaq Deflers in Bull. Soc. Bot. France 43: 122 (1896). Type: Yemen, Schoukra, Bilad Fodhli, J. Areys 30 k E.N.E. [Schughra], *Deflers* 377 (K!, P 3 sheets, iso.)
  S. scindicum Prain in J. Asiat. Soc. Bengal, Pt. 2, Nat. Hist. 65: 542 (1896). Type: India, Rann of Kutch, *Stolizcka* s.n. (CAL, K, syn.) & Pakistan, Sind, *Stocks* s.n. (CAL, K, syn.) & Sind *Cooke* s.n. (CAL, K, syn.) & Rajasthan State, Rajputana, Jessole, *King* s.n. (CAL, K, syn.)
  S. albicaule Kotschy var. *parvifrons* Bitter in F.R. Beih. 16: 102 (1923). Type: Sudan, N Darfur, El Fasher, *Pfund* 407 (no specimens found)

NOTE. *Solanum forskalii* can be recognised by its dense white indumentum of porrect-stellate trichomes on the young stems, and abundant prickles. Most representatives of *S. forskalii* are easy to identify but variability within this widespread species is considerable. *Solanum forskalii* is frequently confused with the largely sympatric *S. cordatum* and can be distinguished by its porrect trichomes with 6–10 rays over 0.15 mm long on the young stems (versus multangulate trichomes with 12–18 rays under 0.15 mm long on the young stems of *S. cordatum*), 1–20 flowers per inflorescence (versus 1–2 flowers per inflorescence in *S. cordatum*), and anthers 4.5–7 mm long (versus anthers 3–5 mm long in *S. cordatum*).

32. **Solanum hastifolium** *Dunal* in DC., Prodr. 13(1): 284 (1852); C.H. Wright in F.T.A. 4, 2: 226 (1906); T.T.C.L.: 580 (1949); K.T.S.L.: 580 (1994); U.K.W.F., 2nd ed.: 243 (1994); Friis in Fl. Eth. 5: 132, fig. 158.12.1–2 (2006) & in Fl. Somalia 3: 214 (2006). Type: Sudan, Mt Cordofan, Arasch-Cool, *Kotschy* 393 (P00344017! lecto, designated here [see note]; B, BM!, GOET!, K!, M, P!, W, WRSL, isolecto.)

Erect or sometimes scandent woody subshrub, 0.3–1(–1.5) m, young stems sparsely to densely stellate-pubescent, trichomes porrect, translucent, sessile or shortly stalked, stalks up to 0.1 mm, rays 5–8, 0.1–0.3 mm, midpoints same length as rays or shorter; prickles hooked to almost straight, reflexed or curved downwards, 1–4 mm long, 0.5–3.5 mm wide at base. Leaf blades drying concolorous, yellow-green, ovate to lanceolate, 2.5–6.5 × 0.8–2.5 cm, 2–4 times longer than wide, base obtuse to cuneate(truncate), usually equal, margin lobed, the broadly rounded lobes 2–4 on each side, 0–10 mm long, extending up to ²/₃ of the distance to the midvein, apex acute to obtuse; moderately to sparsely stellate-pubescent, trichomes on abaxial surface porrect, sessile or shortly stalked, stalks to 0.1 mm, rays 4–8, 0.1–0.4 mm, midpoints reduced or same length as rays, adaxially with fewer shorter rays; primary veins 3–4(–5) pairs; petiole 0.6–1.2 cm (–20 cm on basal leaves) ¹/₃–¹/₄ of the leaf length. Inflorescences not branched, 1–3 cm long, with 3–5 flowers; peduncle absent; rachis 0.2–1.5 cm long; peduncle and rachis unarmed; pedicels 0.6–1.5 cm long in flower, in fruit 1.1–1.9 cm long, unarmed. Flowers perfect, 4-merous or 5-merous. Calyx 4–6 mm long, the lobes deltate, 2–4.5 mm long, acute to acuminate, unarmed. Corolla mauve to purple (white), 1.4–2 cm in diameter, lobed for ± ¹/₅ of its length, the lobes lanceolate to oblong, 6–10 × 1.8–2.5 mm. Stamens equal; anthers 5–8 mm. Ovary glabrous or with trichomes towards the apex; style 10–13 mm long. Berries 1–4 per infructescence, striped when young, red at maturity, spherical, 6–8 mm in diameter. Seeds 2.8–3.3 × 2–2.8 mm.

UGANDA. Karamoja District: near Rupa, Sep. 1958, *J. Wilson* 585!
KENYA. Northern Frontier District: 30 km south of Wamba, 21 Dec. 1971, *Bally & Smith* 14745!;
W Suk District: N of Marich Pass at the foot of the Kaimat Escarpment, 27 Oct. 1977, *Carter & Stannard* 59!; Masai District: near Olorgesailie, 24 Oct. 1955, *Milne-Redhead & Taylor* 7144!
TANZANIA. Shinyanga District: Seseku aerodrome, 7 Jan. 1932, *B.D. Burtt* 3518!; Masai District:
Kitumbeine [Ketumbane], 7 Jan. 1936, *Greenway* 4288!
DISTR. **U** 1; **K** 1–4, 6–7; **T** 1–3; South Sudan, Ethiopia, Somalia
HAB. Dry *Acacia* scrub, open places, disturbed vegetation and roadsides on sand, clay, loam and black cotton soil; 200–1700 m

SYN. *S. longestamineum* Dammer in E.J. 38: 58 (1905). Type: Ethiopia, Dagaga, *Ellenbeck* 1000
(GOET!, lecto., designated by Vorontsova & Knapp in Taxon 59: 1597, 2010)
S. hastifolium Dunal subsp. *velutinellum* Bitter in F.R. Beih. 16: 130 (1923). Type: Tanzania,
Pare District: between Same and Makanya [Makandja], *Winkler* 3780 (WRSL, holo.)
S. cynanchoides Chiov., Fl. Somala 1: 333, fig. 190 (1932). Type: Somalia, Oltregiuba, Obe,
Senni 444 (FT, holo.; K!, iso., fragment)

NOTE. *Solanum hastifolium* is a widespread polymorphic weedy species recognised by its lobed leaves, retrorse or recurved prickles, and inflorescences with more than 2 flowers. *Solanum hastifolium* is closely affiliated with the more local *S. taitense* and *S. setaceum.*
The protologue of *Solanum hastifolium* cites a specimen in „h. Delile": this may be the sheet P00344017 now held in Paris and annotated with the Prodromus species number; no duplicates of *Kotschy* 393 were found at MPU and the other specimen at P, P00344018, originates from Hb. Cosson. P00344017 is chosen as the lectotype to fix the application of this name.
It is difficult to distinguish *Solanum taitense* from *S. hastifolium* reliably as no single character separates all the specimens, but *S. hastifolium* has the majority of the following: lobed leaves 2.6–6.5 cm long (versus entire to subentire leaves 1.2–3(–3.5) cm long in *S. taitense*), abundant prickles 1–4 mm long, spaced 3–10 mm apart (versus inconspicuous prickles up to 1(–2) mm long in *S. taitense*), 3–5 flowers per inflorescence (versus 1–2(–3) flowers per inflorescence in *S. taitense*), plant sparsely to moderately pubescent (versus *S. taitense* densely pubescent), and trichomes sessile to shortly stalked (versus stalked trichomes on all vegetative surfaces with inflated trichome stalks on the adaxial side of the leaves in *S. taitense*). *Solanum setaceum* can be distinguished by its prominent stem bristles.

33. **Solanum inaequiradians** *Werderm.* in N.B.G.B. 12: 90 (1934); T.T.C.L.: 581 (1949). Type: Tanzania, Morogoro District, Uluguru Mts, NW side, *Schlieben* 2707 (B!, holo.; BM!, BR!, LISC!, M, P!, S, Z, iso.)

Scandent or climbing subshrub to 2 m, armed; young stems densely stellate-pubescent, trichomes porrect, reddish-translucent, mostly subsessile, stalks up to 0.15 mm, rays ± 8, 0.1–0.2 mm, midpoints (0.5–)1.5–2.5(–3) mm; prickles curved, 2–4 mm long, 0.8–1(–3) mm wide at base, round to slightly flattened. Leaf blades drying discolorous, yellow-green, ovate, 7–10(–13) × 3.5–4.5(–6) cm, ± 2 times longer than wide, base truncate to cuneate, usually equal, margin lobed to almost entire, the lobes 2–4 on each side, deltate to rounded, up to 10 mm long, apically rounded to acute, extending up to $^1/_3(-^1/_2)$ of the distance to the midvein; apex acute or somewhat long-acuminate; stellate-pubescent on both sides, trichomes on abaxial surface porrect, sessile or shortly stalked, stalks to 0.15 mm, rays 6–8, 0.2–0.35 mm, midpoints 1–2 mm, adaxially with bulbous bases, rays usually absent, midpoints 1.5–2.5 mm long; primary veins 4–5 pairs; petiole 1–2.5 cm, $^1/_4$–$^1/_5$ of the leaf length. Inflorescences not branched, 3–5 cm long, with 1–4 flowers; peduncle absent; rachis ± 1.5 cm long, with 0–2 prickles; pedicels ± 0.8 cm long in flower, in fruit 2–2.5 cm long, with 0–15 prickles. Flowers largely perfect, 5-merous. Calyx 7–9 mm long, lobes narrowly deltate, 4–6 mm long, thin-acuminate, with 0–10 prickles. Corolla mauve to purple, 1.7–2 cm in diameter, lobed for ± $^4/_5$ of its length, lobes oblong, ± 10 × 2–1.5 mm. Stamens equal; anthers ± 7 mm. Ovary not known; style ± 12 mm long. Berries 1–2 per infructescence, striped when young, red at maturity, often elongate when young, 12–14 mm in diameter; fruiting calyx 10–12 mm long. Seeds 2.8–3.2 × 2.5–2.8 mm.

TANZANIA. Morogoro District: Uluguru Mts, Morningside, 14 Apr. 1968, *Harris* 1602! & Uluguru Mts, Kitundu, 22 Nov. 1934, *E.M. Bruce* 204! & Uluguru Mts, NW side, 18 Sep. 1932, *Schlieben* 2707!
DISTR. **T** 6; endemic to Uluguru Mts
HAB. Forest understorey; 1100–2000 m

NOTES. *Solanum inaequiradians* is a rare high altitude forest endemic of the Uluguru Mountains, recognised by its long-attenuate calyx lobes exceeding the corolla in bud and growing to 7–9 mm long in flower and 10–12 mm long in fruit, ovate discolorous yellow-green leaves with deltate lobes and attenuate apices, and dense trichomes with elongated midpoints 1.5–3 mm (rarely 0.5 mm) long on all surfaces of the plant.

34. **Solanum lamprocarpum** *Bitter* in F.R. Beih. 16: 107 (1923); T.T.C.L.: 583 (1949). Type: Tanzania, Lindi District, E of Lake Lutamba, *Busse III* 2498 (P!, lecto., designated by Vorontsova & Knapp in Taxon 59: 1597, 2010; EA!, GOET!, isolecto.)

Shrub, 0.5–2 m, erect, unarmed or heavily armed; young stems densely stellate-pubescent, trichomes porrect, translucent to orange-brown, sessile to shortly stalked, stalks up to 0.1 mm, rays ± 8, 0.2–0.4 mm, midpoints 0.7–1 mm; prickles straight, 2–6 mm long, 0.2–0.8 mm wide at base. Leaf blades drying concolorous to discolorous, red-green or yellow-green, obovate to elliptic, 5–7.5 × 2.5–4 cm, ± 2 times longer than wide, base cuneate, often unequal or oblique, margin lobed, the lobes 3–4 on each side, rounded to oblong, up to 10 mm long, apically rounded to obtuse, extending up to $\frac{1}{2}(\frac{2}{3})$ of the distance to the midvein; apex obtuse to acute; densely stellate-pubescent abaxially and sometimes glabrescent adaxially, trichomes on abaxial surface porrect, sessile or shortly stalked, stalks to 0.1 mm, rays 6–8, 0.2–0.3 mm, midpoints 0.7–1 mm long, adaxially with 0–5 rays, midpoints 0.5–1.5 mm long; primary veins 4–6 pairs; petiole 0.7–2 cm, $\frac{1}{3}$–$\frac{1}{4}$ of leaf length. Inflorescences not branched or branched once, 2–3 cm long, with 2–4 flowers; peduncle 0–3 mm long; rachis 0–1 cm long; peduncle and rachis unarmed; pedicels 0.4–0.8 cm long in flower, in fruit 1.5–2.4 cm long, usually unarmed. Flowers largely perfect, 5-merous. Calyx 4–6 mm long, lobes oblong, 2–3 mm long, obtuse, with 0–10 prickles. Corolla mauve, ± 1.7 cm in diameter, lobed for ± $\frac{4}{5}$ of its length, lobes narrowly ovate to narrowly deltate or oblong, 7–9 × ± 2 mm. Stamens equal; anthers 5–6 mm. Ovary glabrous; style ± 9 mm long. Berries 2–4 per infructescence, red at maturity, spherical, 10–11 mm in diameter; fruiting calyx unarmed. Seeds 2.7–3 × 2.2–2.3 mm.

TANZANIA. Uzaramo District: Kazimzumbwi Forest, south of Kisarawe, Mar. 1991, *Frontier-Tanzania* 1928!; Kilwa District: Libungani, Selous Game Reserve, 19 Feb. 1971, *Ludanga* 1251!; Mikindani District: Mikindani–Lindi road, 11 Mar. 1963, *Richards* 17829!
DISTR. **T** 6, 8; Mozambique
HAB. Shrubland and disturbed areas; 0–300 m

NOTES. *Solanum lamprocarpum* is a rare species recognised by its foliaceous oblong calyx lobes with obtuse apices, reddish elliptic or obovate lobed leaves, straight prickles, erect woody pedicels and trichomes with midpoints elongated to 0.7–1 mm long. Like several other endemic spiny *Solanum* species superficially similar to the widespread *S. anguivi*, very few collections exist and its geographical extent and morphological variability are not fully known.

35. **Solanum lanzae** *J.-P. Lebrun & Stork* in Candollea 50: 217 (1995); Friis in Fl. Eth. 5: 135, fig. 158.12.10–12 (2006), *nom. nov.* based on *Solanum angustifolium* Lanza, *non* Lam., 1794. Type: Ethiopia, Borana, Javello, near airfield, *Cufodontis* 499 (FT, holo.; FT, GE, W, iso.)

Woody herb, 0.3–1 m, erect, unarmed; young stems stellate-pubescent, trichomes porrect, translucent, sessile or stalked, stalks up to 0.1 mm, rays ± 8, 0.1–0.25 mm, midpoints shorter than rays. Leaf blades drying concolorous to discolorous, dull grey-green, narrowly elliptic to lanceolate, 2–9 × 0.3–1.5 cm, 4–8 times longer than wide, base narrowly cuneate to attenuate, usually equal, margin entire, apex acute to obtuse; stellate-pubescent on both sides, trichomes on abaxial surface porrect, sessile or shortly stalked, stalks to 0.1 mm, rays 7–8, 0.15–0.3 mm, midpoints same length as rays or shorter, adaxially with fewer shorter rays; primary veins 6–10 pairs; petiole absent or up to 0.4 cm, less than $^1/_6$ of the leaf length. Inflorescences not branched, 2–4 cm long, with 3–6 flowers; peduncle 0–2 mm long; rachis 0.2–0.5 cm long; pedicels 0.6–1 cm long in flower, fruiting pedicels 0.8–1.8 cm long. Flowers largely perfect, 5-merous. Calyx 1.5–2 mm long, lobes broadly deltate, ± 0.5 mm long, cuspidate. Corolla mauve to purple, 0.9–1.5 cm in diameter, lobed for ± $^1/_2$ of its length, lobes deltate, 3–4.5 × 2–3 mm. Stamens equal; anthers 3.5–4.2 mm. Ovary stellate-pubescent in the upper $^1/_5$; style 4.5–6 mm long. Berries 2–5 per infructescence, evenly green when young, red at maturity, spherical, 6–9 mm in diameter. Seeds 2.4–2.8 × 2–2.3 mm.

UGANDA. Karamoja District: Kangole, Aug. 1957, *J. Wilson* 386!
KENYA. Laikipia District: Ngobit, 22 April 1952, *Bally* 8191!; Meru District: Timau–Meru road, 8 Dec. 1971, *Bally & Smith* 14441!; Masai District: Aitong, 13 June 1961, *Glover et al.* 1818!
TANZANIA. Musoma District: W side of Titushi and Mab Mbalangeti River junction, 20 April 1961, *Greenway & Turner* 10076!; Moshi District: Ushira, about 15 km ESE of Moshi, 8 Feb. 1971, *Pedersen* 242!
DISTR. **U** 1; **K** 1, 3–7; **T** 1–2; Ethiopia
HAB. Disturbed areas, bushland, thickets, and stony slopes, on black cotton soil; 1200–2100 m

SYN. *S. angustifolium* Lanza, Miss. Biol. Borana, Racc. Bot.: 191, fig. 54 (1939), *nom. illeg.*, later homonym of *Solanum angustifolium* Mill., 1768

NOTES. *Solanum lanzae* is a common and distinctive endemic of East African black cotton soils. Its greyish, long-elliptic, almost sessile leaves, complete lack of prickles, and small flowers set it apart from other *Solanum* in the region. It has been suggested that it is not a native species, even though no similar species are known outside Africa. Greyish green leaves and unusually small flowers and fruits suggest an affiliation with the southern African *S. catombelense* Peyr.

36. **Solanum malindiense** *Voronts.* in Syst. Bot. 35: 904 (2010). Type: Kenya, Tana River District, Nairobi Ranch, Ras Wanawali Sabaa, *Festo & Luke* 2337 (EA!, holo.; MO!, NHT, iso.)

Scandent or sometimes erect shrub to 2 m, moderately armed; young stems densely stellate-pubescent; trichomes porrect, translucent, sessile or stalked, stalks up to 0.1 mm, rays 6–8, 0.1–0.2 mm, midpoints same length as rays or shorter, often reduced to globular glands; prickles curved, 1–3 mm long, 1–1.5 mm wide at base. Leaf blades drying concolorous, yellow-green, ovate, 6–8 × 3–6 cm, ± 1.5 times longer than wide, base cordate, often oblique, margin sinuate, the lobes 2–4 on each side, broadly rounded, up to 5 mm long, extending up to $^1/_4$ of the distance to the midvein, apex rounded; densely stellate-pubescent on both sides, trichomes on abaxial surface porrect, subsessile, stalks less than 0.1 mm, rays ± 8, 0.15–0.2 mm, midpoints same length as rays or shorter, adaxially glabrescent; primary veins 5–6 pairs; petiole 1.5–2 cm, $^1/_3$–$^1/_4$ of the leaf length. Inflorescences not branched, 3–4 cm long, with 3–7 flowers; peduncle 1–4 mm long; rachis 0.3–1 cm long; peduncle and rachis usually unarmed; pedicels 0.5–1 cm long in flower, in fruit ± 1.5 cm long, usually unarmed. Flowers heterostylous, only 1–2 long-styled, 5-merous. Calyx 5–8 mm long, lobes long-deltate, 3.5–5 mm long, long-acuminate, unarmed or with up to 20 prickles. Corolla mauve, 2.8–3.7 cm in diameter, lobed for $^2/_3$–$^3/_4$ of its length, lobes long-deltate, ± 12 × 5 mm. Stamens equal; anthers 8.5–9.5 mm. Ovary stellate-pubescent in upper $^1/_2$; style 13–15 mm long on long-styled flowers. Berries 1–2 per infructescence, yellow at maturity, spherical, ± 15 mm in diameter; fruiting calyx with 10–30 prickles. Seeds 2.5–3.5 × 2–2.5 mm.

KENYA. Lamu District: Kitwa Pembe Hill, 15 July 1974, *Faden & Faden* 74/1072! & Nairobi Ranch, Ras Wanawali Sabaa, 18 Oct. 2004, *Luke & Luke* 10326!; Kilifi District: 6 km N of Malindi, 3 Nov. 1961, *Polhill & Paulo* 709!
DISTR. **K** 7; not known elsewhere
HAB. Coastal bush, dunes and sand, often on coral; 0–50 m

NOTE. *Solanum malindiense* has undulate leaves of consistent shape, long-acuminate calyx lobes, and curved prickles covering the calyx from late bud onwards. *Solanum malindiense* is similar to the more southern *S. usaramense* populations in the coastal areas of Mozambique, Tanzania and southern Kenya, with wider and more lobed leaves, larger yellow berries, and sparse indumentum. Observations suggest that it is comparatively rare and occurs sporadically.

37. **Solanum mauense** *Bitter* in F.R. Beih. 16: 42 (1923); K.T.S.L.: 581 (1994); U.K.W.F., 2nd ed.: 243 (1994). Type: Kenya, Nakuru District: Mau plateau, between Sandiani and Njoro, alt. 2300–3000 m., Oct., *Baker* 44 (B†, holo.); neotype: Kenya, Kericho District: Mau area, *Glover, Gwynne & Samuel* 938 (EA!, neo., designated here; K!, isoneo.)

Shrub, 0.5–1.5 m, erect, armed; young stems densely stellate-pubescent, trichomes porrect, orange-translucent, stalked, stalks up to 0.6 mm, rays 7–8, 0.1–0.25 mm, midpoints shorter than rays, sometimes to 1.5 mm; prickles curved, 2–4 mm long, 1–3 mm wide at base, flattened. Leaf blades drying strongly discolorous, yellow-green, ovate or sometimes elliptic, 4–15(–25) × 1–4.5(–7) cm, 2.5–3(–2) times longer than wide, base cuneate to rounded, often unequal, margin subentire (rarely lobed on mature leaves), apex obtuse to rounded(acute); densely stellate-pubescent on both sides, trichomes on abaxial surface porrect, stalked, stalks 0.15–0.4 mm, rays 7–8, 0.2–0.3 mm, midpoints shorter than rays (1 mm), adaxially with thick stalks 0.1–0.2 mm long, reduced rays and midpoints (midpoints to 1–2 mm); primary veins 5–7 pairs; petiole 0.5–2 cm, ⅕–⅛ of the leaf length. Inflorescences not branched, 3–4(–6) cm long, with 6–20 flowers, peduncle 1–4 mm long; rachis 0.6–4 cm long; peduncle and rachis unarmed or with 1–2 prickles; pedicels 0.4–0.7 cm long in flower, in fruit 0.7–1.2 cm long, unarmed or with up to 5 prickles. Flowers hermaphrodite, 5-merous. Calyx 3–6 mm long, lobes deltate, 1–2 mm long, acute, unarmed or with 1–4 prickles. Corolla white to mauve, 1–1.4 cm in diameter, lobed for ± ⅔ of its length, lobes deltate, 3.5–4.5 × 2–3 mm. Stamens equal; anthers ± 3.5 mm. Ovary stellate-pubescent in upper ⅕; style 5–7 mm long. Berries 6–13 per infructescence, evenly green when young, red at maturity, spherical, 6–9 mm in diameter; fruiting calyx usually unarmed. Seeds 2.5–3 × 1.8–2.5 mm.

KENYA. Mt Elgon, Dec. 1931, *Jack* 186!; Kiambu District: Limuru, Feb. 1915, *Dummer* 1650!; Masai District: Orengitok, 20 km from Narok on road to Olokurto, 17 May 1961, *Glover et al.* 1211!
TANZANIA. Masai/Mbulu District: Ngorongoro crater, 31 May 1973, *Frame* 143!
DISTR. **K** 3–4, 6; **T** 2; not known elsewhere
HAB. Forest edges, secondary bushland, grassland, commonly found along roadsides; 1800–3000 m

NOTES. *Solanum mauense* is an often abundant weedy highland shrub with yellowish, densely hairy entire leaves, numerous small curved yellow prickles, and numerous bright red-orange fruits. This species is easy to recognise and has been accepted in the present circumscription by all treatments published since its discovery. *Solanum mauense* has in the past been considered part of the diverse *S. anguivi* sensu lato, previously known as *S. indicum* L. *Solanum anguivi* is a variable widespread species present throughout the African highlands and in the Kenyan highlands *S. anguivi* is replaced by *S. mauense*, although the two species are partly sympatric. The main distinguishing character of *S. mauense* is the largely entire leaves.

38. **Solanum polhillii** *Voronts.* in Syst. Bot. 35: 902 (2010). Type: Kenya, Masai District, Ewaso Ngiro–Loliondo road where it crosses the Masan River, *Verdcourt* 3838 (EA!, holo.; K!, iso.)

Erect to semi-scandent shrub, 1–2 m, armed or unarmed; young stems sparsely to densely stellate-pubescent; trichomes porrect, translucent, stalked, stalks up to 0.2 mm, rays 7–9, 0.1–0.2 mm, midpoints same length as rays or shorter, often reduced to globular glands; prickles straight (curved), 2–3(–6) mm long, 0.5–1.5 mm wide at base. Leaf blades drying discolorous, yellow-green, ovate, 2–6 × 1–4 cm, 1.5–2 times longer than wide, base cordate (rounded or cuneate), often oblique, margin entire to weakly lobed, the lobes if present 1–2 on each side, broadly rounded and up to 3 mm long, extending up to $\frac{1}{4}$ of the distance to the midvein; apex rounded or obtuse; densely stellate-pubescent on both sides, trichomes on abaxial surface porrect, stalked, stalks to 0.2 mm, rays ± 8, 0.1–0.3 mm, midpoints same length as rays or shorter, adaxially with reduced rays pointed upwards; primary veins 3–4 pairs; petiole 0.3–2.5 cm, $\frac{1}{3}$–$\frac{2}{3}$ of the leaf length. Inflorescences not branched, 2–4.5 cm long, with 1–3(–4) flowers; peduncle 0–5 mm long; rachis 0–2 cm long; peduncle and rachis unarmed or with 1–3 prickles; pedicels 1–1.5 cm long in flower, in fruit 0.8–4 cm long, unarmed or with up to 15 prickles. Flowers heterostylous, 5-merous, the basal 1–2 long-styled. Calyx 8–16 mm long, lobes long-deltate to oblong, 4.5–12 mm long, long-acuminate, usually unarmed. Corolla mauve to purple, 2–4.2 cm in diameter, lobed for $\frac{2}{3}$–$\frac{3}{4}$ of its length, lobes ovate to elliptic, 7–15 × 4.5–10 mm. Stamens equal; anthers 4.5–8 mm. Ovary densely stellate-pubescent; style 8–12 mm long on long-styled flowers. Berries 1(–2) per infructescence, striped when young, yellow to orange at maturity, spherical, 13–20 mm in diameter; fruiting calyx usually unarmed. Seeds 2.8–3 × 2–2.5 mm.

KENYA. Laikipia District: Uaso Narok River on Kisima Farm, 40 km N of Rumuruti, 13 Nov. 1977, *Carter & Stannard* 333!; Machakos District: Lukenya Rocks by Nairobi–Mombasa Road, 2 June 1980, *Gilbert* 5961!; Masai District: 59 km from Nairobi on Magadi road, 12 April 1960, *Verdcourt et al.* 2672!

TANZANIA. Masai District: Serengeti Central Plains, 3 km W of the E Boundary, 30 May 1962, *Greenway & Watson* 10677! & Soitayai, 29 Nov. 1956, *Greenway* 9086!

DISTR. **K** 3–4, 6; **T** 1–2; not known elsewhere

HAB. Savanna, rocky hillsides, bushland and scrub, on granite, volcanic rocks or red sandy soil; 1800–2200 m

SYN. "*Solanum* sp. = Greenway 9086" sensu Polhill, *Solanum* in E & NE Africa: 34 (ined., 1961)
Solanum sp. G sensu Agnew, U.K.W.F.: 243 (1994) & K.T.S.L.: 583 (1994)

NOTE. Morphology, including leaf size, petiole length and prickliness, varies with environmental conditions such as aridity, nutrient availability, and herbivory. Particularly remarkable is the variation in flower size, with the corolla 2–4.2 cm wide and the anthers 4.5–8 mm long, smaller than the more southern *S. richardii* but larger than most other species in the region. Limited populations of *Solanum polhillii* remain in some dry upland areas but severe habitat loss has been documented, partly due to grazing (Vorontsova *et al.* in Syst. Bot. 35: 894–906. 2010).

Solanum polhillii is intermediate between the morphologically defined sections *Oliganthes* and *Melongena* due to its limited andromonoecy and medium-sized fruits. In spite of vegetative similarity to members of section *Oliganthes*, its fruits are yellow rather than orange to red and its flowers are too large to fit comfortably into that group.

39. **Solanum ruvu** *Voronts.* in J. E. Afr. Nat. Hist. 99: 230 (2011). Type: Tanzania, Morogoro District, Ruvu Forest Reserve, *Mhoro UMBCP* 113 (K!, holo.; MO!, iso.)

Prostrate or climbing subshrub, ± 1 m, armed; young stems almost glabrous, with minute simple hairs, the stellate trichomes present on the youngest parts only, porrect, translucent, sessile, rays 2–4, ± 0.05 mm, midpoints ± same length as rays; bristles straight, 4–6 mm long, ± 0.2 mm wide at base. Leaf blades drying concolorous, brown-green, elliptic, 9–12 × 2.5–4.5 cm, 2.5–4 times longer than wide, base cuneate, usually equal, margin almost entire, rarely with shallow lobes, apex long-acuminate; almost glabrous, with minute simple hairs and stellate trichomes on the youngest parts only, trichomes on abaxial surface porrect, sessile, rays 2–4, ± 0.05 mm, midpoints ± same length as rays; primary veins 6–8 pairs; petiole 0.5–1.5 cm, $^1/_5$–$^1/_{10}$ of the leaf length. Inflorescences not branched, 5–9 cm long, with 10–15 flowers; peduncle 10–15 mm long; rachis 5–8 cm long; peduncle and rachis densely armed, the prickles like those on stems evenly covering the rachis; pedicels 0.7–1.2 cm long. Flowers perfect, 5-merous. Calyx 2.5–5 mm long, lobes deltate, 1.5–3.5 mm long, long-acuminate, often with numerous prickles. Corolla 1.2–1.8 cm in diameter, colour not known, lobed for ± $^4/_5$ of its length, lobes long-deltate, 6–8 × 1.5–2 mm. Stamens equal; anthers 5–6 mm. Ovary with minute glandular hairs; style ± 8 mm long. Berries not known.

TANZANIA. Morogoro District, Ruvu Forest Reserve, 17 July 2000, *Mhoro UMBCP* 113!
DISTR. **T** 6; only known from the type
HAB. Moist coastal forest understorey; ± 200 m

NOTE. The unusually long filiform inflorescences with a dense covering of long straight prickles on the rachis have no parallels among the known *Solanum* species in Africa and Madagascar. The affinities of *S. ruvu* are most likely with *S. zanzibarense*, the only other scandent and sometimes subglabrous East African coastal forest species with subentire leaves, thin stems, and prickles that are sometimes straight. Failure to recollect this species suggests it may be extinct.

A similarly dense covering of long and flexible prickles or bristles is found on two other East African species that are not directly related to *S. ruvu*: the savanna shrub *S. setaceum* in section *Oliganthes* and the montane forest shrub *S. schumannianum*.

40. **Solanum setaceum** *Dammer* in P.O.A. C: 353 (1895); T.T.C.L.: 580 (1949); U.K.W.F., 2nd ed.: 243 (1994). Type: Tanzania, Kilimanjaro, Dehu, Kahe, *Volkens* 2202 (B†, holo.; GOET!, lecto., designated by Vorontsova & Knapp in Taxon 59: 1599, 2010; GOET!, isolecto.)

Scandent or climbing subshrub to 1 m, armed; young stems stellate-pubescent, trichomes porrect, translucent, stalked, stalks 0.1–0.2 mm, elongating to form the bristles, rays 6–8, 0.2–0.3 mm, midpoints shorter or longer than rays; bristles curved, (2.5–)3–5 mm long, 0.1–0.5 mm wide at base. Leaf blades drying concolorous to weakly discolorous, yellow-green, ovate, 1.5–5(–7) × 0.7–2(–4) cm, 2–4 times longer than wide, base cordate to obtuse, usually equal, margin subentire or lobed, the lobes 1–3 on each side, 0–5 mm long, broadly rounded, extending up to $^1/_3$ of the distance to the midvein; apex obtuse; stellate-pubescent on both sides, trichomes on abaxial surface porrect (multangulate), stalked, stalks 0.1–0.2 mm, rays 6–8(–11), 0.1–0.3(–0.4) mm, midpoints ± same length as rays, adaxially with fewer shorter rays; primary veins 3–5 pairs; petiole 0.5–1 cm, ($^1/_2$–)$^1/_3$–$^1/_4$ of the leaf length. Inflorescences not branched, 2–3 cm long, with 1–3(–4) flowers; peduncle 0(–8) mm long; rachis 0–1 cm long; peduncle and rachis unarmed; pedicels 0.6–1.2 cm long in flower, in fruit 0.9–1.6 cm long, unarmed or with 1–10 bristles. Flowers perfect, 5-merous. Calyx 3.5–7 mm long, lobes long-deltate or oblong, 3–5 mm long, acute to acuminate, unarmed or with a few bristles. Corolla white to mauve, 1.4–2 cm in diameter, lobed for ± $^4/_5$ of its length, lobes lanceolate to oblong, 6.5–8 × 2–3 mm. Stamens equal; anthers 5–8 mm. Ovary stellate-pubescent in the upper $^1/_3$–$^1/_4$; style 8.5–14 mm long. Berries 1–4 per infructescence, 6–9 mm in diameter, spherical, evenly green when young, red at maturity. Seeds 2.8–3 × 2.1–2.3 mm. Fig. 21, p. 181.

KENYA. Masai District: Namanga, 17 km up road to Meto (Ngito Hills) from Junction with main Nairobi–Namanga road just S of Bissel, 22 Dec. 1963, *Verdcourt* 3861!
TANZANIA. Musoma/Maswa District: Seronera River, 25 Apr. 1958, *Paulo* 381!; Masai District: Olkarien, 20 Dec. 1962, *Newbould* 6402!; Singida District: track off Singida–Babati, 30 March 1965, *Richards* 20362!
DISTR. **K** 6; **T** 1–2, 5; not known elsewhere
HAB. *Acacia* bushland, thickets and grassland on sandy loam or black cotton soil; 1000–1500 m

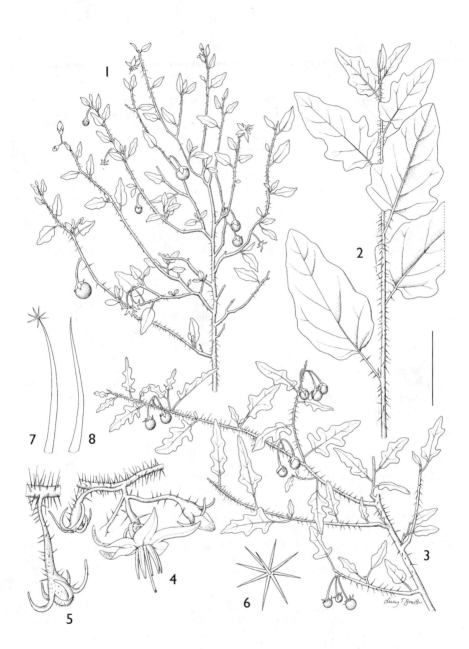

FIG. 21. *SOLANUM SETACEUM* — **1**, fruiting habit in dry environment; **2**, juvenile habit in wet environment; **3**, fruiting habit with lobed leaves; **4**, inflorescence with a long-styled flower; **5**, developing fruit; **6**, abaxial leaf surface trichome; **7**, stalked trichome developing into a bristle, the young bristle with a stellate trichome attached; **8**, fully developed bristle. 1, 6–8 from *Richards* 25706; 2 from *Verdcourt* 3851; 3 from *Richards* 25506; 4–5 from *Vorontsova et al.* 167, field photograph. Drawn by Lucy T. Smith. Scale bar: 1, 2, 3, = 3 cm; 4 = 1 cm; 5 = 0.4; 7, 8 = 0.8 mm.

SYN. *S. setaceum* Dammer var. *irakuanum* Bitter in F.R. Beih. 16: 138 (1923); T.T.C.L.: 580 (1949).
    Type: Tanzania, Mbulu District, Mbugwe and Iraku, *Merker* 297 (GOET!, lecto.,
    designated by Vorontsova & Knapp in Taxon 59: 1599, 2010)

NOTES. *Solanum setaceum* is instantly recognisable by the long bristles on its younger stems. It is
a frequent and fairly uniform species with shallowly lobed ovate yellowish leaves and
trichome stalks 0.1–0.2 mm long on its vegetative parts. Stalks of some trichomes on the
young stems lengthen and lignify to form the bristles, and stellate trichome rays are often
visible on the apices of developing bristles. The bristles vary from long, dense, pale yellowish,
filiform structures perpendicular to the stem to thicker, recurved, dark brown to grey
structures that are further apart and similar to the prickles found in related species.
    Solanum setaceum lies within the distribution area of the variable *S. hastifolium* and has
limited range overlap with *S. taitense*. There are no consistent morphological differences with
S. hastifolium outside the bristles. The uniform nature of *S. setaceum* and its geographical
coherence justify maintaining it as a distinct species.

41. **Solanum stipitatostellatum** *Dammer*, Abh. Königl. Akad. Wiss. Berlin 1894: 63
(1894); C.H. Wright in F.T.A. 4, 2: 227 (1906); T.T.C.L.: 580 (1949). Type: Tanzania,
Lushoto District, Usambara, Kwa Mshuza, *Holst* 9121 (GOET!, lecto., designated by
Vorontsova & Knapp in Taxon 59: 1599, 2010; K!, isolecto.)

Climbing subshrub, 1–3 m, armed or unarmed; young stems densely stellate-
pubescent, trichomes porrect, translucent to orange-brown or brown, stalked,
stalks 0.1–0.4 mm, rays 7–8, 0.1–0.3 mm, midpoints shorter than rays or up to 0.7 mm;
prickles curved downwards or straight, 1–2 mm long, 0.4–0.8 mm wide at base,
inconspicuous. Leaf blades drying somewhat discolorous, red-green or red-yellow-
green, ovate, 6–13 × 3–7 cm, ± 2 times longer than wide, base truncate to cuneate,
often unequal or oblique, margin entire to obscurely lobed, the lobes up to 3 on
each side, up to 5 mm long, broadly rounded, apically rounded to obtuse and
extending up to $^1/_4$ of the distance to the midvein; apex obtuse to acute; densely
stellate-pubescent abaxially, trichomes on abaxial surface porrect, stalked, stalks
0.1–0.3(–0.4) mm, rays ± 8, 0.15–0.3 mm, midpoints variable (–0.4 mm), adaxially
glabrescent, rays reduced, midpoints reduced(0.8 mm); primary veins 4–5 pairs;
petiole 0.5–2 cm, $^1/_5$–$^1/_6$ of the leaf length. Inflorescences not branched or
branched once, 3.5–7 cm long, with 3–10 flowers; peduncle (2–)6–30 mm long;
rachis 0–3 cm long; peduncle and rachis unarmed or with 1 prickle; pedicels
0.8–1 cm long, in fruit 1.5–2.2 cm long, usually unarmed. Flowers heterostylous,
5-merous, the basal 2–5 long-styled. Calyx 4–6 mm long, lobes deltate, 2–3 mm
long, acute to acuminate, unarmed. Corolla mauve to purple, 1.8–3 cm in
diameter, lobed for ± $^2/_3$ of its length, lobes broadly deltate, 8–12 × 3–6 mm.
Stamens equal; anthers 6–8 mm. Ovary with sparse simple hairs; style 10–15 mm
long on long-styled flowers. Berries 2–5 per infructescence, striped when young,
red at maturity, usually elongate when young, 10–13 mm in diameter. Seeds 2–2.7
× 1.7–2.2 mm.

KENYA. Kwale District: Shimba Hills, 14 Jan. 1964, *Verdcourt* 3931!
TANZANIA. Lushoto District: Amani, 17 Apr. 1922, *Soleman* 7250!; Handeni District:
    Kwamarukanga Forest Reserve, 3 Feb. 1971, *Shabani* 638!; Morogoro District: Mkungwe
    Forest Reserve, 11 July 2000, *Mhoro* UMBCP 62!
DISTR. **K** 7; **T** 3, 6; not known elsewhere
HAB. Forest understorey, open forest, forest edges or disturbed ground, 300–2000 m

SYN. *S. glochidiatum* Dammer in P.O.A. C: 354 (1895); C.H. Wright in F.T.A. 4, 2: 230 (1906).
    Type: Tanzania, Lushoto District, Usambara, Mashewa, Bumko, *Holst* 8834 (B†, holo.;
    GOET!, lecto., designated by Vorontsova & Knapp in Taxon 59: 1597, 2010)
    S. kitivuense Dammer in P.O.A. C: 353 (1895); C.H. Wright in F.T.A. 4, 2: 243 (1906); T.T.C.L.:
    583 (1949). Type: Tanzania, Lushoto District, Usambara, Kitivo, *Holst* 276 (GOET!, lecto.,
    designated by Vorontsova & Knapp in Taxon 59: 1597, 2010; GOET!, isolecto.)

S. englerianum Dammer in P.O.A. C: 353 (1895); C.H. Wright in F.T.A. 4, 2: 237 (1906).
   Type: Tanzania, Lushoto District, Gonja, *Holst* 4231 (W!, lecto., designated here [best
   material]; LE!, M, P!, isolecto.)
S. scheffleri Dammer in E.J. 38: 191 (1906). Type: Tanzania, Lushoto District, Usambara
   Mountains, Ngwelo, *Scheffler* 12 (E!, lecto., designated by Vorontsova & Knapp in Taxon
   59: 1599, 2010; BM!, GOET!, isolecto.)
S. kitivuense Dammer var. *glochidiatum* (Dammer) Bitter in F.R. Beih. 16: 111 (1923);
   T.T.C.L.: 583 (1949)
S. kitivuense Dammer subsp. *englerianum* (Dammer) Bitter in F.R. Beih. 16: 112 (1923);
   T.T.C.L.: 583 (1949)
S. kitivuense Dammer var. *scheffleri* (Dammer) Bitter in F.R. Beih. 16: 114 (1923); T.T.C.L.:
   583 (1949)

NOTES. *Solanum stipitatostellatum* includes variable montane populations distinguished by their
almost entire reddish leaves on fertile branches, large inflorescences, prickles small or absent,
and trichomes with stalks 0.1–0.4 mm long. The fruit is markedly elongate during
development and becomes spherical at maturity. The corolla size ranges between 1.8 and 3 cm
in diameter; the majority of specimens have a pedunculate inflorescence with 4–10 flowers.
   Care must be taken with identification of this species. Most *S. stipitatostellatum* herbarium
material is currently held under *S. kitivuense*, *S. englerianum* and *S. glochidiatum*. Specimens
annotated as "sp. near *stipitatostellatum*" are often attributable to the smaller-leaved savanna
species *S. taitense*. *S. stipitatostellatum* has some distribution overlap with *S. taitense* and both taxa
have stalked trichomes, but *S. stipitatostellatum* has mature leaves 6–13 × 3–7 cm, and > 3 (up
to 10) flowers per inflorescence, while *S. taitense* has smaller leaves and fewer flowers per
inflorescence. *Solanum stipitatostellatum* is a larger plant than *S. zanzibarense*, with larger flowers
and long-stalked trichomes, and occurs at higher altitudes than the coastal *S. zanzibarense*.

42. **Solanum taitense** *Vatke* in Linnaea 43: 327 (1882); C.H. Wright in F.T.A. 4, 2: 226
(1906); T.T.C.L.: 580 (1949); Blundell, Wild Fl. E. Africa: 191 (1992); K.T.S.L.: 582
(1994). Type: Kenya, Teita District, between Ndi and Tsavo River, *Hildebrandt* 2605
(B†, holo.; GOET!, lecto., designated by Vorontsova & Knapp in Taxon 59: 1600, 2010)

Erect or scandent subshrub, 0.3–1.5 m, usually unarmed; young stems densely
stellate-pubescent, trichomes porrect, reddish-translucent or yellow-translucent,
stalked, stalks up to 0.1 mm, rays ± 8, 0.1–0.2 mm, midpoints shorter than rays;
prickles curved, up to 1(–2) mm long, 0.2–0.5 mm wide at base. Leaf blades drying
concolorous to weakly discolorous, yellow-green to red-green, ovate to elliptic,
1.2–3(–3.5) × 0.6–1.3 cm, 2–3 times longer than wide, base rounded to obtuse,
sometimes cordate, usually equal, margin entire to subentire, apex rounded; densely
stellate-pubescent on both sides, trichomes on abaxial surface porrect, stalked, stalks
0.1–0.15 mm, rays ± 8, 0.1–0.25 mm, midpoints ± same length as rays or shorter,
adaxially with thick stalks and reduced rays; primary veins 3–5 pairs; petiole 0.3–0.8 cm,
$^1/_4$–$^1/_6$ of the leaf length. Inflorescences not branched, 1.2–2 cm long, with 1–2(–3)
flowers; peduncle absent; rachis absent; peduncle and rachis unarmed; pedicels
0.7–1 cm long, in fruit 1.3–1.6 cm long, unarmed. Flowers perfect, 5-merous. Calyx
4–5 mm long, lobes deltate, 2–3.5 mm long, acute to acuminate, unarmed. Corolla
white to mauve, 1–1.7 cm in diameter, lobed for ± $^1/_5$ of its length, lobes lanceolate to
oblong, 4.5–7 × 1.5–2 mm. Stamens equal; anthers 4–6 mm. Ovary densely stellate-
pubescent in upper $^1/_5$; style 6–10 mm long. Berries 1–2 per infructescence, red at
maturity, spherical, 6–10 mm in diameter. Seeds ± 3 × 2.6 mm.

KENYA. Kitui District: Mutha camp, 24 Jan. 1942, *Bally* 1623!; Masai District: Chyulu Plains,
   Nongiyiaa Kopjes–Kuku, 23 Apr. 2000, *Luke & Luke* 6216!; Tana River District: Kurawa, 48 km
   south of Garsen, 23 Sep. 1961, *Polhill & Paulo* 541!
TANZANIA. Mwanza District: a few miles from Nyliakunga on Mwanza road, 23 Jun. 1945,
   Rensburg 42!; Pare District: Mkomazi Game Reserve, Kamakota to Kifukua, 13 Jun. 1996,
   Abdallah et al. 96/216!
DISTR. **K** 3–4, 6–7; **T** 1, 3; not known elsewhere
HAB. Bushland and grassland on sandy or black clay soil; 0–1500 m

NOTES. *Solanum taitense* is an unremarkable shrub recognised by its apically rounded subentire leaves, sparse inconspicuous prickles under 1(–2) mm long, frequent branching, rarely more than 2 flowers per inflorescence, and relatively small corollas and short anthers. Trichomes on the adaxial sides of the leaf have short, dilated stalks, and rays are reduced in size and point upwards.

Solanum taitense occurs within the distribution range of *S. hastifolium* and identification can be difficult due to the variable nature *of S. hastifolium*. No single morphological character separates all the specimens, but *Solanum taitense* has the majority of the following: entire to subentire leaves, inconspicuous prickles up to 1(–2) mm long, 1–2(–3) flowers per inflorescence, corolla 1–1.7 cm in diameter.

43. **Solanum usambarense** *Bitter & Dammer* in F.R. Beih. 16: 40 (1923); T.T.C.L.: 584 (1949). Type: Tanzania, Lushoto District: Handei near Kwa Mshuza, *Holst* 8925*a* (K!, lecto., designated here [best material]; W, Z, isolecto.)

Shrub, 1–2.5 m, erect, armed; young stems densely stellate-pubescent, trichomes porrect, translucent, sessile, rays 7–8, 0.1–0.2 mm, midpoints 1–2 mm, occasionally reduced to a gland; prickles straight, up to 4 mm long, 1–2 (to 4 on older stems) mm wide at base, round or flattened, inconspicuous. Leaf blades drying strongly discolorous, the youngest shoots a distinctive orange, ovate, 9–22 × 7–16 cm, ± 1.5 times longer than wide, base cuneate to almost cordate, usually unequal or oblique, margin lobed, the lobes 2–5 on each side, broadly deltate, 1–2 cm long, apically rounded, extending $\frac{1}{4}$–$\frac{1}{3}$($\frac{1}{2}$) of the distance to the midvein; apex acute; densely stellate-pubescent on both sides; trichomes on abaxial surface porrect, sessile, rays ± 8, 0.15–0.3 mm, midpoints 0.5–1.5 mm long, adaxially with reduced rays, midpoints to 1–1.8 mm; primary veins 4–5 pairs; petiole 1.5–4.5 cm, $\frac{1}{5}$–$\frac{1}{6}$ of the leaf length. Inflorescences branched more than once, 5.5–8 cm long, with 20–60 flowers, peduncle 10–30 mm long; rachis 3–6 cm long; peduncle and rachis unarmed; pedicels 0.7–1.1 cm long, in fruit 0.8–1.3 cm long, unarmed or with up to 10 prickles. Flowers perfect, 5-merous. Calyx 3–4 mm long, lobes deltate, 0.5–1.5 mm long, acute, unarmed or with 1–4 prickles. Corolla white to mauve, 0.9–1.2 cm in diameter, lobed for ± $\frac{2}{3}$ of its length, lobes deltate, 4–5 × ± 2 mm. Stamens equal; anthers ± 3.5 mm. Ovary with a few stellate trichomes; style 4.5–6 mm long. Berries 15–40 per infructescence, evenly green when young, red at maturity, spherical, 5–8 mm in diameter; fruiting calyx usually unarmed. Seeds 2.2–3 × 2–2.5 mm.

KENYA. Northern Frontier District: Mount Nyiru, Mbarta Forest Zone, 29 Mar. 1995, *Bytebier et al.* 75!; Nakuru District: 6 km E of Londiani along the Kericho road, 17 Nov. 1967, *Perdue & Kibuwa* 9111!; Kiambu District: Muguga, 4 Sep. 1965, *Kokwaro & Kabuye* 334!
TANZANIA. Mbulu District: Olodare Northern Highlands Forest Reserve, 30 Dec. 1962, *Newbould* 6479!; Lushoto District: Shagai forest, 17 Mar. 1954, *Willan* 116!
DISTR. **K** 1, 3–4, 6; **T** 2–3; not known elsewhere
HAB. Forest understorey, forest edges or clearings; 1800–2200 m

NOTES. *Solanum usambarense* is a hirsute mountain shrub with dense many-branched inflorescences with many flowers; it can be distinguished from *S. anguivi* by its inflorescence axis branching more than once (versus not branching or branching only once in *S. anguivi*), more than 20 flowers per inflorescence (versus less than 20 in *S. anguivi*), and recurved, visibly hirsute pedicels (versus recurved or mostly straight, with or without visible indumentum in *S. anguivi*). The two species are sympatric throughout the distribution range of *S. usambarense*.

Plants with forked inflorescences and trichomes with long midpoints occur in higher altitude populations across Africa, but multiple branching inflorescences with numerous flowers are clearly associated with north Tanzanian and Kenyan mountains. Collections from other Eastern Arc mountains display a continuous range of variation with no identifiable cutoff point between *S. usambarense* and *S. anguivi*. Further information on the boundary between *S. usambarense* and *S. anguivi* is impossible to obtain from herbarium sheets that only preserve one branch as it is possible that inflorescence branching and flower number varies within individuals and between individuals growing under different environmental conditions.

44. **Solanum usaramense** *Dammer* in P.O.A. C: 353 (1895); T.T.C.L.: 581 (1949); Gonçalves in F.Z. 8(4): 100 (2005). Type: Tanzania, Usaramo, Bunha, *Stuhlmann* 7066 (B†, holo.; GOET!, lecto., designated by Vorontsova & Knapp in Taxon 59: 1600, 2010)

Prostrate or climbing subshrub to 3 m, armed; young stems densely stellate-pubescent, trichomes porrect, white-translucent, stalked, stalks (0.1–)0.2–0.3(0.4) mm, rays (5–)6–8, 0.1–0.4 mm, midpoints short or up to 0.2(–0.5) mm; prickles ± 1(–3) mm long, 0.2–0.5 mm wide at base, curved, flattened. Leaf blades drying strongly discolorous, green-brown or yellow-green, ovate to elliptic, 3–8 × 1.5–4 cm, ± 2 times longer than wide, base cuneate to rounded, often unequal, margin almost entire to weakly lobed, the lobes 1–3 on each side, up to 5 mm long, broadly rounded, apically rounded, extending up to ¹/₃ of the distance to the midvein; apex rounded to acute; densely stellate-pubescent on both sides, trichomes on abaxial surface porrect, stalked, stalks 0.1–0.3 mm, rays ± 8, 0.15–0.3 mm, midpoints up to 0.1 mm long, adaxially with thicker stalks and reduced rays and midpoints; primary veins 3–5(–8) pairs; petiole 0.5–2 cm, ¹/₃–¹/₅ of the leaf length. Inflorescences not branched, 3–5 cm long, with 2–10 flowers; peduncle 0–5(–10) mm long; rachis 0.5–4.5 cm long; peduncle and rachis with 2–20 prickles; pedicels 0.6–0.8(–1.5) cm long, in fruit 1.2–1.7(–2.5) cm long, with up to 20 prickles. Flowers perfect, (4–)5-merous. Calyx 3–9 mm long, lobes deltate, 1–6 mm long, acute to acuminate, with up to 30 prickles. Corolla mauve to purple, 2.5–3 cm in diameter, lobed for ± ¹/₅ of its length, lobes narrowly ovate to deltate, 7–15 × 2–5 mm. Stamens equal; anthers 6–10 mm. Ovary glabrous; style 10–17 mm long. Berries 2–5 per infructescence, red at maturity, spherical, 8–11 mm in diameter. Seeds 2.5–3.5 × 2.5–3 mm.

KENYA. Kwale District: Diani Forest, 11–13 July 1972, *Gillett & Kibuwa* 19892!
TANZANIA. Tanga District: Bomalandani [Bomandani] 13 km S of Moa, 5 Aug. 1953, *Drummond & Hemsley* 3661!; District unclear, Ngambaula Forest Reserve, 22 Aug. 2000, *Mhoro* UMBCP 414!; Zanzibar: Kisim Kazi, 12 Jan. 1931, *Vaughan* 1833!
DISTR. **K** 7; **T** 3, 6; Mozambique
HAB. Coastal bushland, thickets, disturbed places; 0–500 m

SYN. *S. filicaule* Dammer in E.J. 48: 259 (1912). Type: Mozambique, Delagoa Bay, *Schlechter* 12168 (GOET!, lecto., designated by Vorontsova & Knapp in Taxon 59: 1596, 2010)

NOTES. *Solanum usaramense* is a distinctive climber, easily recognised by its dense covering of small uniform hooked prickles, strongly discolorous leaves, fairly big flowers, and trichomes with stalks 0.2–0.3 mm long. It spends the dry season as unremarkable-looking spiny twigs with almost no leaves or flowers. Anecdotal evidence suggests it is one of the first species to be eaten by goats. The name *Solanum monotanthum* Dammer has been erroneously applied to populations of *Solanum usaramense* in coastal Kenya and Tanzania. Type material of *S. monotanthum* found in GOET demonstrates the name *S. monotanthum* is actually a synonym of *S. zanzibarense* Vatke.

45. **Solanum zanzibarense** *Vatke* in Linnaea 43: 326 (1882); C.H. Wright in F.T.A. 4, 2: 230 (1906); T.T.C.L.: 584 (1949); K.T.S.L.: 583 (1994); Gonçalves in F.Z. 8(4): 99 (2005). Type: Tanzania, Zanzibar, Kidoti, *Hildebrandt* 988 (BM!, lecto., designated here [best material]; K!, W, isolecto.)

Prostrate or climbing subshrub to 4 m, armed; young stems stellate-pubescent, trichomes porrect, translucent to orange-brown, sessile or stalked, stalks up to 0.1 mm, rays 7–8, 0.1–0.15 mm, midpoints short or reduced to glands; prickles straight or curved, 2–4 mm long, 0.8–2 mm wide at base, round or flattened. Leaf blades drying usually concolorous, yellow-green to red-green, ovate to lanceolate, 3–14 × 1.5–7.5 cm, 2.5–3 times longer than wide, base cuneate, often unequal or oblique, margin lobed, sometimes entire, lobes 2–3 on each side, up to 15 mm long, rounded to deltate (obovate), apically rounded to obtuse, extending up to ¹/₂(–²/₃) of the distance to the midvein; apex acute to obtuse; stellate-pubescent to glabrescent on both sides, trichomes on abaxial surface porrect, stalked, stalks to 0.1 mm, rays (4–)6–8, 0.1–0.2 mm, midpoints short or reduced to glands, adaxially glabrescent, rays reduced to (2–)4–8, often less than 0.1 mm long, sometimes rays almost invisible with only brown globular glands seen; primary veins 4–6 pairs; petiole 0.5–4 cm, $\frac{1}{3}$–$\frac{1}{6}$ of the leaf length. Inflorescences not branched, 2–5 cm long, with 2–10 flowers; peduncle 0–5 mm long; rachis 0–2.5 cm long; peduncle and rachis usually unarmed; pedicels 0.3–0.7 cm long, in fruit 1.2–1.8 cm long, usually unarmed. Flowers perfect, 4–5-merous. Calyx 2–4 mm long, lobes deltate, 0.5–1.5 mm long, acute to acuminate, usually unarmed. Corolla white to mauve, 1.5–2 cm in diameter, lobed for ± $\frac{1}{5}$ of its length, lobes narrowly ovate to narrowly deltate or oblong, 6–8(–12) × 1.5–2 mm. Stamens equal; anthers 4–6.5 mm. Ovary glabrous or with minute simple hairs; style 8–12 mm long. Berries 1–4(–10) per infructescence, red at maturity, spherical, 8–14 mm in diameter. Seeds 1.8–2.5 × 1.5–2.3 mm. Fig. 22, p. 187.

KENYA. Kilifi District: Arabuko-Sokoke Forest Reserve, Jilore, 25 Nov. 1961, *Polhill & Paulo* 850!
TANZANIA. Bagamoyo District: Kikoka Forest Reserve, Apr. 1964, *Semsei* 3801!; Masasi District: 8 km NE of Masasi, Masasi Hill, 15 March 1991, *Bidgood et al.* 2010!; Zanzibar: NE of Mkunduchi, 27 Nov. 1930, *Greenway* 2592!
DISTR. **K** 7; **T** 6, 8; **Z**; Mozambique
HAB. Moist or dry forest, forest edges, rocky outcrops, on sand or sandy loam; 0–700 m

SYN. *S. vagans* C.H. Wright in K.B. 1894: 128 (1894). Type: Tanzania, Uzaramo District: Dar es Salaam, May 1879, *Kirk* s.n. (K! [K000413974], lecto., designated here [best material]; K!, isolecto.)
    S. monotanthum Dammer in E.J. 28: 474 (1900); C.H. Wright in F.T.A. 4, 2: 241 (1906); T.T.C.L.: 584 (1949). Type: Tanzania, Ulanga District: Ukutu, *Goetze* 113 (GOET!, lecto., designated by Vorontsova & Knapp in Taxon 59: 1598, 2010)
    S. alloiophyllum Dammer in E.J. 48: 247 (1912); T.T.C.L.: 581 (1949). Type: Tanzania, Lindi District, Kitunda, opposite Lindi, *Busse* 2393 (BM!, lecto., designated by Vorontsova & Knapp in Taxon 59: 1598, 2010; BR, EA!, GOET!, P!, isolecto.)
    S. praematurum Dammer in E.J. 48: 258 (1912). Type: Tanzania, Zanzibar, Jambiani, *Stuhlmann* 124 (B†, holo.)
    S. zanzibarense Vatke var. *vagans* (C.H. Wright) Bitter in F.R. Beih. 16: 116 (1923); T.T.C.L.: 584 (1949); Gonçalves in F.Z. 8(4): 99 (2005)
    S. zanzibarense Vatke subsp. *praematurum* (Dammer) Bitter in F.R. Beih. 16: 117 (1923); T.T.C.L.: 584 (1949)
    S. zanzibarense Vatke var. *abbreviatum* Bitter in F.R. Beih. 16: 118 (1923); T.T.C.L.: 584 (1949). Type: Tanzania, Uzaram, Bunbe (loc. not found), *Stuhlmann* 7026 (no specimens found)
    S. alloiophyllum Dammer subsp. *machisuguense* Bitter in F.R. Beih. 16: 120 (1923). Type: Mozambique, Machisugu, *Schlechter* 12119 (GOET!, lecto., designated by Vorontsova & Knapp in Taxon 59: 1594, 2010)

NOTE. *Solanum zanzibarense* can be recognised by its small flowers with long narrow corolla lobes, scrambling or climbing habit, slightly recurved or sometimes straight prickles over 2 mm long, and trichomes with very short stalks and short midpoints. This species includes a wide range of interbreeding populations across coastal East Africa from Kenya to Mozambique, encompassing a variety of leaf shapes and prickle morphologies, stalked-trichome lobed-leaved populations from Zanzibar (*S. zanzibarense* sensu stricto) and glabrescent large-leaved populations from Kenyan and Tanzanian coast (often annotated as *S. vagans*). None of the distinguishing characters can alone separate all specimens, but the combination of leaf size and shape, trichome stalk length, calyx lobe shape and flower size correlates with geographical distribution. Like many similar climbing spiny *Solanum* species treated here, *S. zanzibarense* occurs sporadically over a large geographical area but can be difficult to find outside its flowering season.

46. **Solanum jubae** *Bitter* in E.J. 54: 501 (1917); Chiovenda in Bull. Soc. Bot. Ital.: 106 (1925) & in Fl. Somal.: 238 (1929) & in Fl. Somal. 2: 333 (1932); Polhill, *Solanum* in E & NE Africa: 31 (ined., 1961); E.P.A.: 870 (1963); Jaeger, Syst. stud. *Solanum* in Africa: 364 (1985, ined.); K.T.S.L.: 581 (1994); Friis in Fl. Somalia 3: 212 (2006) & in Fl. Eth. 5: 128 (2006). Type: Somalia, "Djuba Steppe, Elmeged," *Keller* s.n. (Z, syn.) & Webi Suabeli (?Schebeli), *Keller* s.n. (Z, syn.)

FIG. 22. *SOLANUM ZANZIBARENSE* — habits showing variation: **1**, weakly lobed leaves and curved prickles; **2**, subentire leaves and curved prickles; **3**, lobed leaves and straight prickles; **4**, long-styled flower; **5**, abaxial leaf surface sessile trichome; **6**, abaxial leaf surface stalked trichome. 1, 4–5 from *Drummond & Hemsley* 3897; 2 from *Torre* 6287; 3 from *Richards* 17958; 6 from *Greenway* 2592. Drawn by Lucy T. Smith. Scale bar: 1, 2, 3 = 3 cm; 4 = 1.5; 5, 6 = 0.4 mm.

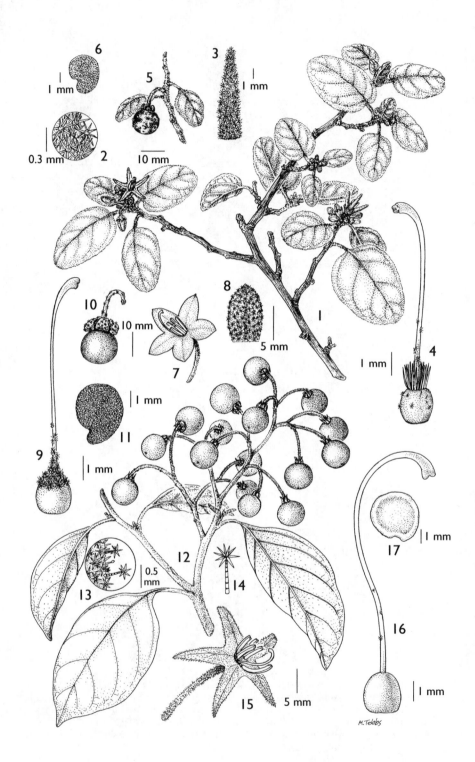

Shrub, spreading, straggling or much branched from the base to 3 m high, unarmed; young stems smooth, reddish-brown, stout at maturity, often lenticellate and bearing prominent leaf scars, with dense yellowish tomentose indumentum of 8-equal-rayed eglandular stellate hairs when young, glabrescent. Leaves often clustered around short shoots on the stem and opposite in pairs or threes, rough, yellowish to light green, orbicular to ovate, 1–2.4(–4.5) × 0.5–2.7 cm, bases cuneate, margins entire, apices rounded to obtuse, rarely acute; both surfaces with tomentose indumentum of stellate hairs larger than those on stems but with shorter erect central rays, bulbous basally, hairs densely interlocking on lower surfaces, on upper surfaces lamina visible between hairs; petioles 2–10 mm long. Inflorescences terminal up to 8-flowered umbellate cymes, or lateral when on short or long shoots and solitary or 2–3-flowered cymes; peduncles absent or vestigial; pedicels erect and 2–11 mm long in flower, recurved and 2.5–12 mm in fruit, stellate-tomentose. Calyx campanulate, 2–4(–5) mm long, stellate-tomentose externally; calyx lobes triangular (ovate), 1–2(–3.5) × 1–1.5(–2) mm, acute to apiculate (obtuse); adherent becoming reflexed in fruit, (1–)2.5–4 × 1.2–3 mm, sometimes unequal. Corolla pale lilac to purple and cream, stellate, 16–28 mm diameter; tube 1.5–2 mm long, glabrous externally; lobes narrowly lanceolate, 5–11 × 1.6–3 mm, densely stellate-pilose externally, glabrous internally, strongly reflexed exposing androecium after anthesis. Stamens usually equal and connate; filaments free for ± 0.5 mm, glabrous; anthers yellow to brownish, poricidal, often with orange pores, 4.6–7.6 × 0.6–1.4 mm, free and often spreading at maturity. Ovary brownish, 1.2–2 × 1–1.4 mm, glabrous to stellate-pubescent, occasionally mixed with simple hairs; style often curved apically, 6–11 × 0.2–0.5 mm, hairy on lower half with a dense collar of mainly simple hairs at junction with ovary, always exserted 2–5 mm; stigma capitate, 0.4–1 mm diameter. Berries mottled green and white maturing to yellow/orange, globose to ovoid, 4–9 × 5–11 mm, with scattered stellate hairs, sometimes with unpleasant smell. Seeds 7–22 per berry, light brown to golden yellow, ovoid, discoid or reniform, 2.1–3 × 1.8–2.8 mm, foveolate; sclerotic granules absent. Fig. 23/1–6, p. 188.

KENYA. Northern Frontier District: NE of El Wak, 11 Dec. 1971, *Bally & Smith* 14552! & 40 km on the El Wak–Wajir Road, 29 Apr. 1978, *Gilbert & Thulin* 1189! & 30 km on Ramu–Malka Mari Road, 8 May 1978, *Gilbert & Thulin* 1544!
DISTR. **K** 1; Ethiopia, Somalia
HAB. *Acacia-Commiphora* woodland or mixed bushland, often on steep slopes, limestone ridges and valleys; 350–500 m

NOTE. Specimens typically exhibit open flowers in one-, two- or three-flowered cymes. However, many juvenile terminal inflorescences have up to 8 flower buds, and it is possible that not all of these develop to maturity. Herbarium material of this species is generally poor, often consisting of a few leaf fragments on bare woody stems; these are however characteristically tough and multi-angular with smooth reddish bark. Thulin (2006) noted that the syntypes of *S. jubae* cited by Keller are from present day Ethiopia and not Somalia. This species is grazed by goats in Somalia.

FIG. 23. *SOLANUM JUBAE* — **1**, flowering habit; **2**, leaf indument; **3**, sepal indument; **4**, gynoecium. **5**, fruit on recurved pedicel; **6**, seed. *S. PAMPANINII* — **7**, open flower; **8**, sepal indument; **9**, gynoecium; **10**, fruit; **11**, seed. *S. SOMALENSE* — **12**, infructescence habit; **13**, young stem indument; **14**, multicellular stalked stellate hair; **15**, complete flower; **16**, gynoecium; **17**, seed. 1–4 from *Gilbert & Thulin* 1544; 5–6 from *Gilbert & Thulin* 1189; 7–9 from *Gilbert* 24912; 10–11 from *Gillespie* 120; 12–14 from *Faden & Faden* 74/797; 13–16 from *Makin* 13054; 17 from *Kirika* 365. Drawn by Margaret Tebbs.

47. **Solanum pampaninii** *Chiov.*, Result. Sc. Miss. Stefan.-Paoli, Coll. Bot.: 128 (1916); Chiovenda in Bull. Bot. Ital.: 107 (1925) & in Fl. Somal. 2: 331 (1932); Polhill, *Solanum* in E & NE Africa: 30 (1961, ined.); E.P.A.: 875 (1963 & 1974); Jaeger, Syst. stud. *Solanum* in Africa: 366 (1985, ined.); K.T.S.L.: 582 (1994); Friis in Fl. Somalia 3: 212 (2006). Type: hills of Giumbo, *Paoli* 166 (FT, syn.; photo.!); thickets near Mogadishu, *Paoli* 72 (FT, syn.; photo.!) & 91 (FT, syn.; photo!)

Shrub, scrambling or climbing to 2 m high or creeping, unarmed; stems often lenticellate, young stems flexuose, yellowish and densely pubescent with short stellate hairs of 8 equal eglandular rays to 0.3 mm long with a central ray of equal size arising from bulbous centre, glabrescent. Leaves often alternate, membranaceous, greyish- to dark green, ovate, 2–5.8 × 1.4–4 cm, bases cordate, subcordate or cuneate when sometimes oblique, margins entire, sometimes slightly sinuate below, apices obtuse; densely pubescent on both surfaces when young, becoming moderate to sparse with maturity, hairs as on stems; petioles 1–2.5 cm long. Inflorescences terminal to subterminal, sometimes on short shoots, 1–2-flowered; peduncles 0–4 mm; pedicels axillary if peduncles absent, erect and 13–25 mm long in flower, often strongly recurved and 15–22 mm long in fruit; axes stellate-pubescent. Calyx campanulate, 6–13 mm long, stellate-pubescent externally and internally in flower and in fruit; lobes ovate to obovate with prominent median veins, 4.5–9.5 × 3–5 mm, acute to mucronate; adherent becoming reflexed in fruit when 5–6.5 × 3–4.5 mm. Corolla blue, purple or mauve and yellow centrally, campanulate/stellate, 3–5.8 cm diameter; tube to 1–1.5 mm long; lobes broadly ovate, 1–2 × 0.6–1.6 cm, acute, stellate-pubescent externally, hairs confined to veins internally, spreading after anthesis. Stamens slightly unequal; filaments free for 0.8–1.25 mm, glabrous; anthers yellow to brown, 6.1–7.8 × 1.2–1.7 mm. Ovary 1.6 × 1.4–1.6 mm, upper part with stellate hairs forming a dense apical collar, glabrous below, bilocular; style curved apically, with a few scattered stellate hairs towards base, 9–11.5 × 0.2–0.5 mm, exserted 3.8–5 mm; stigma bilobed to clavate, 0.4–0.8 mm diameter. Berries smooth, orange to red, glossy, globose to ovoid, 1.3–1.5 × 1.1–1.5 cm, sometimes with scattered stellate hairs. Seeds 15–25 per berry, yellow to light brown, ovoid to orbicular, 4–5 × 3.2–3.8 mm, foveolate; sclerotic granules absent. Fig. 23/7–11, p. 188.

KENYA. Lamu District: Kiunga Point, 88 km NE Lamu, 3 July 1961, *Gillespie* 120! & Kui Island, June 1956, *Rawlins* 25! & Mkokoni sand dunes, Sept. 1956, *Rawlins* 156!
DISTR. **K** 7; Somalia
HAB. Coastal sand dunes and shores, cliff tops amongst scrub, may be locally common; sea-level to 9 m

SYN. ?*S. mesadenium* Bitter in E.J. 54: 501 (1917). Type: Somalia, Djuba Prov., Gobwin, *Ellenbeck* 2355 (B†, holo.)

NOTE. All East African specimens identified as this species are characterised by a dense indumentum of short equal-rayed eglandular stellate hairs, which Friis (2006) also pointed out. However, Chiovenda's (1916) protologue of *S. pampaninii* described the stem indumentum as being composed of stellate hairs with longish terminal branches mixed with simple hairs. Later however, Chiovenda (1925) specifically described this species (with which he then synonymised *S. mesadenium*) as having short glandular median-rayed hairs. I cannot comment on this as I have not seen the type material of *S. pampaninii*.
    Chiovenda (1925, 1932), Polhill (1961), Jaeger (1985) and Friis (2006) all regarded *S. mesadenium* as a synonym of *S. pampanini*. According to Jaeger (1985) Bitter was unaware that Chiovenda had already described *S. pampaninii* when he described *S. mesadenium*. However, Bitter's protologue of the latter described the median ray of the stellate hairs as being glandular though not elongated, whereas those found in *S. pampaninii* sensu stricto are eglandular. Clearly the location and examination of duplicate type material is necessary to clarify the synonymy of this species.
    There are a few Somalian species which other authors have variously cited as being synonyms or possible synonyms of *S. pampaninii*. These include:

• *S. benadirense* Chiov., Result. Sc. Miss. Stefan.-Paoli, Coll. Bot.: 126 (1916) & in Bull. Soc. Bot. Ital.:106 (1925); Friis in Fl. Somalia 3: 211 (2006); Friis in Fl. Eth. 5: 128 (2006); Jaeger,

Syst. stud. *Solanum* in Africa: 366 (1985, ined.). Type: Boscaglia near Mogadishu, *Paoli* 68 (FT, syn., photo!); dunes of Mogadishu, *Paoli* 41 (FT, syn., photo!)

Polhill (1961) and Jaeger (1985) thought that *S. benadirense* was a synonym of *S. pampaninii*, but Friis (2006) considered this to be a distinct species and he will have examined the type material. Vegetative and floral measurements taken from a photograph of the syntype *Paoli* 68 indicate that this species is closely allied to *S. pampaninii*. The protologue of *S. benadirense* described a stellate-tomentose indumentum of many- and short- rayed dense hairs and Chiovenda (1925) in his later key to the Somalian species placed both *S. benadirense* and *S. cicatricosum* (see below) in the eglandular-haired species group. However, Friis described the stellate indumentum of this species as having long central rays – presumably after examining the type material. Moreover, Friis also recorded subsessile anthers for this taxon though the protologue cited glabrous filaments of 3–4 mm. Clearly further examination of the type material is necessary to determine whether *S. benadirense* is a distinct species, a variant, or a synonym of *S. pampaninii*.

- *S. cicatricosum* Chiov. in Bull. Soc. Bot. Ital.; 106 (1925); Chiov., Fl. Somal.: 238 (1932); Jaeger, Syst. stud. *Solanum* in Africa: 367 (1985, ined.). Type: [fide Fl. Somal. 1932] Somalia, Garbauen–Durgale, [*Stefanini & Puccioni*] 420 (FT, syn.) & Scermarca-Hassan Tobungab, near Obbia, [*Stefanini & Puccioni*] 578 (FT, syn., fragment!)

Chiovenda's protologue is brief and lacks useful diagnostic characters though the key described the indumentum as being shortly stellate with subulate and acute rays and glandular hairs rarely present. Chiovenda did not cite any type specimens in 1925, but in 1932 cited two syntypes without collectors (see above). Jaeger (1985) thought that this species should probably be included in *S. pampaninii*. However, a leaf and small stem fragment of *Puccioni* & *Stefanini* 578 [633] had a dense tomentose indumentum of stellate hairs in which some long glandular headed central rays are visible. If verified by future examination of the complete type material, *S. cicatricosum* would be synonymised with the distinct species *S. robecchii* Bitter & Dammer (see below).

- *S. cicatricosum* Chiov. var. *gorinii* Chiov. *nom. invalid.*, based on *Gorini* 71 (FT; photos.!) and *Gorini* 88 (FT; photos.!) both from Kisimayu [Chisimayo], Somalia (cf. Friis 2006)

These specimens together with *Gorini* 80 (FT, photo!) seem to have been determined by Chiovenda, with a type label being attached only to *Gorini* 71. Friis considered this a synonym of *S. pampaninii* in Fl. Somalia and gave it as a synonym of *S. benadirense* Chiov. in Fl. Eth. Again examination of the actual type material is necessary before this variety of *S. cicatricosum* can be correctly placed.

- *S. robecchii* Bitter & Dammer in E.J. 54: 502 (1917); Chiovenda in Bull. Soc. Bot. Ital.: 107 (1925); Polhill, *Solanum* in E & NE Africa: 30 (ined., 1961); Jaeger, Syst. stud. *Solanum* in Africa: 365 (1985, ined.); Friis in Fl. Somalia 3: 211 (2006). Type: Somalia, Webi, *Robecchi-Bricchetti* s.n. (Herb. ROM., holo.).

This seems to be a good species which was recognised as such by Polhill (1961) and which is found in the Ogaden region of Ethiopia and Somalia. Plants identified as this species have a distinct stellate-haired indumentum in which the central ray is long (–2 mm), several-celled and usually terminates in a small gland. The juvenile parts appear densely tomentose with the long rays clearly visible to the naked eye. The floral parts are all smaller than those in *S. pampanini*. Friis (2006) thought this species a possible synonym of *S. benadirense*, but Jaeger (1985), thought that *S. robecchii* showed affinity to *S. jubae*.

The boundaries between these taxa are clearly very difficult to define; Friis (2006) noted that there is much variation in the indumentum and flower size in *S. pampanini* and *S. benadirense* and indeed the specimens that he cited in the Flora of Somalia exhibit a mixed indumentum. Herbarium collections of all of these taxa are relatively sparse and their definitive specific boundaries and correct taxonomic recognition require field work and more extensive collecting.

The 'cherry'-sized red fruits of *S. pampaninii* are eaten locally in **K** 7.

48. **Solanum robustum** *Wendl.* in Flora 27: 784 (1844); Sendtner in Fl. Bras. 10: Sp. 72, t. 5, fig. 38–48 (1846); Dunal in DC., Prodr. 13(1): 257 (1852); Bitter in F.R. 16: 182 (1923); T.T.C.L.: 587 (1949); Polhill, *Solanum* in E & NE Africa: 41 (ined., 1961); Seithe in E.J. 81: 325 (1962); Whalen in Gentes Herb. 12: 241 (1984); Jaeger, Syst. studies *Solanum* in Africa: 485 (1985, ined.); RHS Gard. Dict. 4: 319 (1997). Type: Brazil; type specimen not specified

Shrub or perennial herb to 5 m high; all stems ferrugineous with a dense indumentum of intertwined stalked stellate hairs to 1.5 mm diameter, up to 9-rayed with long central rays, stalks 0–2 mm long, mixed with sessile brown glands; main stems and branches with scattered stout sharp yellow pyramidal prickles to 1.5 × 1.8 mm. Leaves alternate or opposite, often thick and soft, dark green above, yellow to brown below with orange to brown veins and midrib, ovate, 12–27 × 9–24 cm, bases cordate and oblique, margins deeply sinuate-dentate with 2–4 broadly triangular acute lobes to 6 cm deep, apices acute; upper surfaces stellate-pubescent, hairs thicker and interwoven on midrib and main veins, lower surfaces densely stellate-pubescent, the interwoven hairs with unequal rays to 1.3 mm long; scattered prickles present on midribs and upper main veins; petioles 3–20 cm long, decurrent and with wings up to 1.5 cm wide often extending down to next node, with scattered prickles to 2 cm long below. Inflorescences subterminal to lateral, simple, few- to 20-flowered, lax, helicoid cymes; flowers ?andromonoecious, 5-merous, becoming lax in fruit; axes densely ferrugineous, hairs on stems mixed with small glandular hairs; peduncles erect and 1.2–4.5 cm long, with prickles; pedicels usually erect, sometimes recurved apically and 0.5–1.4 mm long in flower, spreading and 1–1.8 cm in fruit. Calyx green, campanulate/cupulate, 5–9 mm long, stellate-pubescent and with sessile glands externally and internally; lobes lanceolate, 5–9 × 1.8–3.4 m, acute, reflexed between corolla lobes, appressed becoming reflexed in fruit and 5–10 × 2–4.5 mm. Corolla white, with a contrasting basal star and median veins, stellate, 1.5–2.1 cm radius; tube ± 1 mm long, glabrous; lobes lanceolate to narrowly lanceolate, 7–15 × 2–3.5 mm, stellate-pubescent externally, acute with apical tufts of small hairs, glabrous internally except for occasional hairs on veins, corolla reflexed exposing androecium after anthesis. Stamens equal or unequal; filaments free for 1–3.5 mm, often varying in same flower, glabrous; anthers yellow to brownish, poricidal, 5–7 × 1–1.8 mm, connivent. Ovary 1.2–2.6 × 1.4–2.5 mm, pubescent with long appressed silky hairs, bilocular; style straight, *either* enclosed at base of staminal tube when 1.2–3.5 × 0.25–1 mm, *or* 9–10 × 0.4–0.6 mm when exserted up to 4 mm, pubescent with scattered stellate and small simple glandular hairs in lower half; stigma capitate, 0.2–1 mm diameter. Berries dark green becoming black, globose, 1.2–2 cm diameter, rusty tomentose with simple silky appressed hairs, glabrescent apically. Seeds > 100, dark brown, obovoid, reniform to orbicular, 1.7–2 × 1.2–1.4 mm, rounded not flattened, reticulate; sclerotic granules absent.

TANZANIA. Lushoto District: Western Usambaras, Magamba–Mkuzi road, 7 June 1953, *Drummond & Hemsley* 2874!; Morogoro District: Uluguru Mountains, Bunduki, 25 Jan. 1969, *Batty* 357! & same, Oct. 1930, *Haarer* 1868!
DISTR. **T** 3, 6; cultivated and now a successful escape which has become locally common; native to Brazil, Paraguay and Argentina, also recorded from Reunion and Java
HAB. Secondary vegetation, old cultivation, shamba edges, moist forest margins and clearings, plantations, roadsides; 500–2150 m

NOTE. Nee (in Nee *et al.* (eds), Solanaceae IV: 321 (1991)) commented that species belonging to the group/section *Erythrotrichum* constitute the most difficult of those in the subgenus *Leptostemonum*, with its species being difficult to differentiate from those in the section *Micracantha*.

Solanum robustum has been introduced into Africa from S America and is commonly known as the White Potato. Although cultivated as an ornamental this species has become a troublesome weed as an escape from cultivation in **T** 3 and **T** 6. Solitary plants of *S. robustum* have been reported but it often grows in groups which can form impenetrable masses of viciously spiny vegetation. These spines and the dense ferrugineous pubescence composed of varied and complex hairs characterise this species. Not only do stalks and rays of the stellate hairs vary considerably in length, but stalks can also be invested with small spreading multicellular glandular hairs. Plants of this species are reportedly used to treat gonorrhoea and worms in Lushoto.

49. **Solanum wrightii** *Benth.*, Fl. Hongkong: 243 (1861); Wright in K.B. 1914: 304 (1914); Heine in K.B. 14: 248 (1960) & in F.T.W.A., ed. 2: 335 (1963); Polhill, *Solanum* in E & NE Africa: 41 (ined., 1961); Seithe in E.J. 81: 330 (1962); F.F.N.R.: 377 (1962); Gbile in Biol. & Tax. Solanaceae: 119 (1979); Whalen in Gentes Herb. 12: 247 (1984); Jaeger, Syst. stud. *Solanum* in Africa: 456 (1985, ined.); Troupin, Fl. Rwanda 3: 382 (1985); Bukenya & Hall in Bothalia 18: 85 (1988); RHS Gard. Dict. 4: 320 (1997); Whalen in Nee *et al.*, Solanaceae IV: 321 (1999); Mansfeld, Encycl. Ag. & Hort. Crops 4: 1831 (2001); Gonçalves in F.Z. 8, 4: 71 (2005). Type: China, Hong Kong, *Wright* 489 (K!, holo.)

Large shrub or small tree to 12 m high with large flat crown; trunk becoming thick, dark brown, girth to 70 cm with smooth pale grey-brown bark, glabrescent; younger stems lanate/pubescent with spreading stalked (to 1.6 mm) stellate hairs mixed with scattered short simple eglandular hairs; shortly-stalked club-glands also present throughout; stems and branches with scattered to dense sharp stout yellow and sometimes recurved acute prickles to 10.5 × 4 mm, sometimes unarmed. Leaves alternate or opposite, often coriaceous, dark green above, yellowish below, ovate to ovate-lanceolate, 11–30(–60) × 6–25(–40) cm, bases usually cordate and distinctly unequal, margins entire, sinuate or deeply sinuate-dentate often on the same plant, with 0–4 acute or obtuse antrorse lobes up to 9 cm deep, apices acute; with scattered small prickles on lower midribs; upper surfaces strigose with simple bulbous-based eglandular hairs to 1 mm mixed with scattered club-glands especially on midrib where mixed with stalked stellate hairs; lower surfaces lanate with long-stalked stellate hairs to 1.25 mm diameter with 4–8 spreading rays with a short central ray; petioles 1.5–7.5 cm long, stellate-strigose; occasional prickles sometimes present. Inflorescences subterminal to lateral simple and few- to 50-flowered lax, helicoid cymes, with extended rachides in fruit, flowers andromonoecious, 5-merous; axes strigose with eglandular and glandular hairs mixed with long-stalked (–2.5 mm) stellate-hairs and stalked club-glands, becoming woody and glabrescent in fruit; peduncles erect and 1.4–6 cm long in flower and fruit; pedicel erect to recurved and (7–)11–21 mm long in flower, 10–26 mm in fruit. Calyx pale green, campanulate/cupulate, 1.1–1.8 cm long, densely glandular- and stellate-strigose, glabrous internally; lobes narrowly triangular, 9.5–15 × 2–3.5 mm, with membranous margins, subulate-acute and recurved; calyx tube enlarging, becoming thickened and raised in fruit forming a woody rim around the berry base surmounted by adherent triangular calyx lobes. Corolla blue, purple or lilac and fading with age often to white, usually with a contrasting basal star and median veins, stellate, 2.5–4.4 cm radius; tube 1–2 mm long, glabrous; lobes broadly triangular, 1.3–2.8 × 1.1–2.5 cm, acute with dense tufts of small hairs, with dense central bands of small sessile stellate hairs externally, ± glabrous internally except for median veins, corolla reflexed exposing androecium after anthesis. Stamens usually equal within each flower, occasionally unequal; filaments free for 2–3.5 mm, sometimes thick and tubular, glabrous; anthers yellow to brownish, poricidal, 10–16 × 1.2–2.2 mm long, connivent. Ovary 1.5–4 × 1.8–3 mm, glabrous, 4-locular; style straight *either* enclosed at base of staminal tube when 1.5–3 × 0.4–1 mm, *or* 6.5–18 × 0.7–1 mm and exserted up to 2 mm, white, glabrous but with small stipitate glands especially towards the base, sometimes tubular; stigma green, capitate, 0.3–1.5 mm diameter. Berries smooth, yellow to greenish-yellow (? to red), spherical, 4–5 cm diameter, glabrous glossy, leathery with thick pericarp. Seeds numerous (> 100), reddish- to dark brown, ovoid, obovoid or orbicular, 2.5–3.3 × 2–2.7 mm, rounded not flattened, deeply reticulate; sclerotic granules absent.

KENYA. NW Nairobi, 24 Jan. 1965, *Gillett* 16596! & Nairobi, Bell Stud (?), 3 July 1921, *Russell* 51541 from seed collected by *Shantz* 1141! & Nairobi Arboretum, 17 Mar. 1952, *Williams* 372! TANZANIA. Moshi District: Rau Forest Reserve near Rau river, SE of Moshi town, 23 Nov. 1983, *Macha* 376!; Arusha District, Tengeru, 4 Oct. 1960, *Kanywa* 30!; Lushoto District: Amani, 17 July 1930, *Greenway* 2277!

DISTR. **U** (cult., fide Bitter, 1923); **K** 4; **T** 1–3; probably native to the northern Andes of South America, but now widely cultivated throughout the tropics; Sierra Leone, Ghana, Nigeria, Cameroon, Gabon, Congo-Kinshasa, Rwanda, Malawi, Zambia and Zimbabwe

HAB. Widely cultivated in gardens and arboreta, an escape in riverine and evergreen forest; 800–1700 m

SYN. *S. macranthum* Carr. in Rev. Hort.: 132 (1867), *non* Dunal. Type: supposedly of Asiatic orgin, cultivated in Cairo, no specimen cited; Plate in Rev. Hort.!, lecto., designated here [The beautiful plate leaves little doubt that Carrière's new species is synonymous with *S. wrightii*]
  S. grandiflorum auctt., e.g. Bitter in F.R. 16: 180 (1923); T.T.C.L.: 586 (1949), *non* Ruiz & Pav. [Bitter's name based on Egypt, Alexandria, *Gaillardot* 365 (C); Tanzania, Uzaramo, *Holtz* 2524 (B†)]

NOTE. This species was originally described by Bentham (1861) from Hong Kong, but its native country is Bolivia. Bentham did note in his protologue that "no authentically wild specimens have ever been seen in China". *Solanum wrightii* is now frequently cultivated as an ornamental tree, largely for its showy flowers, in tropical regions of the world, where it often escapes becoming naturalised; it is commonly known as the "Potato-tree" or the Brazilean Potato Tree. It is also occasionally used for shade in Mexico. Bitter incorrectly united this species with *S. grandiflorum* Ruiz & Pav. without examining the type of the latter. This is a very different species characterised by a floccose indumentum, absence of prickles and deeply lobed corolla (White in F.F.N.R.: 377, 1962). The considerable confusion surrounding the synonymy of this species was discussed by Heine (1960).

Bentham (1861) originally mentioned that his new species seemed to belong to the *Melongena* group. Later authors agreed or followed this treatment including Jaeger (1985) & Bukenya & Hall (1988). Bitter (1923) included it in his series *Acanthocalyx* Bitter; Seithe (in E.J. 81: 330, 1962) tentatively included it in the section *Stellatipilum* while Whalen (1984) placed it in his *Solanum crinitum* group, later including it in the section *Crinitum* (Whalen) Childs (Whalen 1999). Gonçalves (2005) subsequently followed this infrageneric categorisation.

The species is reportedly strongly andromonecious with reduced and sterile gynoecia. All herbarium material examined exhibited large anthers and the majority had only short styles embedded within the staminal column, though the ovaries all appeared to bear numerous ovules. Occasional specimens exhibited long exserted styles. Only one or two fruits reportedly develop from each inflorescence, and these have been variously described as yellow or red but no annotations were found on herbarium specimens to clarify its berry colour. Symon (in J. Adelaide Bot. Gard. 8: 158 (1985)) recorded up to 258 seeds in one New Guinean fruit examined, though he followed other authors by identifying it as *S. grandiflorum* Ruiz & Pav.

Solanum wrightii is densely pubescent, exhibiting a mixture of stalked stellate hairs in which the reduced central ray may or may not be glandular; both long and short multicellular simple hairs which can be spreading or appressed and have glandular or eglandular heads, together with scattered stalked club glands. The stalked stellate hairs are often visible to the naked eye with the tubular and striated stalks varying in length up to 2 mm. As with other species, prickles can be present or absent on plants of *S. wrightii*; when present they are usually found on the stems, petioles and lower midribs.

Apart from its widespread ornamental use, this species is cultivated as a nematode resistant rootstock for *S. quitoense* in Ecuador, Colombia and Florida, with the fruits commonly being used as soap for washing clothes and medicinally for treating tumours (cf. Mansfeld, 2001). Its charcoal has been used as gunpowder when mixed with 'chora' in Tanzania (*Kanywa* 30).

50. **Solanum mammosum** *L.*, Sp. Pl. 1: 187 (1753); Don, Gen. Hist. Dichlam. Pl. 4: 435 (1837); Dunal, Synopsis: 41 (1816) & in DC., Prodr. 13(1): 250 (1852); Seithe in E.J. 81: 301, 321 (1962); D'Arcy in Ann. Miss. Bot. Gard. 60: 712 (1973); Nee in Biol. & Tax. Solanaceae: 576 (1979); Symon in J. Adelaide Bot. Gard. 4: 103 (1981); Whalen in Gentes Herb. 12: 253 (1984); Jaeger, Syst. stud. *Solanum* in Africa: 479 (1985, ined.); Bukenya & Hall in Bothalia 18: 83 (1988); Nee in Solanaceae III: 260, 264 (1991); Bukenya & Carasco in Bothalia 25: 51 (1995); RHS Gard. Dict. 4: 319 (1997); Mansfeld, Encycl. Ag. & Hort. Crops 4: 1832 (2001); Gonçalves in F.Z. 8, 4: 70 (2005). Type: "Habitat in Virginia, Barbados" *Solanum Barbadense spinosum, foliis villosis, fructu aureo rotundiore, Pyri parvi inversiforma et magnitude* in Plukenet, *Phytographia, pars tertia.* t. 226, fig 1 (1692), (lecto., designated by Knapp & Jarvis in J.L.S. 104: 344, 365 (1990) based on observations by Jaeger (1985)

Annual or short-lived perennial bushy herb or shrub to 2 m high, loosely branched; stems to 3 cm diameter, densely hirsute with simple spreading predominantly long and eglandular hairs to 2.5 mm, mixed with shorter glandular- and multi-headed hairs, occasionally mixed with shortly stalked stellate hairs and short-stalked club glands; stems and branches with scattered robust slightly recurved pyramidal yellow prickles to 1.5 × 6 mm, occasionally absent. Leaves usually alternate, darkish green, velvety, broadly ovate, 6–25 × 9–26 cm, bases cordate, margins deeply sinuate-dentate with 2–7 main broadly triangular antrorse acute lobes to 4.5 cm deep some of which biserrate, apices acute; straight acicular prickles to 2 cm long on midribs and main veins; both surfaces densely lanate-hirsute with appressed eglandular and erect glandular hairs as stems, denser below and on midribs and veins; petioles 3–16 cm long with scattered straight prickles to 2.5 cm long. Inflorescences lateral, simple, 1–5(–7)-flowered lax cymes. Flowers 5-merous; axes densely viscid/hirsute; peduncles erect, 0–1.5 cm long, becoming woody in fruit; pedicels erect to recurved and 5–11 mm long in flower, recurved and up to 2.2 cm in fruit when stout, glabrescent, with or without small yellow prickles to 2 mm long × 0.3 mm basally. Calyx dark green, campanulate, 6–8 mm long, densely hirsute externally, mixed with small prickles, glabrous internally; lobes usually equal, narrowly triangular becoming ligulate, (3.5–)5–8 × 0.5–2 mm, acute/acuminate, with scattered small prickles; adherent becoming reflexed in fruit, 6–11 × 2–5 mm. Corolla blue to purple, often with a yellowish internal star, stellate, 1.2–2 cm radius; tube ± 0.5 mm long, glabrous; lobes narrowly lanceolate, 12–16(–20) × 2.3–5 mm, acute, hirsute with long hairs externally becoming denser towards the apices, glabrescent internally, strongly reflexed after anthesis. Stamens usually equal; filaments free for 1.2–1.6 mm, glabrous; anthers poricidal, yellow to orange, 8.5–11 × 1.2–1.6 mm, connivent. Ovary brownish, 1.1–3.6 × 1–3.1 mm, ± glabrous, bilocular; style straight, 1–12 × 0.4–0.8 mm, glabrous, included within the staminal column; stigma capitate, sometimes bilobed, 0.6–0.8 mm diameter. Berries pendulous on recurved pedicels; only one or two developing on each inflorescence; smooth, yellow to orange, obovoid, 4.5–8.5 × 3.5–6.5 cm, dull with glabrous leathery pericarp and a characteristic nipple-like apex with up to 5 basal mammillae or protuberances when cultivated. Seeds numerous, smooth, greyish- to dark brown, obovoid to reniform, 2.5–3.8 × 2–3 mm, not winged or flattened, minutely punctuate to reticulate; sclerotic granules absent.

UGANDA. Teso District: Serere, Dec.1931, *Chandler* 203!
KENYA. Nairobi, 18 Mar. 1958, *Imbert* H64/58!
DISTR. U 3–4; **K** 4; native to the West Indies; cultivated as an ornamental and found in Ghana, Burundi, Angola, Malawi, Mozambique, Zimbabwe and South Africa; Panama, Guyana, Surinam, French Guiana, Colombia, Venezuela, Ecuador, Bolivia, Brazil and Peru; also cultivated in the US, India, Hong Kong, and Australia
HAB. Rare as a cultivar or on unclaimed land; 1050–1700 m

SYN. *S. cornigerum* Andre, Rev. Hortic.: 33 (1868). Type: cultivated at Hyères, France by Huber from seed believed to come from Africa (?K, holo., not found; BM! ?lecto in Herb. Shuttleworth)
   S. corniculatum Andre, Rev. Hort.: 33 (1868), *nom. nud.*
   S. platanifolium Sims in Bot. Mag. 53, t. 2618 (1926), *non* sensu Sendtner in Martius Fl. Bras., 10: 58 (1846). Type: Native of northern parts of South America; communicated in flower by *Mrs Walker* (type not located)
   Solanum sp. [Imbert H64/58] in Polhill, *Solanum* in E & NE Africa: 37 (ined., 1961)

NOTE. Native to the Caribbean where it is found on most of the West Indian islands, *S. mammosum* is now widely cultivated as an ornamental in many parts of tropical America where it is known as the Nipple Fruit. The fruits of *S. mammosum* are among the largest found in *Solanum*, and are unique through their characteristic protuberances. According to Nee (1979, 1991) only the lower flower in each inflorescence is fertile with its pedicel becoming reflexed. Certainly the stylar length varied from very short to long on the specimens examined though all were included within the stylar column.

The illustration of *S. platanifolium* leaves little doubt that this species is a synonym of *S. mammosum*; the fruits are immature and mottled green and white though the text described them as ripening yellow. Referring to its characteristic leaf shape, Hooker named his new species the Plane-tree Nightshade.

Andre included differences in anther and corolla morphology, together with the lack of fruit protuberances on Linnaeus' species *S. mammosum* as justification for describing his new species *S. cornigerum*. However, the illustration and description of the latter leave little doubt that these two species are synonymous.

The history and biogeography of this species together with its unique fruit shape development was fully discussed by Nee (1979). Though typically invested with sharp acicular spines, some plants without spines have been reported (cf. Nee 1979, 1991). The widespread use of *S. mammosum* as a cockroach poison is well-documented and according to Nee (1979) could account for its spread by tribes through tropical America. It is only rarely found in Africa where it has failed to become naturalised; the fruits are said to be poisonous in Ghana (Bukenya & Hall in Bothalia 18: 83, 1988). Mansfeld (2001) reported that the plant was a potential source of solasodine for medicinal use, and of glycoalkaloids for schistosomiasis control.

51. **Solanum aculeatissimum** *Jacq.*, Icon. Pl. Rar. 1: 5, t. 41 (1786) & in Collectanea 1: 100 (1787); Dunal, Synopsis: 41 (1816); Don, Gen. Hist. Dichlam. Pl. 4: 434 (1837); Sendtner in Fl. Bras. 10: Sp. 59, t. 10, fig. 10–16 (1846); Dunal in DC., Prodr. 13(1): 244 (1852) saltem pro parte excl. syn.; Clarke in Fl. Brit. Ind. 4: 237 (1885); P.O.A. C: 352 (1895); Wright in Fl. Cap. 4: 97 (1904) pro parte excl. syn. & in F.T.A. 4, 2: 228 (1906); Dammer in E.J. 38: 189 (1906); Bitter in F.R. 16: 148 (1923); Cat. Vasc. Pl. S. Tomé: 252 (1944); Robyns in F.P.N.A. 2: 214 (1947); T.T.C.L.: 580 (1949); Polhill, *Solanum* in E & NE Africa: 37 (ined., 1961); Seithe in E.J. 81: 301, 308 (1962); Watt & Breyer-Brandwijk, Med. & Poisonous Pl. S and E Africa, ed. 2: 990 (1962); Heine in F.W.T.A., ed. 2: 334 (1963); Verdcourt & Trump, Common Pois. Pl. E. Africa: 169, 171 (1969); Gbile in Biol. & Tax. Solanaceae: 115 (1979); Whalen in Gentes Herb. 12: 252 (1984); Jaeger, Syst. stud. *Solanum* in Africa: 478 (1985, ined.); Jaeger & Hepper in Solanaceae, Biol. & Syst.: 48 (1986); Bukenya & Hall in Bothalia 18: 83 (1988); Nee in Solanaceae III: 261, 265 (1991); Blundell, Wild Flowers E. Afr.: 189 (1992); U.K.W.F., 2nd ed.: 243 (1994); Bukenya & Carasco in Bothalia 25: 51 (1995); RHS Gard. Dict. 4: 317 (1997); Gonçalves in F.Z. 8, 4: 118 (2005); Friis in Fl. Eth. 5: 145 (2006). Type: cultivated in Vienna, *Jacquin* s.n. (W, holo., iso., photos!) fide D'Arcy in Ann. Miss. Bot. Gard. 60: 711 (1973) in adnot.

Annual or short-lived perennial herb or shrub to 3(–5) m high, branches often spreading, sometimes procumbent; mature stems brown, younger stems villous with spreading simple eglandular hairs to 2 mm mixed with short and long spreading purple glandular-headed hairs *and* shortly-stalked brown glands scattered throughout, glabrescent apart from scattered shortly-stalked glands; always covered with dense spreading or recurved acicular prickles to 6 mm long, often blackish. Leaves alternate, sometimes opposite above, membranaceous, dark green above, pale green below and often with prominent brown veins, broadly ovate to palmatifid, usually wider than long, 6–25 × 6–20 cm, bases cuneate to cordate, often oblique, margins deeply sinuate-dentate with 3–4 antrose acute biserrate lobes to 6 cm deep on either side, apices acute; both surfaces with green or purple acicular prickles on midribs and main veins, conspicuously hispid with long simple spreading eglandular hairs, denser on the midribs and veins, often mixed with glandular hairs and sometimes also with sessile stellate hairs; petioles 2–12 cm long, with scattered acicular prickles to 2 cm. Inflorescences lateral, extra-axillary, sessile, 2–4(–6)-flowered umbellate cymes; pedicels erect to spreading, 1–2.2 cm long in flower, with glandular- and eglandular hairs mixed with sessile glands, woody, recurved and 1–3 cm long in fruit, with scattered acicular prickles to 4.5 mm. Calyx green, campanulate to cupular, 4–7.5 mm long, densely sericeus externally with long eglandular and shorter glandular hairs, sessile glands and acicular prickles to 4.5 mm long, glabrescent internally; lobes narrowly triangular, (2.5–)3.5–6 × 1–2.5 mm, acute to acuminate, adherent becoming reflexed and 4–10 × 3–5 mm in fruit. Corolla white, cream, pale mauve to purple, often with purple veining and lobe tips, stellate, 1.6–2.6(–4) cm diameter, tube 1–2 mm long, glabrous; lobes lanceolate, 8–10(–14) × 3–5 mm, acute, glabrous to pubescent externally with eglandular and glandular hairs, glabrous internally, reflexed after anthesis. Stamens usually equal, sometimes slightly unequal; filaments white to pale yellow, free for 1–2.2 mm, glabrous; anthers yellow to brown, poricidal, 5.5–7 × 1.2–2 mm, usually connivent. Ovary brownish, 1.2–2 × 1.7–2 mm, covered with small glands, becoming glabrous, bilocular; style white, straight, linear, glabrous, 3 mm and included *or* 7–9.5 mm and exserted up to 2 mm; stigma green, capitate, occasionally clavate, 0.5–0.8 mm diameter. Berries smooth, usually conspicuously veined with light to dark green stripes when young, becoming uniformly cream to yellow when mature, globose to ovoid, 1.2–4 cm diameter, glabrous, pericarp leathery. Seeds often > 100 per berry, brown, obovoid, orbicular, 2–3.2(–4.5) × 1.8–3(–3.6) mm, not winged but rounded (not flattened) when fully viable, reticulate-foveolate; sclerotic granules absent.

UGANDA. Kigezi District: Kachwekano Farm, Dec. 1949, *Purseglove* 3155!; Acholi/Karamoja District, Imatong Mountains, Langia, Apr. 1943, *Purseglove* 1413!; NTB region (???), Mar. 1923, *Maitland* 681! (prickle-less variant)
KENYA. Northern Frontier District: Mt Nyiro, July 1960, *Kerfoot* 1987!; Elgeyo District: Kapsowar, Kipkunurr Forest Reserve, E end of Cherangani Hills, 14 Mar. 1975, *Hepper & Field* 4985!; N Nyeri District: off Naro Moru Road up to Mt Kenya, 16 Dec. 1979, *Jaeger* 2!
TANZANIA. Kilimanjaro, Narnuai Farm, 12 June 1994, *Grimshaw* 94/570!; Iringa District: near Kigogo River, Mufindi, 4 May 1968, *Renvoize & Abdallah* 1925!; Njombe District: Elton Plateau, 7 Jan. 1957, *Richards* 7550!
DISTR. U 1, 2; K 1–7; T 2–4, 6, 7; Sierra Leone, Liberia, Ivory Coast, Ghana, Nigeria, Cameroon, Equatorial Guinea, Congo-Kinshasa, Rwanda, Sudan, Ethiopia, Zambia, Malawi, Mozambique, Zimbabwe and South Africa; India, Singapore and Malaysia, tropical South America, southern USA, West Indies
HAB. Moist and evergreen dry forest (*Olea-Teclea*, *Podocarpus-Juniperus*, *Tabernaemontana-Neoboutonia*) including margins and clearings, *Acacia*-woodland, swamp edges, stream-sides, thickets, bushland, grassland, pasture; 1350–2750 m

SYN. *S. aculeatissimum* Jacq. var. *hispidissimum* Dunal in DC., Prodr. 13(1): 244 (1852); Dammer in P.O.A. C: 352 (1895), as *hispidissima* & in Z.A.E: 285 (1914) & in E.J. 38: 189 (1906); De Wildeman in Pl. Bequaert.: 419 (1922). Type: Cape of Good Hope, *Masson* s.n. in herb. Banks (BM!, holo.)
  S. angustispinosum De Wild. in B.J.B.B. 4: 397 (1914). Type: Congo-Kinshasa, Bolanda, *Nannan* 21 (BR, syn., photo!); Bomputu, *Seret* 1025 (BR, syn., photo!) & *Seret* 1099 (BR, syn., photo!: K!, isosyn.); between Buta and Bima, *Seret* 129 (BR, syn., photo !)
  S. aculeatissimum Jacq. var. *purpureum* A. Chev. in Expl. Bot. Afr. Occ. France: 460 (1920), *nom. nud.* Type: Ivory Coast, Cavally Basin, between Taté and Tabou, *Chevalier* 19830 (?P, holo.)
  S. aculeatissimum Jacq. var. *dolichoplum* Bitter in F.R. 16: 151 (1923). Type: Rwanda, Rugege Forest, Rukarara, *Mildbraed* 909 (?B†, holo.)

NOTE. Bitter (1923) placed *S. aculeatissimum*, along with three other species, into his new section *Simplicipilum* Bitter which he differentiated from species in the section *Leptostemonum* by their simple, long erect hairs [though these are often interspersed with scattered stellate hairs in specimens of *S. aculeatissimum*]. This species is thought to be native to Brazil (cf. Whalen (1984) and Nee (1991)). However, although Don (1837) recorded it as native to tropical America (and Ceylon) and Dunal (1852) to Mexico and Ceylon, Bitter (1923) along with many other authors (e.g. Polhill (1961), Heine (1963), Bukenya & Hall (1988), Gonçalves (2005) and Friis (2006)) considered the species to be native to southern Africa from where it has become widely distributed throughout tropical and southern Africa. Jaeger (1985) initially suggested that the widespread occurrence of *S. aculeatissimum* in Africa and its comparative scarcity in the New World supported this, though he went on to propose that its occurrence as a weed of secondary vegetation in Africa was suggestive of its introduction via trade routes from the New World. Jaeger & Hepper (1986) then included *S. aculeatissimum* under African introductions reporting that it had been present in Africa for a long time, achieving a very wide distribution in Afromontane habitats throughout the Continent. The species is often found in forest clearings, in rough or rocky ground and in secondary vegetation, sometimes associated with damp or marshy habitats.

This species apparently exhibits considerable morphological variability throughout its native distribution. It is characterised by a covering of vicious acicular green or purple prickles, which are dense on all stems and branches – even to the plant base – and scattered on the petioles, leaf laminas and calyces. Those on the stems are of variable lengths but much shorter than those found on the leaves. However, two *Maitland* specimens from Uganda completely lacked prickles but are morphologically similar in all other characters. The striking pubescence is often visible to the naked eye especially on the leaves which appear 'silky' above. The hairs themselves vary considerably in their density, lengths and glandulosity between different parts of the plants, and the glands themselves result in the stems, leaves and inflorescence parts appearing purple. Variations in prickle abundance and hair lengths have also been reported between lowland and montane plants (cf. Bukenya & Hall (1988)). Leaves on Kenyan specimens seem to be smaller than those found in Tanzania. The plants are said to have an extremely unpleasant odour. Only one to three fruits mature on each infructescence. The seeds are generally small (2–3 × 2 mm), rounded and not winged; however, two Tanzanian specimens (from **T** 2 and **T** 7) had larger seeds (4–4.5 × 3.1–3.6 mm), but were otherwise similar to *S. aculeatissimum*. The flowers of this species seem to be heterostylous; Polhill (*in sched.*) noted that only one flower opens at a time and that the styles in basal flowers were longer. Many East African herbarium specimens have been identified as *S. capsicoides*; this is a distinct species which only rarely occurs in East Africa and is distinguishable from *S. aculeatissimum* through its winged seeds and orange fruits.

Dunal differentiated var. *hispidissimum* from *S. aculeatissimum* purely on the basis of its indumentum being hispid rather than villose; such a distinction was not evident from the holotype. However, the holotype specimen comprises an apical fragment lacking fully open mature leaves. Although covered with a dense pubescence its composition together with the overall morphology of the specimen conforms to the variability exhibited by *S. aculeatissimum*.

Bitter used prickle posture and length, indumentum density and hair colour, and leaf lobing to differentiate his Rwandan variety *dolichoplum* from *S. aculeatissimum*. These are all characters which are extremely variable in this species, and although type material has not yet been located, this variety is considered synonymous with the latter.

The protologue, the Kew isosyntype and digital images of the syntypes leave little doubt that *S. angustispinosum* is a synonym of *S. aculeatissimum*. De Wildeman (1914) discussed the similarity of his new species with the latter but, referring to Wrights' (1906) description, considered the two to be sufficiently different to warrant the description of a new species.

The roots and leaves of *S. aculeatissimum* are reportedly toxic to horses and cattle (Verdcourt & Trump, 1969).

52. **Solanum aculeastrum** *Dunal* in DC., Prodr. 13(1): 366 (1852); C.H. Wright in F.T.A. 4, 2: 243 (1906); T.T.C.L.: 585 (1949); I.T.U.: 413 (195); Heine in F.W.T.A. 2nd ed., 2: 332 (19632); Blundell, Wild Fl. E. Afr.: 189 (1992); U.K.W.F., 2nd ed.: 244 (1994); K.T.S.L.: 579 (1994); Gonçalves in F.Z. 8(4): 107 (2005); Friis in Fl. Eth. 5: 138 (2006). Type: South Africa, Cape of Good Hope, near Morleg, *Drège* s.n. (G-DC, holo.; AD, BM!, K!, P!, possible iso.)

Shrub or tree, 2–6 m, erect, usually armed; young stems densely stellate-pubescent, trichomes porrect and multangulate, white-translucent or yellowish, densely matted, sessile or stalked, stalks up to 2 mm, rays 8–12, 0.1–0.5 mm, midpoints shorter than rays or up to 1 mm; prickles straight becoming curved, 8–17 mm long, 3–10 mm wide at base, flattened, prominent. Leaf blades drying strongly discolorous, red-green to yellow-green above, white-grey underneath, ovate, sometimes elliptic, 8–15 × 6–12 cm, 1.5–2 times longer than wide, base cordate to truncate(auriculate), usually equal, margin lobed, the broadly deltate lobes 2–3 on each side, 1.5–4.5 cm long, often with secondary lobing, apically obtuse (rounded), extending $\frac{1}{4}$–$\frac{3}{4}$ of the distance to the midvein; apex acute; densely stellate-pubescent abaxially, trichomes on abaxial surface porrect, sessile or stalked, stalks to 0.1 mm, rays 8–12, 0.15–0.8 mm, midpoints same length as rays or up to 1 mm, adaxially glabrescent, the trichomes reduced; primary veins 3–6 pairs; petiole 1.2–3 cm, ± $\frac{1}{6}$ of the leaf length. Inflorescences rarely branched, 3–6 cm long, with 4–12 flowers; peduncle 0–35 mm long; rachis 0.5–2.5 cm

Fig. 24. *SOLANUM ACULEASTRUM* — **1**, flowering habit with secondary leaf lobes; **2**, fruiting habit with simple leaf lobes; **3**, leaf with shallow lobing; **4**, inflorescence; **5**, mature fruit in section; **6**, prickles on mature stem; **7**, multangulate trichome from the young stem and porrect trichome from the abaxial surface of the leaf. 1, 6–7 from *Scheepers* 1043; 2 from *Richards* 24651; 3, 5 from *Thulin & Mhoro* 3162; 4, 6 from *Vorontsova et al.* 173, field photograph. Drawn by Lucy T. Smith. Scale bar: 1, 2, 3, 4 = 4 cm; 5 = 3 cm; 6 = 1 cm; 7 = 0.4 mm

long; peduncle and rachis unarmed; pedicels 1–3.5 cm long, in fruit 2.5–4.5 cm long, unarmed or with 1–5 prickles. Flowers heterostylous, 5-merous, only the basal 1(–2) long-styled. Calyx 3–15 mm long, lobes long-deltate to broadly deltate, 2–7 mm long, acute to apiculate, usually with 2–10 prickles in long-styled flowers. Corolla white (mauve), 1.5–3.2 cm in diameter, lobed for $^1\!/_2$–$^1\!/_5$ of its length, the lobes deltate or long-deltate, 6–15 × 2–5 mm. Stamens equal; anthers 4.5–6.5 mm. Ovary stellate-pubescent in the upper $^1\!/_5$; style 7–10 mm long on long-styled flowers. Berries 1(–3) per infructescence, evenly green when young, yellow at maturity, 3–5 × 2–4.5 cm, apically rounded to apiculate or acute; fruiting calyx usually unarmed. Seeds 4–5 × 2.7–4 mm. Fig. 24, p. 199.

UGANDA. Kigezi District: Mabungo and many parts of Bufumbira county, 25 Oct. 1929, *Snowden* 1612!; District unclear: 1 km east of Simba Hills, 14 June 1971, *Lye & Katende* 6243!
KENYA. Naivasha District: Naivasha, 22 Oct. 1979, *Kangai Mulu* 3!; Machakos District: Ol Donyo Sabuk National Reserve, Kisukioni, about 1 km to the summit, 9 March 2007, *Mbale et al.* *NMK*835!; N Kavirondo District: W Kakamega Forest Reserve, 13 July 1960, *Paulo* 550!
TANZANIA. Arusha District: Mt Meru, Sura Forest, 10 Nov. 1969, *Richards* 24668!; Kilosa District: Rubeho Mountains, Ukwiva Forest Reserve, between Mwega and Digitove, 8 km SE of Reding'ombe Village, 2 June 2005, *Mwangoka et al.* 3893!; Mbeya District: Mbeya Range, 15 March 1960, *Kerfoot* 1538!
DISTR. U 2, 4; K 3–7; T 1–3, 6–7; Nigeria, Cameroon, Congo-Kinshasa, Rwanda, Burundi, Sudan, Angola, Malawi, Mozambique, Zimbabwe, Swaziland, South Africa
HAB. Forest margins, grassland, scrub, and disturbed open places; 1200–2500 m

SYN. *S. albifolium* C.H. Wright in K.B. 1894: 127 (1894). Type: Angola, District Colungo Alto, *Welwitsch* 6095*b* (BM000778216!, lecto., designated here [best material]; BM!, K!, isolecto.)
   S. sapiaceum Dammer in E.J. 38: 60 (1905). Type: Tanzania, Songea District, Matengo, Kwa Djimula, *Busse* 917 (EA!, lecto., designated here [best material; no material of the other cited collection known]; EA!, isolecto.)
   S. conraui Dammer in E.J. 48: 242 (1912). Type: Cameroon, Bangwe Station, *Conrau* 255 (E!, lecto., designated here [only extant material known])
   S. rugulosum De Wild., Pl. Bequaert. 1: 433 (1922). Type: Congo-Kinshasa, Irumu, *Bequaert* 2797 (BR!, holo.; BR!, iso.)
   S. subhastatum De Wild., Pl. Bequaert. 1: 437 (1922). Type: Congo-Kinshasa, Runwenzori, Vallée de la Lamina, *Bequaert* 4340 (BR!, holo.)
   S. aculeastrum Dunal var. *exarmatum* Bitter in F.R. Beih. 16: 169 (1923); T.T.C.L.: 585 (1949). Type: Tanzania, Buoba District: Karagwe, *Stuhlmann* 1733 (no specimens found)
   S. aculeastrum Dunal var. *parceaculeatum* Bitter in F.R. Beih. 16: 169 (1923). Type: Congo-Kinshasa, Lunda-Kassai-Urua-Zone, Lusambo to Nyangwe, *Laurent* s.n. (BR!, holo.)
   S. aculeastrum Dunal var. *albifolium* (C.H. Wright) Bitter in F.R. Beih. 16: 170 (1923); Heine in F.W.T.A. 2ⁿᵈ ed., 2: 332 (1963)
   S. aculeastrum Dunal var. *conraui* (Dammer) Bitter in F.R. Beih. 16: 170 (1923)
   S. aculeastrum Dunal subsp. *sapiaceum* (Dammer) Bitter in F.R. Beih. 16: 171 (1923); T.T.C.L.: 585 (1949)
   S. aculeastrum Dunal subsp. *pachychlamys* Bitter in F.R. Beih. 16: 173 (1923); T.T.C.L.: 585 (1949). Type: Tanzania, Mbulu District: Wanege highlands, Iraku, Mama Isara's country, *Jaeger* 212 (no specimens found)

NOTES. *Solanum aculeastrum* is a widespread and variable species found across the African highlands. It is easy to recognise by its markedly discolorous lobed leaves that are almost white below and by its prominent orange downwardly hooked prickles, although these are sometimes absent. The fruit is frequently compared to lemons: big, almost always single, yellow, round or apiculate or elongated, with a tendency to develop warts. *Solanum aculeastrum* is a distinctive species unlikely to be misidentified, except for possible confusion with the closely related *S. phoxocarpum* and *S. thomsonii*, which differ in fruit characters.

53. **Solanum campylacanthum** A. Rich., Tent. Fl. Abyss. 2: 102 (1850); C.H. Wright in F.T.A. 4, 2: 239 (1906); T.T.C.L.: 589 (1949); F.P.U.: 129 (1962); Friis in Fl. Eth. 5: 140, fig. 158.15.6–8 (2006& in Fl. Somalia 3: 217 (2006). Type: Ethiopia, Adoa, *Schimper* I 1082 (P, lecto., designated by Lester 1997: 283; BM!, BR!, G, GH!, K!, P!, W, isolecto.)

Erect herb or shrub, 0.2–1.5(–4) m, armed or unarmed; young stems stellate-pubescent, trichomes porrect, translucent to orange-translucent, sessile or stalked, stalks up to 1.5(–2) mm, rays 6–12, 0.2–0.4 mm, midpoints ± same length as rays or reduced or elongated to 0.8 mm; prickles straight or curved, 0–5 mm long, 0.2–2 mm wide at base, round or flattened. Leaf blades drying concolorous to weakly discolorous, yellow-green to red-brown, ovate to elliptic or lanceolate, 3.5–17(–40) × 0.6–10(–19) cm, 1.5–4 times longer than wide, base rounded to cordate, usually equal, margin usually entire, sometimes lobed, the broadly rounded lobes 2–5 on each side, apically rounded (obtuse or acute)up to 1(–3) cm long, extending up to $\frac{1}{4}(-\frac{1}{2})$ of the distance to the midvein; apex rounded to acute; moderately to densely stellate-pubescent, trichomes on abaxial surface porrect, sessile or stalked, stalks to 1 mm, rays 6–8, 0.2–0.6 mm, midpoints reduced or to 0.8 mm, trichomes of adaxial surface smaller; primary veins 4–7 pairs; petiole 0.5–4(–11) cm, $\frac{1}{6}-\frac{1}{4}(-\frac{1}{3})$ of the leaf length. Inflorescences simple or occasionally branched once, 2–11 cm long, with (2–)5–10(–40) flowers; peduncle 0–10 mm long; rachis 0.5–8 cm long; peduncle and rachis unarmed; pedicels 0.7–3 cm on long-styled flowers, 0.5–2 cm long, in fruit 1.5–3 cm long, with 0–10 prickles. Flowers heterostylous, (4–)5(–6)-merous, the basal 1–3(–5) long-styled. Calyx 7–15 mm on long-styled flowers, 5–10 mm long on short-styled flowers, lobes deltate to long-deltate, sometimes foliaceous on long-styled flowers, 5–10 mm on long-styled flowers, 1.5–5 mm long on short-styled flowers, acute to obtuse or acuminate, unarmed or with up to 20 prickles. Corolla mauve to purple, 2.5–4.5 cm in diameter on long-styled flowers, 1.8–4 cm in diameter on short-styled flowers, lobed for $\frac{1}{4}-\frac{1}{3}$ of its length in long-styled flowers, lobed for $\frac{1}{4}-\frac{1}{2}$ of its length in short-styled flowers, the lobes broadly deltate, 7–15 × 7–13 mm on long-styled flowers, 5–9 × 4–11 mm on short-styled flowers. Stamens equal; anthers 6.5–9 mm on long-styled flowers, 4–9 mm on short-styled flowers. Ovary glabrous or stellate-pubescent in the upper $\frac{1}{4}$; style 10–15 mm long on long-styled flowers. Berries 1–2(–4) per infructescence, striped when young, yellow-orange at maturity, spherical, 1.5–3 cm in diameter; fruiting calyx with 0–30 prickles. Seeds 2.7–3.2 × 1.9–2.6 mm. Fig. 25, p. 202.

UGANDA. Toro District: Toro Game Reserve, 14 Dec. 1962, *Buechner* 12!; Busoga District: Busoga villages, 13 Dec. 1898, *Whyte s.n.*!; Mengo District: Busuju, Kasa forest, 17 Nov. 1949, *Dawkins* 454!
KENYA. Northern Frontier District: Mount Kulal, Narangani, 8 June 1960, *Oteke* 60!; N Kavirondo District: Elgon Nyanza, Feb. 1960, *Templer H*1!; Lamu District: 88 km NE of Lamu, 23 July 1961, *Gillespie* 24!
TANZANIA. Tanga District: Sawa, 8 Sept. 1965, *Faulkner* 3642!; Kigoma District: Gombe Stream Reserve, from Gombe stream to Missonge, 30 Dec. 1963, *Pirozynski* 128!; Iringa District: Ruaha National Park near Mbagi camp, 14 Jan. 1966, *Richards* 20992!; Zanzibar: Oct. 1873, *Hildebrandt* 987!
DISTR. U 1–4; K 1–7; T 1–8; Z, P; Congo-Kinshasa, Rwanda, Burundi, Sudan, Djibouti, Eritrea, Ethiopia, Somalia, Angola, Zambia, Malawi, Mozambique, Zimbabwe, Namibia, Botswana, Swaziland, Lesotho, South Africa
HAB. Ubiquitous weed of roadsides, abandoned cultivation, wooded grassland, bushland, dunes, forest edges; 0–2000(–2300) m

SYN. *S. bojeri* Dunal in DC., Prodr. 13(1): 344 (1852); C.H. Wright in F.T.A. 4, 2: 240 (1906). Type: Africa, "Ex ora orientali Africae australis, in Mozambico insula, in Madagascaria Zanzibar", 1839, *Bojer s.n.* (G00131445!, lecto., designated here [label matches protologue exactly]; G!, isolecto.)
  S. bojeri Dunal var. *sinuatorepandum* Dunal in DC., Prodr. 13(1): 345 (1852). Type: Mauritius, 1839, *Bouton s.n.* (G!, holo.)
  S. panduriforme Dunal in DC., Prodr. 13(1): 370 (1852), as *panduraeforme*; C.H. Wright in F.T.A. 4, 2: 214 (1906); T.T.C.L.: 587 (1949). Type: South Africa, Cape of Good Hope, Bashee, 1000 ft, *Drège s.n.* (G!, holo.; K!, K!, MO! probable iso.)
  S. delagoense Dunal in DC., Prodr. 13(1): 349 (1852); T.T.C.L.: 589 (1949). Type: Mozambique, Delagoa Bay, 1822, *Forbes s.n.* (G, holo.; K!, 5 iso.)
  S. benguelense Peyr. in Sitz. K. Preuss. Akad. Wiss. Berlin 38: 575 (1860); C.H. Wright in F.T.A. 4, 2: 215 (1906). Type: Angola, Benguela, *Wawra* 289 (W!, holo.; LE!, iso.)

FIG. 25. *SOLANUM CAMPYLACANTHUM* — habits showing variation: **1**, elliptic leaves; **2**, lobed leaves and a branched inflorescence; **3**, no prickles and ovate leaves; **4**, lanceolate leaves; **5**, dense prickles; **6**, porrect stalked trichome from leaf abaxial surface. 1, 6 from *Drummond and Hemsley* 2421; 2 from *Symes* 715; 3 from *Richards* 25663; 4 from *Richards* 23546; 5 from *Anderson* 1169. Drawn by Lucy T. Smith. Scale bar: 1, 2, 3, 4, 5 = 4 cm; 6 = 1 mm.

S. pharmacum Klotzsch in Peters, Reise Mossamb.: 234 (1861); C.H. Wright in F.T.A. 4, 2: 230 (1906). Type: Mozambique, 'Allenthalben auf der Insel und dem Festlande', *Peters* s.n. (B†, holo.)

S. phoricum Klotzsch in Peters, Reise Mossamb.: 234 (1861). Type: Mozambique, Ríos de Sena, *Peters* s.n. (B†, holo.)

S. tomentellum Klotzsch in Peters, Reise Mossamb.: 236 (1861). Type: Mozambique, Island of Mozambique, Río Sena and Río Querimba, *Peters* s.n. (B†, holo.)

S. mossambicensis Klotzsch in Peters, Reise Mossamb. 1: 235 (1861). Type: Mozambique, *Peters* s.n. (B†, holo.)

S. fischeri Dammer in P.O.A. C: 353 (1895); C.H. Wright in F.T.A. 4, 2: 236 (1906). Type: Tanzania, Mwanza District, Kagehi, *Fischer* 406 (B†, holo.)

S. trepidans C.H.Wright in K.B. 1894: 128 (1894). Type: Malawi, Shiré Valley, near Bishop Mackenzie's house, *Scott* s.n. (K!, lecto., designated here, [best flowering material])

S. psilostylum Dammer in P.O.A. C: 354 (1895); C.H. Wright in F.T.A. 4, 2: 215 (1906); T.T.C.L.: 588 (1949). Type: Tanzania, Mwanza District, Mwanza [Muansa], *Stuhlmann* 4160 (B†, holo.)

S. volkensii Dammer in P.O.A. C: 354 (1895); C.H. Wright in F.T.A. 4, 2: 215 (1906); T.T.C.L.: 588 (1949). Type: Tanzania, Moshi District, Marangu, Habari, *Volkens* 2144 (WU!, lecto., designated here [best material; only collection known from amongst syntypes]; BR!, isolecto.)

S. antidotum Dammer in P.O.A. C: 355 (1895); C.H. Wright in F.T.A. 4, 2: 242 (1906). Type: Tanzania, Kilimanjaro, Pare Moëta, *Fischer* 314 (B†, holo.)

S. obliquum Dammer in P.O.A. C: 354 (1895). Type: Tanzania, Dar es Salaam, *Stuhlmann* 7576 [Bitter (1923) says the number is 7536] (B†, holo.)

S. urbanianum Dammer in P.O.A. C: 355 (1895). Type: Mozambique, Zambezi region, Gorungosa, *de Carvalho* s.n. (COI, holo.)

S. astrochlaenoides Dammer in E.J. 28: 476 (1900). Type: Tanzania, Kilosa District, between Khutu and Uhehe, Kidodi, base of Vidunda Mts, shore of the Ruhembe, *Goetze* 384a (K!, lecto., designated here [best flowering material]; BR!, isolecto.)

S. magdalenae Dammer in E.J. 38: 194 (1906); T.T.C.L.: 588 (1949). Type: Tanzania, Iringa District, Uhehe near Iringa, *Magdalene Prince* s.n. (GOET!, lecto., designated by Vorontsova & Knapp 2010: 1597)

S. lachneion Dammer in E.J. 38: 194 (1906); T.T.C.L.: 590 (1949). Type: Tanzania, Lushoto District, Usambara, Kwai, *Eick* 423 (GOET!, lecto., designated by Vorontsova & Knapp in Taxon 59: 1597, 2010)

S. neumannii Dammer in E.J. 38: 195 (1906). Type: Ethiopia, Alata, Sidamo, *Neumann* 6 (GOET!, lecto., designated by Vorontsova & Knapp in Taxon 59: 1598, 2010)

S. bussei Dammer in E.J. 48: 240 (1912). Type: Tanzania, Kilwa District: near Kwa Matumola, Mtama shamba, *Busse* 504 (EA!, lecto., designated here [best flowering material]; EA!, isolecto.)

S. merkeri Dammer in E.J. 48: 254 (1912); T.T.C.L.: 590 (1949). Type: Tanzania, Kilimanjaro, *Merker* s.n. (B†, holo.)

S. deckenii Dammer in E.J. 48: 257 (1912). Type: Comoros, Kitanda–Mdjini, N on road to Mrigini, May 1864, *Kersten* s.n. (B†, syn.); Tanzania, Zanzibar, *Schmidt* 6 (B†, syn.).

S. delpierrei De Wild. in B.J.B.B. 4: 398 (1914). Type: Congo-Kinshasa, Uele, 1904, *Delpierre* s.n. (BR!, holo.)

S. macrosepalum Dammer in E.J. 53: 331 (1915). Type: Uganda, Mengo District, Kampala, *Nägele* 16 (B†, holo.)

S. iodes Dammer in E.J. 53: 332 (1915). Type: Kenya, Machakos District: Kibwezi, *Scheffler* 466 (E!, lecto., designated here [best material]; BM!, K!, isolecto.)

S. secedens Dammer in E.J. 53: 345 (1915). Type: Tanzania, Kilimanjaro, *Endlich* 306 (M!, lecto., designated here [only known extant material])

S. endlichii Dammer in E.J. 53: 347 (1915). Type: Tanzania, Kilimanjaro, Kikafu, *Endlich* 572 (M!, lecto., designated here [only material known])

S. himatacanthum Dammer in E.J. 53: 348 (1915). Type: Tanzania, Kilimanjaro, Kikafu, Kibo, *Endlich* 306a (M!, lecto., designated here [only material known])

S. tabacicolor Dammer in E.J. 53: 349 (1915); T.T.C.L.: 588 (1949). Type: Tanzania, Kilimanjaro, between Kibo and Kikafu, *Endlich* 297 (M, lecto., designated by Vorontsova & Knapp in Taxon 59: 1599, 2010)

S. omahekense Dammer in E.J. 53: 350 (1915). Type: Namibia, Omaheke, Owinauanaua, *Seiner* 436 (B†, holo.)

S. omitiomirense Dammer in E.J. 53: 351 (1915). Type: Namibia, Omitiomire, Black Nossob River, *Seiner* 42 (B†, holo.)

S. pentheri Gand. in Bull. Soc. Bot. France 65: 61 (1918). Type: South Africa, Cape Province, Colossa, *Krook* in *Penther* 1842 (LY, holo.; M, W, iso.)

S. beniense de Wild., Bequaert 1: 419 (1922). Type: Congo-Kinshasa, P. N. Albert, Vieux Beni, 4 Apr. 1914, *Bequaert* 3402 (BR!, holo.; BR! K! iso.).

S. tuntula De Wild., Pl. Bequaert. 1: 438 (1922). Type: Congo-Kinshasa, Kabare, *Bequaert* 5492 (BR!, lecto., designated here [best material])

S. bojeri Dunal var. *deckenii* (Dammer) Bitter in F.R. Beih. 16: 239 (1923)

S. bojeri Dunal var. *houyanum* Bitter in F.R. Beih. 16: 242 (1923); T.T.C.L.: 589 (1949). Type: Tanzania, Mpwapwa District: S Ussagara, Kisuira Mt, *Houy* in *Meyer* IV, 1201 (no specimens found)

S. bojeri Dunal var. *integrum* Bitter in F.R. Beih. 16: 239 (1923). Type: Tanzania, Zanzibar, Sep. 1843, *Peters* s.n. (B†, holo.)

S. campylacanthum A. Rich. subsp. *cordifrons* Bitter in F.R. Beih. 16: 214 (1923). Type: Tanzania, Lushoto District, Usambara, Mission station Mlalo on the Umba River, *Holst* 4050 (B†, holo.)

S. campylacanthum A. Rich. var. *ellipsoideum* Bitter in F.R. Beih. 16: 213 (1923). Type: Ethiopia, Scholloda near Adua, *Schimper* 6 pro parte (most likely B†, holo.)

S. campylacanthum A. Rich. subsp. *kondeense* Bitter in F.R. Beih. 16: 213 (1923). Type: Tanzania, Rungwe District: Kyimbila Station, Bach Kale, N of Lake Nyassa, *Stolz* 1580 (B†, holo.; K000414114!, lecto., designated here [best material]; A!, BM!, K!, isolecto.)

S. delagoense Dunal var. *tomentellum* (Klotzsch) Bitter in F.R. Beih. 16: 254 (1923)

S. delagoense Dunal var. *munitius* Bitter in F.R. Beih. 16: 255 (1923). Type: Mozambique, Coast and Rios de Sena, *Peters* s.n. (no specimens found)

S. delagoense Dunal var. *brachyastrotrichum* Bitter in F.R. Beih. 16: 256 (1923); T.T.C.L.: 590 (1949). Type: Tanzania, Lushoto District: Usambara, Djuani, *Braun* 3658 (EA!, iso.)

S. delagoense Dunal var. *astrochlaenoides* (Dammer) Bitter in F.R. Beih. 16: 257 (1923); T.T.C.L.: 589 (1949)

S. delagoense Dunal var. *fischeri* (Dammer) Bitter in F.R. Beih. 16: 258 (1923); T.T.C.L.: 590 (1949)

S. delagoense Dunal var. *karagweanum* Bitter in F.R. Beih. 16: 260 (1923); T.T.C.L.: 590 (1949). Type: Tanzania, Bukoba District: Kafuro in Karagwe, *Stuhlmann* 1771 (no specimens found)

S. delagoense Dunal var. *obliquum* (Dammer) Bitter in F.R. Beih. 16: 256 (1923)

S. delagoense Dunal subsp. *pliomorphum* Bitter in F.R. Beih. 16: 261 (1923). Type: Zambia, Bwana Mkubwa, *Fries* 473 (UPS, holo.)

S. delagoense Dunal subsp. *punctatistellatum* Bitter in F.R. Beih. 16: 262 (1923). Type: Kenya, District unclear, Tanaland, Ngai, *F. Thomas* 141 (B†, holo.)

S. delagoense Dunal subsp. *epacanthastrum* Bitter in F.R. Beih. 16: 263 (1923). Type: Zimbabwe, Chirinda, *Synnerton* 387 (BM!, lecto., designated here [best material]; K!, isolecto.)

S. delagoense Dunal var. *benguelense* (Peyr.) Bitter in F.R. Beih. 16: 263 (1923)

S. delagoense Dunal subsp. *baumii* Bitter in F.R. Beih. 16: 264 (1923). Type: Angola, Habungu, *Baum* 469 (lectotype, BM!, lecto., designated here [best flowering material]; K!, E, W, isolecto.)

S. delagoense Dunal subsp. *omahekense* (Dammer) Bitter in F.R. Beih. 16: 264 (1923)

S. delagoense Dunal subsp. *transvaalense* Bitter in F.R. Beih. 16: 267 (1923). Type: South Africa, Transvaal, Lijdenburg, *Wilms* 1019 (GOET!, lecto., designated here [annotated by Bitter] JE!, G!, M!, W!, isolecto.)

S. incanum L. var. *kavirondoense* Bitter in F.R. Beih. 16: 276 (1923). Type: Kenya, N Kavirondo District: between Nandi and Mumias, Nov.–Dec., *Whyte* s.n. (K! [K000414123], lecto., designated here [best material]; K!, isolecto.)

S. malacochlamys Bitter in F.R. Beih. 16: 215 (1923); T.T.C.L.: 587 (1949). Type: Tanzania, Tanga, *Winkler* 3525 (WRSL, holo.; GOET!, iso.)

S. malacochlamys Bitter var. *transgrediens* Bitter in F.R. Beih. 16: 217 (1923); T.T.C.L.: 587 (1949). Type: Tanzania, Tanga District, Amboni, *Holst* 2678 (K!, lecto., designated here [best material]; WRSL, isolecto.)

S. neumannii Dammer var. *schoense* Bitter in F.R. Beih. 16: 218 (1923). Type: Ethiopia, Schoa, near Akaki, *Ellenbeck* 1595 (B†, holo.)

S. repandifrons Bitter in F.R. Beih. 16: 219 (1923); T.T.C.L.: 588 (1949). Type: Tanzania, ?Morogoro District: between Ruaha, Rufiji and Ruwu, E of Kitondwe, *Stuhlmann* 8273 (GOET!, tracing of type)

S. merkeri Dammer var. *tobleri* Bitter in F.R. Beih. 16: 223 (1923); T.T.C.L.: 591 (1949). Type: Tanzania, Lushoto District, Usambara, near Amani, 1000 m, *Grote* s.n. (no specimens found)

S. merkeri Dammer var. *mediidominans* Bitter in F.R. Beih. 16: 223 (1923). Type: Tanzania, Lushoto District: Usambaras, *Holst* 51 (B†, holo.)

S. merkeri Dammer subsp. *militans* Bitter in F.R. Beih. 16: 223 (1923); T.T.C.L.: 591 (1949). Type: Tanzania, Lushoto District, Usambara, Kwai, *Eick* 201 pro parte (GOET!, lecto., designated by Vorontsova & Knapp in Taxon 59: 1598, 2010)

S. merkeri Dammer var. *endastrophorum* Bitter in F.R. Beih. 16: 224 (1923); T.T.C.L.: 591 (1949). Type: Tanzania, Dodoma District: Wembere-, Ugogo- and Ussangu-Steppe, Kilimatinde, *Claus* 20 (GOET!, neo., designated by Vorontsova & Knapp in Taxon 59: 1598, 2010)

S. merkeri Dammer var. *intermontanum* Bitter in F.R. Beih. 16: 225 (1923); T.T.C.L.: 591 (1949). Type: Tanzania, Moshi District: between Kilimanjaro and Meru, *Merker* s.n. (B†, holo.)

S. merkeri Dammer var. *ruandense* Bitter in F.R. Beih. 16: 226 (1923). Type: Rwanda, East Rwanda, *Meyer* 604 (GOET!, lecto., designated by Vorontsova & Knapp in Taxon 59: 1598, 2010)

S. merkeri Dammer forma *subinerme* Bitter in F.R. Beih. 16: 222 (1923). Type: Tanzania, Lushoto District, Usambara, near Amani, 1000 m, Aug., *Grote* s.n. (no specimens found)

S. lachneion Dammer var. *abbreviatum* Bitter in F.R. Beih. 16: 229 (1923); T.T.C.L.: 590 (1949). Type: Tanzania, Lushoto District, W Usambara, Kwambuguland, *Engler* 1328, 1329 (not found)

S. lachneion Dammer var. *intercedens* Bitter in F.R. Beih. 16: 230 (1923); T.T.C.L.: 590 (1949). Type: Tanzania, Arusha District: near Arusha, S of Meru, *Holtz* 3272 (B†, holo.)

S. lachneion Dammer var. *protopyrrhotrichum* Bitter in F.R. Beih. 16: 230 (1923); T.T.C.L.: 590 (1949). Type: Kenya, Nairobi, *Horn* s.n. (W, holo.; GOET!, iso.)

S. maranguense Bitter in F.R. Beih. 16: 234 (1923); T.T.C.L.: 588 (1949). Type: Tanzania, Kilimanjaro, Station Marangu, *Volkens* 617 (GOET!, lecto., designated by Vorontsova & Knapp in Taxon 59: 1598, 2010; GOET!, PH, isolecto.)

S. ochracanthum Bitter in F.R. Beih. 16: 108 (1923). Type: Tanzania?, Mossambique coast, Yangwani near Lindi, 150 m, June, *Busse* 2989 (no specimens found)

S. stellativillosum Bitter in F.R. Beih. 16: 226 (1923). Type: Kenya, Central Province, Kibwezi Station, 28 Jan. 1906, *Scheffler* 94 (Z! lecto., designated here [following unpublished lectotypification by R. Lester]; E!, G!, K!, P!, P!, W!, WAG!, isolecto).

S. stellativillosum Bitter var. *makinduense* Bitter, F.R. Beih. 16: 228 (1923). Type: Kenya, Machakos Distr., Makindu River, 14 Apr., *Kässner* 580 (Z! lecto., designated here [annotated by Bitter]; K!, isolecto.)

S. verbascifrons Bitter in F.R. Beih. 16: 236 (1923); T.T.C.L.: 588 (1949). Type: Tanzania, Masai District, Mbuga Kitwai, *Jaeger* 82 (GOET!, lecto., designated by Vorontsova & Knapp in Taxon 59: 1600, 2010)

S. ukerewense Bitter in F.R. Beih. 16: 242 (1923); T.T.C.L.: 591 (1949). Type: Tanzania, Mwanza District: Ukerewe, Kagunguli, *Conrads* 240 (no specimens found)

S. mesomorphum Bitter in F.R. Beih. 16: 246 (1923). Type: Rwanda, Mountains south of Lake Kivu, *Fries* 1484 (UPS, holo.)

S. pembae Bitter in F.R. Beih. 16: 244 (1923). Type: Tanzania, Pemba, *Voeltzkow* 29 (B†, holo.)

S. volkensii Dammer var. *himatacanthum* (Dammer) Bitter in F.R. Beih. 16: 249 (1923), as *himatiacanthum*; T.T.C.L.: 588 (1949)

S. sennii Chiov., Fl. Somala 2: 334, fig. 191 (1932). Type: Somalia, Colbio, *Senni* 220 (FT, holo.)

S. cufodontii Lanza, Miss. Biol. Borana, Racc. Bot.: 193, fig. 56 (1939). Type: Ethiopia, Borana, Neghelli, affluente del Ganale, *Cufodontis* 51 (FT, holo.; FT, iso.)

S. goniocalyx Lanza, Miss. Biol. Borana, Racc. Bot.: 195, fig. 57 (1939). Type: Ethiopia, Javello, Quota Littorio, *Cufodontis* 581 (FT! [FT003061], lecto., designated here [best material]; FT, W, isolecto.)

S. melongenifolium Lanza, Miss. Biol. Borana, Racc. Bot.: 198, fig. 59 (1939). Type: Ethiopia, Javello, Pozzi Acacie, *Cufodontis* 416 (FT! [FT003063], lecto., designated here [best material]; FT, W, isolecto.)

NOTES. *Solanum campylacanthum* is a familiar species to anybody visiting north-eastern, eastern, or southern Africa, omnipresent in almost any disturbed environment between sea level and 2000 m elevation. A perennial weed flowering and fruiting abundantly throughout the year, it has mauve flowers, big bright yellow leathery fruits, a long taproot, and leaves that are usually entire. Comparing isolated specimens of *S. campylacanthum* creates the impression of numerous species and indeed, many species names have been published. Morphological variation is in fact continuous with similar variants occurring randomly in different parts of the range. *Solanum campylacanthum* does not usually reach above 1.5 m in height, but in the forests of upland Kenya it can grow up to 4 m (these variants have previously been misidentified as "*Solanum richardii*").

Solanum campylacanthum is often referred to as *"Solanum incanum"*. *Solanum incanum* sensu Gonçalves (Flora Zambesiaca 8(4): 1–124. 2005) is a group of closely related weedy species including *S. campylacanthum* in N, E and S Africa, *S. incanum* in northern Africa and Arabia, *S. lichtensteinii* in southern Africa, *S. cerasiferum* Dunal in western and northern Africa, and the cultivated *S. melongena*. Almost all East African members of this group are referable to *S. campylacanthum*, except some populations of *S. incanum* sensu stricto in northern Kenya and rare occurrences of *S. lichtensteinii* in southern Tanzania. The concept of *S. campylacanthum* used in this treatment includes populations from Ethiopia to South Africa, including specimens identified by other authors as *S. panduriforme* and *S. delagoense*.

54. **Solanum dasyphyllum** *Schumach. & Thonn.* in Beskr. Guin. Pl.: 126 [146] (1827); C.H. Wright in F.T.A. 4, 2: 244 (1906); T.T.C.L.: 585 (1949); Heine in F.W.T.A. 2nd ed., 2: 334 (1963); U.K.W.F., 2nd ed.: 243 (1994); Friis in Fl. Eth. 5: 137, fig. 158.13.8–10 (2006) & in Fl. Somalia 3: 216 (2006). Type: 'Guinea', Dyrkes af Negerne, *Thonning* 144 (C!, lecto. [IDC microfiche 2203 102:I.1–2, "362"] designated here [best material]; C, G-DC, isolecto.)

Woody herb, 0.5–1 m, erect, heavily armed; young stems sparsely to densely stellate-pubescent, trichomes porrect, white-translucent or yellowish, stalked, stalks 0.5–2.5 mm, often broad and lignified, rays ± 4, 0.5–1 mm, midpoints ± same length as rays or up to 1 mm; prickles straight, 2–7 mm long, 0.5–1.5 mm wide at base, flattened. Leaf blades drying discolorous to almost concolorous, yellow-green to red-brown, elliptic, 10–35 × 6–20 cm, 1.2–2 times longer than wide, base attenuate (cuneate), usually equal, margin lobed, the lobes (3–)4–5(–6) on each side, 3–9 cm long, usually with extensive secondary lobing, apically acute to obtuse, extending $^1/_2$–$^2/_3$ of the distance to the midvein; apex acute to obtuse; sparsely to densely stellate-pubescent on both sides, trichomes on abaxial surface porrect, stalked, stalks 0.1–0.5 mm, rays 4(–5), 0.4–0.7 mm, midpoints ± same length as rays or up to 0.8 mm, adaxially appearing as simple hairs visible to the naked eye, with bulbous bases and no visible rays, the midpoint 0.5–2 mm; primary veins 5–8 pairs; petiole usually decurrent (0–4 cm), usually less than $^1/_6$ of the leaf length. Inflorescences not branched, 4–7 cm long, with 5–10 flowers; peduncle 0–3 mm long; rachis 1.5–3.5 cm long; peduncle and rachis unarmed or with numerous prickles; pedicels 1.5–3 cm long and stout in long-styled flowers, 0.7–2.5 cm and slender in short-styled flowers, in fruit 1.5–3 cm long, with numerous prickles. Flowers heterostylous, 5-merous, basal one long-styled. Calyx 10–30 mm on long-styled flowers, 8–20 mm long on short-styled flowers, lobes deltate, 12–25 mm on long-styled flowers, 7–10 mm long on short-styled flowers, apically acute to long-acuminate, densely covered in prickles in long-styled flowers. Corolla mauve to purple (white), 3.5–6 cm in diameter on long-styled flowers, 1.5–3.5 cm in diameter on short-styled flowers, lobed for $^1/_4$–$^1/_3$ of its length, lobes broadly deltate, 5–20 × 10–25 mm on long-styled flowers, 2–7 × 7–15 mm on short-styled flowers. Stamens equal; anthers 16.5–7.5 mm on long-styled flowers, 5.5–7.5 mm on short-styled flowers. Ovary with a few trichomes on the upper $^1/_3$; style 8–13 mm long on long-styled flowers. Berries 1(–2) per infructescence, striped when young, yellow at maturity, spherical, 2.5–4 cm in diameter; fruiting calyx with numerous prickles. Seeds 2.8–4.5 × 2–3.5 mm.

UGANDA. District unclear: Nakishenyi, 23 Apr. 1941, *A.S. Thomas* 3811!; Masaka District: Dumu village on Lake Victoria, 18 May 1971, *Lye & Katende* 6122!
KENYA. Machakos District: Kibwezi, 27 Dec. 1971, *Gillett* 19396!; Masai District: Olosendo area, Trans Mara, 19 June 1961, *Glover et al.* 1877!; Teita District: Taita Hills, top ridge ± 5 km south of Bura mission west side of Bura valley, 8 Apr. 1998, *Mwachala et al.* 1267!
TANZANIA. Arusha District: Ngurdoto Crater National Park, Leopard Point, 18 March 1966, *Greenway & Kanuri* 12434!; Kigoma District: Lake Tanganyika, shore near Kigoma, 10 July 1960, *Verdcourt* 2783!; district unclear: river Ruvu, 6 Nov. 1955, *Milne-Redhead & Taylor* 7055!; Zanzibar: May 1879, *Kirk* s.n.!

DISTR. U 1–2, 4; **K** 1, 4–7; **T** 1–7; **Z**; Gambia, Sierra Leone, Guinea Bissau, Liberia, Burkina
Faso, Ivory Coast, Ghana, Nigeria, Cameroon, Gabon, Equatorial Guinea, Central African
Republic, Congo-Kinshasa, Rwanda, Burundi, Sudan, Ethiopia, Somalia, Angola, Zambia,
Mozambique, South Africa

HAB. Mainly a forest species but also in wooded grassland, grassland and wasteland, frequently
near water; (0–)600–1600 m

SYN. *S. afzelii* Dunal in DC., Prodr. 13(1): 363 (1852). Type: Sierra Leone, *Afzelius* s.n. (BM!,
holo.)

 S. acanthoideum Dunal in DC., Prodr. 13(1): 364 (1852). Type: South Africa, Cape of Good
Hope, Natal, *Drège* 4862 (G!, holo.; MPU, P, iso.)

 S. duplosinuatum Klotzsch in Peters, Reise Mossamb.: 233 (1861); C.H. Wright in F.T.A.
4, 2: 243 (1906). Type: Mozambique, 'Festlandes und Insel Mossambique', *Peters* s.n.
(B†, holo.)

 S. kilimandschari Dammer in P.O.A. C: 352 (1895). Type: Tanzania, Moshi District: Himo,
river crossing on road from Taveta to Moschi, *Volkens* 1729 (B†, holo.; BR!, lecto.,
designated here [only extant material known])

 S. macrocarpon L. var. *hirsutum* De Wild. & T. Durand in Ann. Mus. Congo Belge sér. 3, 1: 290
(1901), as *hirsuta*. Type: Congo-Kinshasa, cultivated, *Dewèvre* s.n. (BR, holo., not found)

 S. duplosinuatum Klotzsch var. *semiglabrum* C.H. Wright in F.T.A. 4, 2: 244 (1905). Type:
Nigeria, Nupe, *Barter* 1344 (K!, holo.; W, iso.)

 S. eickii Dammer in E.J. 38: 192 (1906); T.T.C.L.: 586 (1949). Type: Tanzania, Lushoto
District, Usambara, Kwai, *Eick* 30 (GOET!, lecto., designated by Vorontsova & Knapp in
Taxon 59: 1596, 2010)

 S. sapinii De Wild. in Ann. Mus. Congo, sér. 2, Bot., ser. 5, 2: 341 (1908), as *sapini*. Type:
Congo-Kinshasa, Ikongo, on the Sankuru River, Sep. 1906, *Sapin* s.n. (BR!, holo.)

 S. dasyphyllum Schumach. & Thonn. var. *semiglabrum* (C.H. Wright) Bitter in F.R. Beih. 16:
191 (1923)

 S. dasyphyllum Schumach. & Thonn. var. *decaisneanum* Bitter in F.R. Beih. 16: 191 (1923).
Type: Ethiopia, Begemder, near Repp River, *Schimper* 1404 (BM!, lecto., designated here
[best material]; K!, E, isolecto.)

 S. dasyphyllum Schumach. & Thonn. var. *kilimandschari* (Dammer) Bitter in F.R. Beih. 16:
192 (1923); T.T.C.L.: 586 (1949)

 S. dasyphyllum Schumach. & Thonn. var. *inerme* Bitter in F.R. Beih. 16: 192 (1923). Type:
Cameroon, Campo area, Bebai, *Tessmann* 404 pro parte (no specimens found)

 S. dasyphyllum Schumach. & Thonn. var. *brevipedicellatum* Bitter in F.R. Beih. 16: 193 (1923).
Type: Angola, Distr. Loanda, *Gossweiler* 388 (BM!, lecto., designated here [best material];
P!, isolecto.)

 S. dasyphyllum Schumach. & Thonn. var. *natalense* Bitter in F.R. Beih. 16: 193 (1923). Types:
South Africa, Berea near Durban, *Medley Wood* 5530 (BM!, lecto., designated here [best
material]; MO, isolecto.)

 S. dasyphyllum Schumach. & Thonn. var. *transiens* Bitter in F.R. Beih. 16: 193 (1923);
T.T.C.L.: 586 (1949). Type: Tanzania, Lushoto District: Usambara, Rinko,
Wambugeland, *Buchwald* 450 (no specimens found)

 S. crepidotrichum Bitter in F.R. 16: 194 (1923). Type: Cameroon, Bagiri, *Houy* 101 (no
specimens found)

 S. macrocarpon L. subsp. *sapinii* (De Wild.) Bitter in F.R. Beih. 16: 201 (1923)

NOTES. The usually attenuate leaf base and decurrent petiole provide the easiest identification
feature for *S. dasyphyllum* and *S. macrocarpon*. The sizeable stellate hairs on stems and leaves have
a maximum of 4 or 5 rays, another useful identification feature overlooked in other treatments.

 It is now widely accepted that *S. dasyphyllum* is the wild progenitor of the cultivated *S.
macrocarpon*. Less hairy forms of *S. dasyphyllum* with fewer prickles and larger fruit were
selected for cultivation to produce the visibly different *S. macrocarpon*. The distinction
between the cultivated *S. macrocarpon* and the wild *S. dasyphyllum* is largely artificial but
maintained here for practical purposes.

 Several species of African *Solanum* have confusingly similar large leaves, pointed leaf lobes,
straight yellow flattened prickles on both surfaces of the leaf, and elongated midpoints of the
adaxial leaf surface: *S. cerasiferum*, *S. nigriviolaceum*, and *S. aculeatissimum* can all be
superficially similar to *S. dasyphyllum*. *Solanum dasyphyllum* is most frequently confused with
the introduced *S. aculeatissimum* Jacq. and can be distinguished by its lack of distinct petiole
and long-attenuate leaf bases, almost rotate corolla on short-styled flowers, and only 4(–5)
rays on the star hairs on vegetative parts of the plant.

55. **Solanum incanum** *L.*, Sp. Pl.: 188 (1753); C.H. Wright in F.T.A. 4, 2: 238 (1906); Heine in F.W.T.A. 2nd ed., 2: 332 (1963); Cribb & Leedal, Mountain Fl. Tanzania: 111, t. 26a (1982); Blundell, Wild Fl. E. Afr.: 190, t. 591 (1992); K.T.S.L.: 581 (1994); U.K.W.F., 2nd ed.: 244 (1994); Gonçalves in F.Z. 8(4): 112 (2005); Friis in Fl. Eth. 5: 140, fig. 158.15.1–3 (2006) & in Fl. Somalia 3: 216 (2006). Type: 'Africa', Herb. *J. Burser* Vol. 9, no. 20 (UPS, neo., designated by Hepper & Jaeger 1985: 388)

Herb or shrub, 0.4–1.5 m, erect, armed; young stems densely stellate-pubescent, trichomes multangulate, translucent, a mixture of sessile and stalked, stalks up to 1 mm, rays 6–20, 0.1–0.4 mm, midpoints ± same length as rays; prickles straight or curved, 3–9 mm long, 1.5–6 mm wide at base, round or flattened. Leaf blades drying concolorous to weakly discolorous, yellowish, ovate, 6–22 × 4–15 cm, ± 1.5 times longer than wide, base rounded to cordate, often unequal and oblique, margin lobed to almost entire, the broadly rounded lobes (3–)4(–5) on each side, 0.5–2 cm long, apically rounded, extending up to $\frac{1}{4}$ of the distance to the midvein; apex rounded to acute; densely stellate-pubescent abaxially, trichomes on abaxial surface porrect, sessile or stalked, stalks to 0.2 mm, rays 8–12, 0.15–0.5 mm, midpoints ± same length as rays or up to 0.6 mm, trichomes of adaxial surface smaller; primary veins ± 5 pairs; petiole 1–9 cm, $\frac{1}{4}$–$\frac{1}{3}$ of the leaf length. Inflorescences not branched, 3–8 cm long, with 5–10 flowers; peduncle 1–4 mm long; rachis 0.5–5 cm long; peduncle and rachis unarmed; pedicels 0.8–1.5 cm on long-styled flowers, 0.5–0.9 cm long on short-styled flowers, in fruit 1.3–1.8 cm long, with 0–15 prickles. Flowers heterostylous, 5-merous, basal one long-styled flower. Calyx 6–10 mm on long-styled flowers, 4.5–8 mm long on short-styled flowers, lobes long-deltate and sometimes foliaceous on long-styled flowers, broadly deltate on short-styled flowers, 2.5–5 mm on long-styled flowers, 1.5–3 mm long on short-styled flowers, acute to obtuse, with 15–60 prickles on long-styled flowers. Corolla mauve, 2.4–3 cm in diameter on long-styled flowers, 1.5–2.3 cm in diameter on short-styled flowers, lobed for $\frac{1}{3}$–$\frac{1}{2}$ of its length, lobes broadly deltate, 7–10 × 7–10 mm on long-styled flowers, 6–9 × 5–7 mm on short-styled flowers. Stamens equal; anthers 6–7.5 mm on long-styled flowers, 4.2–7 mm on short-styled flowers. Ovary stellate-pubescent in the upper $\frac{1}{4}$; style 10–13 mm long on long-styled flowers. Berries 1(–2) per infructescence, striped when young, yellow at maturity, spherical, 2.5–3.5 cm in diameter; fruiting calyx with 5–30 prickles. Seeds 2.2–2.8 × 1.8–2.3 mm.

KENYA. Northern Frontier District: Dandu, 8 Apr. 1952, *Gillett* 12729! & 89 km SW of Ramu, 16 Dec. 1971, *Bally & Smith* 14628A!
DISTR. **K** 1; Senegal, Benin, Burkina Faso, Mali, Niger, Nigeria, Sudan, Eritrea, Ethiopia, Somalia; Egypt, Arabia, Oman, Israel, Jordan, Iran, Afghanistan, Pakistan, India
HAB. Thickets, bushland, wooded grassland; 500–800 m

SYN. *S. sanctum* L., Sp. Pl. ed. 2: 269 (1762), *nom. illeg. superfl.* for *S. incanum* fide Hepper & Jaeger 1985: 388
    S. unguiculatum A. Rich., Tent. Fl. Abyss. 2: 102 (1850). Type: Ethiopia, Chiré, *Quartin Dillon & Petit* s.n. (P, lecto., designated by Lester 1997: 291; P, isolecto.)
    S. coagulans Forssk. var. *griseum* Dunal in DC., Prodr. 13(1): 369 (1852). Type: Egypt, near Kenne, *Schimper* 951 (G-DC!, lecto., designated here [collection with best duplicate distribution; specimen cited by Dunal]; AV, CAS, E, LE!, MPU, NY, P!, W, isolecto.)
    S. coagulans Forssk. var. *ochraceum* Dunal in DC., Prodr. 13(1): 369 (1852). Type: Saudi Arabia, Wadi Sel, *Schimper* 786 (G-DC!, lecto., designated here [only known material of that cited in protologue]; CAS, E, LE!, M, MPU, P!, W, isolecto.)
    S. hierochuntinum Dunal in DC., Prodr. 13(1): 369 (1852), *nom. nov.* based on *S. sanctum* L.
    S. hierochuntinum Dunal var. *lanuginosum* Dunal in DC., Prodr. 13(1): 369 (1852). Type: Ethiopia, Djeladjeranne, *Schimper* 1574 (G-DC, holo.; BM!, K!, LE!, MO!, P!, W, iso.)
    S. melongena L. var. *incanum* (L.) Kuntze, Revis. Gen. Pl. 2: 454 (1891)
    S. floccosistellatum Bitter in F.R. Beih. 16: 218 (1923). Type: Somalia, without locality, June, *Mrs. Lort Phillips* s.n. (K00441661!, lecto., designated by Vorontsova & Knapp in Taxon 59: 1596, 2010; GOET!, isolecto.)
    S. incanum L. var. *brevitomentosum* Bitter in F.R. Beih. 16: 275 (1923). Type: Eritrea, Mahio, Haddas Valley, *Schweinfurth* 556 (Z, holo.)

S. incanum L. var. *pluribaccatum* Bitter in F.R. Beih. 16: 275 (1923). Type: Sudan, Kordofan, Melbeis, *Pfund, Exped. Colston Bey* 404 (no specimens found)

S. incanum L. var. *integrascens* Bitter in F.R. Beih. 16: 276 (1923). Type: Nigeria, Sokoto Province, *Dalziel* 387 (K!, lecto., designated here [only material known]).

S. incanum L. var. *unguiculatum* (A. Rich.) Bitter in F.R. Beih. 16: 276 (1923)

S. incanum L. subsp. *schoanum* Bitter in F.R. Beih. 16: 277 (1923). Type: Ethiopia, S Schoa, Gennet, 2000 m, Mar., *Rosen* s.n. (WRSL, holo.)

S. incanum L. var. *unguiculatum* (A. Rich.) Abedin, Al-Yahya, Chaudhary & J.S.Mossa in Pakistan J. Bot. 23: 277 (1991)

NOTE. The name "*S. incanum*" has been commonly and incorrectly applied to the widespread weedy species *S. campylacanthum. Solanum incanum* sensu stricto is distinguished by its lobed leaves, yellowish drying colour, adundant long-stalked indumentum, and does not occur any further south than K 1. Herbarium specimens of *Solanum incanum* bear a striking resemblance to the similarly yellowish and densely tomentose southern African *S. lichtensteinii*; the two species are not sympatric and geographical location data can greatly simplify the identification process. The ridged stems of *S. lichtensteinii* differentiate it from *S. incanum.* The majority of yellow-fruited weeds commonly encountered around East Africa are part of *S. campylacanthum.*

Solanum incanum sensu stricto may be the closest African relative of the cultivated aubergine *S. melongena* (Weese & Bohs, Taxon 59: 49–56; 2010).

56. **Solanum lichtensteinii** *Willd.*, Enum. Pl. (Willd.): 238 (1809). Type: cultivated in Berlin, seeds from *Lichtenstein* from interior regions of South Africa, Cape of Good Hope, *Willdenow* s.n. (B-W!, holo.)

Erect herb or shrub, 0.5–2 m, armed; young stems ridged, densely stellate-pubescent, trichomes porrect to multangulate, translucent to orange-translucent, stalked, stalks up to 0.8(–2.5) mm, rays 8–20, 0.1–0.4 mm, midpoints ± same length as rays; prickles curved, 3–5 mm long, 2–5 mm wide at base, round to flattened. Leaf blades drying weakly to strongly discolorous, dirty green-brown adaxially and whitish abaxially, ovate, 8–27 × 3–20 cm, 1.5–2.5 times longer than wide, base cordate(cuneate), often unequal and oblique, margin lobed, the deltate to broadly rounded lobes 3–5 on each side, 0.5–3 cm long, apically rounded, extending $^1/_4$–$^1/_3$(–$^1/_2$) of the distance to the midvein; apex rounded to acute; densely stellate-pubescent abaxially, trichomes on abaxial surface porrect, sessile or stalked, stalks to 0.3 mm, rays ± 8(–15), 0.2–0.4 mm, midpoints ± same length as rays or elongated up to 0.7 mm, trichomes of adaxial surface smaller; primary veins 5–7 pairs; petiole 2–7 cm, $^1/_4$–$^1/_3$ of the leaf length. Inflorescences not branched, 3–8 cm long, with 5–10 flowers; peduncle 1–25 mm long; rachis 1.5–6 cm long; peduncle and rachis usually unarmed; pedicels 0.8–1.5 cm on long-styled flowers, 0.4–0.9 cm long, in fruit 1.5–3.2 cm long, with 3–10 prickles. Flowers heterostylous, (4–)5-merous, with basal one long-styled. Calyx 7–15 mm on long-styled flowers, 4.5–9 mm long on short-styled flowers, lobes long-deltate and often foliaceous and 3.5–6 mm on long-styled flowers, broadly deltate and 2–5 mm long on short-styled flowers, acute to obtuse, with 20–50 prickles on long-styled flowers. Corolla white to mauve, 2.5–3.5 cm in diameter on long-styled flowers, 1.5–2.6 cm in diameter on short-styled flowers, lobed for $^1/_4$–$^1/_2$ of its length, lobes broadly deltate, 7–10 × ± 10 mm on long-styled flowers, 5–9 × 4–10 mm on short-styled flowers. Stamens equal; anthers 5–7 mm on long-styled flowers, 4.2–6.5 mm on short-styled flowers. Ovary stellate-pubescent in the upper $^1/_3$; style 10–12 mm long on long-styled flowers. Berries 1 per infructescence, striped when young, yellow at maturity, spherical, 2.5–4.5 cm in diameter; fruiting calyx with 5–30 prickles. Seeds 2.2–3.2 × 2.2–2.5 mm.

TANZANIA. Mpanda District: Rukwa valley, Tumba, 5 Jan. 1952, *Siame* 68!; Kilwa District: Selous Game Reserve, 43km SW of Kingupira, 5 Aug. 1975, *Vollesen* 2631!; Tunduru District: 95 km from Masasi, at the base of Namakambili rock, 20 March 1963, *Richards* 17984!

DISTR. T 4, 8; Congo-Kinshasa, Angola, Zambia, Mozambique, Zimbabwe, Namibia, Botswana, South Africa

HAB. Dry grassland, woodland, and thickets; 500–700 m

SYN. *S. subexarmatum* Dunal in DC., Prodr. 13(1): 367 (1852). Type: South Africa, Cape of Good
    Hope, Omvamwubo, 500–1000 ft, *Drège* s.n. (G-DC!, holo.)
    S. homblei De Wild. in F.R. 13: 141 (1914). Type: Congo-Kinshasa, Lubumbashi [Elisabethville],
    Homblé 136 (BR! [BR0000008993441], lecto., designated here [annotated by the author];
    BR!, 3 isolecto.)
    S. incanum L. var. *lichtensteinii* (Willd.) Bitter in F.R. Beih. 16: 278 (1923)
    S. incanum L. var. *subexarmatum* (Dunal) Bitter in F.R. Beih. 16: 280 (1923)
    S. incanum L. subsp. *horridescens* Bitter in F.R. Beih. 16: 281 (1923). Type: South Africa,
    Transvaal, Königsberg, *Langenheim* 179 (HBG, holo.)

NOTES. *Solanum lichtensteinii* is the southern African member of the *Solanum incanum* sensu lato
    complex which has suffered from extensive taxonomic confusion. This species is rare in East
    Africa; specimens dry a characteristic yellowish-grey colour and there are pronounced ridges
    on the young stems. *Solanum lichtensteinii* bears a strong resemblance to the similarly yellowish
    and densely tomentose Arabian and NE-African *S. incanum*; the two species are not sympatric
    and geographical location data can greatly simplify the identification process. The
    distinctness of *S. lichtensteinii* and *S. incanum* has been demonstrated by molecular
    phylogenetic work (Weese & Bohs, Taxon 59: 49–56. 2010). The majority of single-fruited
    weeds commonly encountered around East Africa are *S. campylacanthum.*

57. **Solanum macrocarpon** *L.*, Mant. Pl. Altera: 205 (1771); C.H. Wright in F.T.A.
4, 2: 214 (1906); T.T.C.L.: 586 (1949); Heine in F.W.T.A. 2nd ed., 2: 334 (1963);
Gonçalves in F.Z. 8(4): 105 (2005); Friis in Fl. Eth. 5: 137 (2006). Type: *Hort. Uppsala*
s.n. (LINN 248.11!, lecto., designated by Hepper & Jaeger, 1985: 391)

Herb or shrub, 0.3–1 m, erect, unarmed or with a few prickles; young stems
almost glabrous, often with minute orange glands. Leaf blades drying concolorous,
distinctive red-brown, elliptic, 10–35 × 6–20 cm, ± 2 times longer than wide, base
attenuate(cuneate), usually equal, margin lobed to subentire, the lobes 3–5 on each
side, 1–7 cm long, apically obtuse to rounded, extending $\frac{1}{3}$–$\frac{2}{3}$ of the distance to the
midvein, usually with extensive secondary lobing; apex obtuse; primary veins 5–8
pairs; petiole often decurrent (0–2.5 cm), usually less than $\frac{1}{7}$ of the leaf length.
Inflorescences not branched, 3–5 cm long, with 1–10 flowers; peduncle 0–5 mm
long; rachis 0.4–4 cm long; pedicels 1–3 cm long and stout in long-styled flowers,
0.6–1.5 cm and slender in short-styled flowers, in fruit 1–4 cm long. Flowers
heterostylous, 5-merous, basal one long-styled. Calyx 12–35 mm on long-styled
flowers, 6–20 mm long on short-styled flowers, lobes deltate, 6–30 mm on long-
styled flowers, 4–10 mm long on short-styled flowers, acute to long-acuminate,
unarmed or with a few prickles. Corolla white to purple, 3.5–5 cm in diameter on
long-styled flowers, 1.2–2.5 cm in diameter on short-styled flowers, lobed for $\frac{1}{4}$–$\frac{1}{2}$
of its length, lobes broadly deltate, 5–10 × 10–20 mm on long-styled flowers, 4–10 ×
7–10 mm on short-styled flowers. Stamens equal; anthers 4–7 mm. Ovary with
simple hairs on the upper $\frac{1}{2}$; style ± 10 mm long on long-styled flowers. Berries
1(–2) per infructescence, striped when young, yellow to orange (white or almost
black) at maturity, spherical, 4–6 cm in diameter; fruiting calyx with occasional
prickles. Seeds 2.8–3.8 × 2.2–3.2 mm.

UGANDA. Masaka District: Kyotera, Nov. 1945, *Purseglove* 1861!
TANZANIA. Bukoba District: Bulembe Hill, N of Kagera River, 21 Nov. 1999, *Sitoni et al.* 962!;
    Ulanga District: Ifakara, Mbasa, 4 Aug. 1958, *Haerdi* 52/92!
DISTR. U; **K**; T; probably cultivated throughout Africa, but collections are sporadic: Sierra
    Leone, Ghana, Nigeria, Cameroon, Gabon, Equatorial Guinea, Central African Republic,
    Congo-Kinshasa, Ethiopia, Angola; Egypt, Guatemala, Brazil
HAB. Cultivated species: "Gboma eggplant"

SYN. *S. thonningianum* Jacq., Ecl. Pl. Rar. 1: 123, t. 83 (1816). Type: cultivated in Vienna, original
    collection made in Ghana, 1807, *Thonning* s.n. (W?, holo., not found)

S. atropo Schumach. & Thonn. in Beskr. Guin. Pl.: 124 [144] (1827). Type: Ghana, *Thonning* 117 (C!, lecto., designated here [IDC microfiche 101: III.5–6 "522"; best material]; C, LE, isolecto.)

S. macrocarpon L. var. *parcesetosum* Bitter in F.R. Beih. 16: 199 (1923). Type: Equatorial Guinea, Bebai, Akum, *Tessmann* 577 (K! lecto, designated here [only material known])

S. macrocarpon L. var. *primovestitum* Bitter in F.R. Beih. 16: 200 (1923); T.T.C.L.: 586 (1949). Type: Mozambique, Kibata, S side of the Matumbi Mts, *Busse* 3111; Tanzania, Lushoto District, West Usambara, Wilhelmstal, Amani, *Braun* in Herb. Amani 1833, 1 (no specimens of either syntype found)

S. macrocarpon L. var. *setosiciliatum* Bitter in F.R. Beih. 16: 200 (1923). Type: Togo, near Sansugu, W of Basari, *Kersting* 552 (no specimens found)

[Synonyms of non-African material can be found on the Solanaceae Source website: http://www.solanaceaesource.org]

NOTES. Cultivated for its fruits and as a leaf vegetable. The attenuate leaf bases, long calyx lobes covering the developing fruit, and an easily recognisable red-black drying colour are distinguishing features. *Solanum macrocarpon* can be confused with the leaf vegetable varieties of *S. aethiopicum* due to their similar rounded-lobed dark red leaves; it can be distinguished by the absence of clear petioles, attenuate leaf bases and fruit more than 4 cm in diameter.

It is now widely accepted that *S. macrocarpon* is the cultivated form of the wild *S. dasyphyllum*. The distinction between the cultivated *S. macrocarpon* and the wild *S. dasyphyllum* is largely artificial and maintained here for practical purposes. *Solanum macrocarpon* can be distinguished from *S. dasyphyllum* by the absence of prickles and indumentum, and fruit more than 4 cm in diameter. *Solanum macrocarpon* generally has smaller leaves, leaves that are less lobed, and lobes more rounded than *S. dasyphyllum*.

58. **Solanum melongena** *L.*, Sp. Pl.: 186 (1753); C.H. Wright in F.T.A. 4, 2: 242 (1906); Heine in F.W.T.A. 2nd ed., 2: 332 (1963); U.O.P.Z.: 446 (1949); Friis in Fl. Eth. 5: 138, fig. 158.14 (2006) & in Fl. Somalia 3: 216 (2006). Type: Hort. Uppsala, *Anonymous* s.n. (LINN 248.28!, lecto., designated by Schönbeck-Temesy in Fl. Iran. 100: 70, 1972)

Annual or perennial herb, 0.2–0.5 m, erect, unarmed(armed); young stems moderately stellate-pubescent to glabrescent, trichomes porrect, translucent, sessile or stalked, stalks up to 0.2 mm, rays 8–15, 0.3–0.7 mm, midpoints ± same length as rays or to 1 mm; prickles straight, 0–6 mm long, 0.2–1 mm wide at base. Leaf blades drying concolorous to weakly discolorous, green-brown, ovate, 7–23 × 5–17 cm, 1.5–2 times longer than wide, base usually rounded, sometimes obtuse to cordate, often unequal and oblique, margin lobed, the broadly rounded lobes 1–3 on each side, 0.5–2 cm long, apically rounded, extending $^{1}/_{4}$–$^{1}/_{3}$ of the distance to the midvein; apex rounded to acute; moderately stellate-pubescent on both sides, trichomes on abaxial surface porrect, sessile or stalked, stalks to 0.2, rays 5–8, 0.3–1 mm, midpoints ± same length as rays, trichomes of adaxial surface smaller; primary veins 4–5(–7) pairs; petiole 1.5–5(–10) cm, $^{1}/_{4}$–$^{1}/_{3}$(–$^{2}/_{3}$) of the leaf length. Inflorescences not branched, 4–7 cm long, with 1–5 flowers; peduncle 0–2 mm long; rachis 0–4 cm long; peduncle and rachis unarmed; pedicels 2–3 cm on long-styled flowers, 0.8–2 cm long on short-styled flowers, in fruit 2–9 cm long, unarmed or with up to 5 prickles. Flowers heterostylous, 4–8-merous, only the basal one long-styled. Calyx 8–23 mm on long-styled flowers, 7–12 mm long on short-styled flowers, lobes deltate to long-deltate, 6–12 mm long, 3–8 mm long on short-styled flowers, acute to long-acuminate, unarmed or with up to 20 prickles. Corolla white to purple, 2.5–5 cm in diameter on long-styled flowers, 2.4–4 cm in diameter on short-styled flowers, lobed for $^{1}/_{4}$–$^{1}/_{2}$ of its length, lobes broadly deltate, 6–18 × 7–12 mm on long-styled flowers, 5–13 × 6–12 mm on short-styled flowers. Stamens equal; anthers 5.5–7.5 mm on long-styled flowers, 5.5–7 mm on short-styled flowers. Ovary stellate-pubescent in the upper $^{1}/_{4}$; style ± 9 mm long on long-styled flowers. Berries 1 per infructescence, green, sometimes mottled or striped, white, pink, mauve, purple, or black when young, usually white or dark purple at maturity, variously shaped, 3–20 × 3–7 cm; fruiting calyx unarmed or with up to 30 prickles. Seeds 2.9–3.2 × 2.2–2.5 mm.

KENYA. Kiambu District: Muguga, Dec. 1960, *Verdcourt* 12303!
TANZANIA. Lushoto District: Korogwe, Old Ambangulu village, 15 Apr. 1999, *Mwangoka* 433!
DISTR. U; **K**; **T**; probably cultivated throughout, but collections are sporadic: Sierra Leone, Ghana, Togo, Nigeria, Central African Republic, Zambia; South East Asia
HAB. Cultivated species: "Aubergine/eggplant"

SYN. [There are numerous synonyms of the cultivated eggplant, but these are not treated here]

NOTES. The characters distinguishing *S. melongena* from the rest of the *incanum* group are mainly those directly associated with cultivation: larger fruit, altered fruit shape and colour, and lack of prickles. Like the fruit crop cultivars of *S. aethiopicum*, it can be recognised by fasciation in the flowers: increase in the number of flower parts up to 8, inflated ovaries, and straight thick styles not exserted further than 2 mm above the anthers.

The cultivated eggplant has close relationships with *S. campylacanthum* and *S. incanum* sensu stricto but was domesticated in Indo-China (Wang et al. in Ann. Botany 102: 891–897, 2008; Weese & Bohs in Taxon 59: 49–56, 2010). One uniform South East Asian cultivar is commonly grown in East Africa but it does not represent a significant food source.

59. **Solanum nigriviolaceum** *Bitter* in F.R. Beih. 16: 163 (1923). Type: Kenya, Kiambu District, Limuru, *Scheffler* 278 (W!, lecto., designated by Vorontsova & Knapp in Taxon 59: 1599, 2010; BM!, GOET!, K!, isolecto.)

Prostrate or climbing subshrub, 0.3–1(–3) m, heavily armed, young stems moderately stellate-pubescent, trichomes porrect, orange-translucent, usually sessile, rays 4–8, 0.1–0.4 mm, midpoints 1.5–2.5 mm long, visible to the naked eye; prickles straight, 5–9 mm long, 0.3–1.5 mm wide at base. Leaf blades drying slightly discolorous, yellow-green, ovate, 6–12 × 5–11 cm, ± 1.5 times longer than wide, base cordate(cuneate), often oblique, margin lobed, the lobes ± 2 on each side, 1–3 cm long, apically obtuse, extending $^1/_3$–$^2/_3$ of the distance to the midvein, with secondary lobing; apex obtuse; moderately stellate-pubescent on both sides, trichomes on abaxial surface multangulate, sessile or stalked, stalks to 0.2 mm, rays 4–8, 0.2–0.6 mm, midpoints 1.5–2.5 mm long, visible to the naked eye, adaxially with inflated stalks, reduced rays, and midpoints 1–2.5 mm; primary veins 3–5 pairs; petiole 1.5–6 cm, ± $^1/_3$ of the leaf length. Inflorescences not branched, 8–12 cm long, with 3–7 flowers; peduncle 0–35 mm long; rachis 3–10 cm long; peduncle and rachis densely armed; pedicels 2–5.5 cm long on basal flowers, 0.8–1.5 cm on distal flowers, in fruit 2–6 cm long, with numerous prickles. Flowers heterostylous, 5(–6)-merous, the basal one long-styled. Calyx 13–20 mm on long-styled flowers, 7–18 mm long on short-styled flowers, lobes long-deltate, 5–17 mm long, long-acuminate, densely armed on long-styled flowers. Corolla mauve to purple, 4.5–5 cm in diameter on long-styled flowers, 3.5–4.5 cm in diameter on short-styled flowers, lobed for $^1/_2$–$^2/_3$ of its length, the lobes oblong to broadly deltate, 10–22 × 8–18 mm. Stamens equal; anthers 7–9 mm. Ovary with simple hairs on the upper $^1/_2$; style 13–17 mm long on long-styled flowers. Berries 1 per infructescence, striped when young, yellow at maturity, spherical, 2.5–3 cm in diameter; fruiting calyx with numerous prickles. Seeds 2.8–3.2 × 2–2.5 mm.

KENYA. Nakuru District: Eburru Forest Reserve, near camp 269, 17 July 2002, *Luke et al.* 8939!; Kisumu-Londiani District: Mt Londiani western highlands, 5 Dec. 1974, *J.G. Williams* 62!; Masai District: Nasampolai [Enesambulai] Valley, 2 June 1969, *Greenway & Kanuri* 13641!
DISTR. **K** 3–6; not known elsewhere
HAB. Open ground, grassland and forest edges; often locally common; 2500–3000 m

SYN. *S. sessilistellatum* Bitter in F.R. Beih 16: 187 (1923); Blundell, Wild Fl. E. Africa: 191, t. 689 (1992); U.K.W.F., 2nd ed.: 243 (1994). Type: Kenya, "Masai Highlands, Mau Plateau, Pajo Mountain, Leikipia Plateau, Baringo Lake, Naiwascha, Mau-edge near Londiani", *Baker* 39 (B†, holo.)

NOTES. Morphologically uniform Kenyan endemic species commonly confused with *S. aculeatissimum* and *S. dasyphyllum*. *Solanum aculeatissimum* has similar purple stems, a hirsute appearance, abundant long straight prickles, and sometimes similar leaf shape, and but its flowers are much smaller with a more deeply dissected stellate white corolla, and its stem trichomes are simple and have no rays, unlike the stellate trichomes in *S. nigriviolaceum*. *Solanum dasyphyllum* has similar straight prickles and elongated midpoints but its leaves are larger with secondary lobing and attenuate leaf bases, and its flowers have less dissected corollas and finer pedicels.

Herbarium material of *S. nigriviolaceum* is frequently annotated as *S. sessilistellatum*. The two names were published simultaneously and *S. nigriviolaceum* has been chosen as the accepted name because of extant type material.

60. **Solanum phoxocarpum** *Voronts.* in Syst. Bot. 35: 903 (2010). Type: Kenya, Masai District, Lake Naivasha to Nasampolai [Enesambulai] Valley, crest of the Western Rift Wall, *Greenway & Kanuri* 13869 (EA!, holo.; K!, iso.)

Shrub or tree, erect, 1–3(–6) m, armed, young stems densely stellate-pubescent, trichomes multangulate, white-translucent, irregular and densely matted, sessile, rays 11–16, 0.1–0.2 mm, midpoints same length as rays or up to 0.8 mm long; prickles straight becoming gently curved, 6–15 mm long, 2–7 mm wide at base, rounded to flattened. Leaf blades drying strongly discolorous, reddish-green above, white-grey underneath, elliptic, 6–8 × 2.5–4 cm, ± 2.5 times longer than wide, base cuneate, usually equal, margin subentire to weakly lobed, the broadly rounded lobes 1–2(–3) on each side, up to 0.5(–1.5) cm long, extending up to $^1/_3$ of the distance to the midvein, apex acute; densely stellate-pubescent abaxially, trichomes on abaxial surface porrect and multangulate, sessile and stalked, stalks ± 0.1(–0.2) mm, rays 11–16, 0.15–0.3 mm, midpoints same length as rays or up to 0.8 mm, adaxially glabrescent; primary veins (4–)5–6 pairs; petiole 0.5–0.9 cm, ± $^1/_6$ or less of the leaf length. Inflorescences 3–5.5 cm long, not branched, with 1–7 flowers; peduncle 0–6 mm long; rachis 0–0.3 cm long; peduncle and rachis unarmed; pedicels 1.8–3 cm on long-styled flowers, 0.8–1.2 cm on short-styled flowers, in fruit 2.5–4 cm long, usually with 2–10 prickles. Flowers heterostylous, 5(–6)-merous, the basal 1–3 long styled. Calyx ± 15 mm on long-styled flowers, ± 7 mm long on short-styled flowers, lobes long-deltate, 7–10 mm on long-styled flowers, ± 4 mm long on short-styled flowers, acute to apiculate, unarmed or with up to 15 prickles in long-styled flowers. Corolla mauve, ± 3 cm in diameter on long-styled flowers, ± 1.7 cm in diameter on short-styled flowers, lobed for $^1/_2$–$^2/_3$ of its length, lobes deltate, 10–15 × 4–5 mm on long-styled flowers, 6 × 3–4 mm on short-styled flowers. Stamens equal; anthers 3.5–4 mm. Ovary stellate-pubescent in the upper $^1/_5$; style ± 7 mm long on long-styled flowers. Berries 1–3(–5) per infructescence, evenly green when young, yellow at maturity, conical, 2.8–3.7 × 1.8–2.2 cm, acute; fruiting calyx with 0–10 prickles. Seeds 4–4.5 × 3–4 mm.

KENYA. Naivasha District: South Kinangop, 11 July 1965, *Gillett* 16766!; Kiambu District: Limuru, 30 Aug. 1938, *Bally* 7438!; Masai District: Olokurto, Mau Area, 13 May 1961, *Glover et al.* 935!
TANZANIA. Masai District: Ngorongoro Crater, Embagai [Empakai], 16 Aug. 1973, *Frame* 229!; Lushoto District: Gologolo, 9 June 1959, *Mgaza* 172!
DISTR. **K** 3, 4, 6; **T** 2, 3; not known elsewhere
HAB. Woodland, moist forest understorey or secondary scrub; 2100–3000 m

SYN. *S. aculeastrum* Dunal var. 1 sensu Polhill, *Solanum* in E & NE Africa: 42 (ined., 1961)
   S. aculeastrum Dunal subsp. 1 sensu Jaeger, Syst. stud. *Solanum* in Africa: 441 (1985)
   Solanum sp. K sensu U.K.W.F.: 243 (1994)

NOTES. Fully sympatric with the closely related *S. aculeastrum*, *S. phoxocarpum* can be recognised by the unusual cylindrical pointed fruits, subentire leaves on fertile branches, and mauve flowers. *Solanum phoxocarpum* and *S. aculeastrum* frequently grow together but no intermediate individuals have been observed. Morphology of *S. aculeastrum* is reminiscent of typical juvenile *Solanum* morphology with more leaf lobing and abundant prickles, while the morphology of *S. phoxocarpum* is more similar to the typical mature *Solanum* morphology with more entire leaves and fewer prickles.

Udo Dammer also recognised the distinctness of these plants and annotated the sheet *Scheffler* 306 (K) as "*Solanum sepiaceum* Dammer var. *fructile verrucans* spec. nov." in his handwriting, with a printed label "Brit. Uganda. Station Lamuru. Buschiges Hochland. b.c. 3000 m". This name does not seem to have been published and the specimen is not cited in the protologue of *S. sepiaceum* Dammer. "Station Lamuru" refers to Limuru in Kenya.

61. Solanum richardii *Dunal*, in Poir., Encycl., Suppl. 3: 775 (1814); T.T.C.L.: 587 (1949); K.T.S.L.: 582 (1994); Gonçalves in F.Z. 8(4): 109 (2005). Type: Madagascar, *Richard* s.n. (P00352404!, lecto., designated here)

Prostrate or climbing shrub to 2 m, heavily armed; young stems sparsely to densely stellate-pubescent, trichomes porrect, translucent or orange-brown to bright orange, sessile or stalked, stalks 0.1–0.5(–1.5) mm, rays 4–8, 0.1–0.4 mm, midpoints shorter than rays or up to 1.5 mm; prickles curved, 1–10 mm long, 1–5 mm wide at base, flattened. Leaf blades drying discolorous, green-brown to red-green, ovate to elliptic, 7–22 × 5–13 cm, 1.5–2.5 times longer than wide, base cordate (truncate or cuneate), usually oblique, margin lobed or sometimes entire, the lobes (1–)3–6 on each side, broadly rounded to deltate, oblong, or obovate, up to 5 cm long, rounded to acute, extending $^1/_3$–$^3/_4$ of the distance to the midvein, frequently with secondary lobing; apex acuminate to obtuse; moderately to sparsely stellate-pubescent, trichomes on abaxial surface porrect, sessile or stalked, stalks 0.1–0.2(–1.2) mm, rays 6–8, 0.1–0.3 mm, midpoints same length as rays or shorter, adaxially glabrescent, the trichomes reduced; primary veins 5–10 pairs; petiole 1–8 cm, $^1/_2$–$^1/_5$ of the leaf length. Inflorescences not branched or branched once, 6–11 cm long, with 3–10 flowers; peduncle 2–25 mm long; rachis 1–5 cm long; peduncle and rachis usually with numerous prickles; pedicels 1–2.5 cm long, in fruit 2–4 cm long, with numerous prickles. Flowers heterostylous, 5-merous, the basal 1–6 long-styled. Calyx 7–15 mm long, lobes deltate to ovate, 5–12 mm long, acute to somewhat acuminate, densely armed with 10–100 prickles. Corolla mauve to purple, 3.5–6 cm in diameter, lobed for $^1/_2$–$^3/_4$ of its length, lobes broadly deltate, 10–20 × 8–20 mm. Stamens equal; anthers 8.5–11.5 mm. Ovary stellate-pubescent in the upper $^1/_3$ only; style 10–20 mm long on long-styled flowers. Berries 2–6 per infructescence, glaucous green with dark green markings when young, yellow to orange at maturity, often elongate when young, 3–5 cm in diameter; fruiting calyx with 10–100 prickles. Seeds 3–4 × 3–3.7 mm.

TANZANIA. Tanga District: Pongwe, Maweni, 19 Oct. 1965, *Faulkner* 3690!; district unclear: Ngambaula Forest Reserve, 22 Aug. 2000, *Mhoro UMBCP*412!; Lindi District: Rondo Plateau, Mchinjiri, 10 Dec. 1955, *Milne-Redhead & Taylor* 7606!; Pemba: Wete [Weti], 10 Nov. 1929, *Vaughan* 942!
DISTR. **T** 3, 6–8; **Z, P**; Congo-Kinshasa, Zambia, Malawi, Mozambique, Zimbabwe; Comoros, Madagascar
HAB. Disturbed areas, open bushland with grass, open forest, thickets and roadsides; 0–1300 m

SYN. *S. richardii* Dunal var. *pallidum* Dunal in DC., Prodr. 13(1): 326 (1852). Type: Comoro Islands, Anjouan [Johanna], *Bojer* s.n. (G-DC!, holo.)
  S. acanthocalyx Klotzsch in Peters, Reise Mossamb. 6(1): 232 (1861); C.H. Wright in F.T.A. 4, 2: 235 (1906). Type: Comoro Islands, Anjouan [Anjoana], *Peters* s.n. (B†, holo.)
  S. magnusianum Dammer in E.J. 28: 475 (1900); C.H. Wright in F.T.A. 4, 2: 240 (1906). Type: Tanzania, Kilosa District: between Khutu and Uhehe, Kidodi, base of Vidunda Mts, Ruhambe River, *Goetze* 384 (K!, lecto., designated here [only extant material known])
  S. bathocladon Dammer in E.J. 28: 476 (1900); C.H. Wright in F.T.A. 4, 2: 240 (1906). Type: Tanzania, Morogoro District, E Uluguru, Lussegwa, *Stuhlmann* 8731 (B†, holo.)
  S. acutilobatum Dammer in E.J. 53: 338 (1915). Type: Zimbabwe, Chirinda forest, *Swynnerton* 93 (K!, lecto., designated by Vorontsova & Knapp in Taxon 59: 1594, 2010; BM!, GOET!, K!, isolecto.)
  S. burtt-davyi Dunkley in K.B. 1937: 471 (1937). Type: Malawi, Mt Nchisi, *Burtt Davy* 21193 (K!, holo.)

S. richardii Dunal var. *acutilobatum* (Dammer) A.E. Gonç. in Garcia de Orta, Ser. Bot. 11: 72 (1993); Gonçalves in F.Z. 8(4): 110, t. 20 (2005)

S. richardii Dunal var. *burtt-davyi* (Dunkley) A.E. Gonç. in Garcia de Orta, Ser. Bot. 11: 73 (1993)

NOTES. This attractive and commonly collected species with large showy flowers and prominent hooked prickles is morphologically fairly uniform and immediately recognisable across Madagascar, Mozambique, Zambia and Malawi. Tanzanian populations appear to be more diverse and occasional Tanzanian specimens have smaller leaves, smaller flowers, and smaller anthers, intermediate with the higher elevation *S. stipitatostellatum* and the coastal small-flowered *S. zanzibarense*. Individuals with entire leaves and dense orange indumentum on fertile branches have previously been recognised as *S. burtt-davyi* but such morphological forms are occasional and not geographically coherent.

62. **Solanum thomsonii** *C.H. Wright* in F.T.A. 4, 2: 217 (1906); C.H. Wright in F.T.A. 4, 2: 217 (1906); T.T.C.L.: 587 (1949). Type: Tanzania, Rungwe/Njombe District, lower plateau N of Lake Malawi, *Thomson* s.n. (K! [K000414127], holo.)

Erect shrub or tree, 1–4 m, armed; young stems densely stellate-pubescent, trichomes porrect and multangulate, white-translucent to yellow-orange, stalked, stalks 1–3 mm, rays 8–10(–15), 0.1–0.25 mm, midpoints shorter than rays; prickles straight becoming curved, 6–15 mm long, 3–7 mm wide at base, flattened. Leaf blades drying strongly discolorous, dark yellow-green above, yellowish grey underneath, ovate, 8–18 × 7–12.5 cm, ± 1.5 times longer than wide, base cordate to auriculate (truncate), usually equal, margin lobed, the lobes (2–)3–4 on each side, oblong to deltate or obovate, (1.5–)2–4(–5) cm long, apically rounded, extending $^{1}/_{3}$–$^{2}/_{3}$ of the distance to the midvein, often with secondary lobing; apex acute; densely stellate-pubescent abaxially, trichomes on abaxial surface porrect or multangulate, stalked, stalks 0.3–1 mm, rays 8–10, 0.25–0.8 mm, midpoints shorter than rays, adaxially glabrescent, the trichomes reduced; primary veins 3–4(–5) pairs; petiole 1.5–4 cm, $^{1}/_{4}$–$^{1}/_{6}$ of the leaf length. Inflorescences frequently branched, 4–8 cm long, with 6–15 flowers; peduncle 15–40 mm long; rachis 1–4.5 cm long; peduncle and rachis unarmed; pedicels 1–1.5 cm long, in fruit 0.9–2.1 cm long, unarmed. Flowers heterostylous, 5(–6)-merous, the basal 3–7 long-styled. Calyx 10–15 mm long, the lobes long-deltate, 5–8 mm long, acuminate, usually unarmed. Corolla white, 2.3–3 cm in diameter, lobed for $^{1}/_{2}$–$^{2}/_{3}$ of its length, lobes deltate, 7–12 × 4–5 mm. Stamens equal; anthers 4–6 mm. Ovary with stellate trichomes on the upper $^{1}/_{3}$; style 7–11 mm long on long-styled flowers. Berries 4–10 per infructescence, evenly green when young, yellow-orange at maturity, spherical, 1.4–1.7 cm in diameter; fruiting calyx unarmed. Seeds 3.5–4.5 × 3.5–3.8 mm.

TANZANIA. Njombe/Mbeya District: Kitulo [Elton] Plateau, 30 Nov. 1963, *Richards* 18471!; Iringa District: Mufindi, Ifupira, 18 Mar. 1962, *Polhill & Paulo* 1808! & Udzungwa Mountain National Park, pt. 355, 11 Oct. 2002, *Luke et al.* 9074!

DISTR. **T** 7; not known elsewhere

HAB. Forest edges, thickets and roadsides; 1800–2400 m

SYN. *S. protodasypogon* Bitter in F.R. Beih. 16: 173 (1923); T.T.C.L.: 586 (1949). Type: Tanzania, Njobe District: Kitogo [Kidoko], *von Prittwitz* 53 (E!, lecto., designated here [only material known])

NOTES. Previously considered to be a synonym of the variable *S. aculeastrum*, this commonly observed Southern Highland endemic differs from *S. aculeastrum* primarily by its numerous small round fruits, but also by its long yellowish-orange indumentum with stout branches, short internodes and few prickles. It seems to occupy higher altitudes than *S. aculeastrum* in the Southern Highlands and few intermediate individuals have been observed at the boundary. It is particularly frequent in villages and roadsides.

63. **Solanum arundo** *Mattei* in Boll. Reale Orto Bot. Palermo, 7: 188 (1908); Chiovenda, Fl. Somalia: 241 (1929) & Fl. Som 2: 335 (1932); Polhill, *Solanum* in E & NE Africa: 33 (ined., 1961); E.P.A.: 862 (1963); Whalen in Gentes Herb. 12: 263 (1984); Jaeger, Syst. stud. *Solanum* in Africa: 418 (1985, ined.); Bukenya & Hall in Bothalia 18, 1: 84 (1988); K.S.T.L.: 580 (1994); U.K.W.F., 2nd ed.: 243 (1994); Friis in Fl. Somalia 3: 215 (2006). Type: Somalia, Mogadishu, *Macaluso* 82 (PAL, holo.)

Shrub or small tree to 5 m high, often much-branched, with reddish smooth flaking bark; young stems densely whitish-pubescent with sessile or shortly (–0.2 mm) stalked stellate hairs to 0.5 mm diameter with erect median rays and ± 8 equal eglandular rays to 0.2 mm long, glabrescent; simple glandular hairs often also present throughout; main stems and branches with sharp pyramidal downwardly recurved prickles, 5–8(–14) × 6–8 mm basally. Leaves alternate or opposite, usually dark green on both surfaces, shiny, sometimes rough, broadly ovate, rarely lanceolate, 1.1–5(–7) × 0.7–3.8(–6.5) cm, bases cuneate, margins repand or sinuate with 0–3 deep or shallow obtuse lobes on either side, apices usually obtuse; both surfaces stellate-tomentose to glabrescent, the hairs to 0.8 mm diameter with up to 12 equal rays 0.2–0.3 mm long, below usually denser on the midribs and main veins, always with straight acicular yellow prickles to 19 × 1.2–3 mm on both midribs, sometimes smaller ones also present on main veins; petioles 0.1–1 cm long, occasionally with straight acicular prickles. Inflorescences terminal to lateral, few–16-flowered simple or forked to multiply-branched, lax, racemose or scorpioid cymes, often with basal pedicel arising at or near junction with stem. Flowers usually 5-(–6)-merous; axes initially white stellate-tomentose, becoming woody and glabrescent; peduncles erect, 0–13 mm long in flower and fruit; pedicels erect to recurved and 4–11 mm long in flower, strongly recurved and 7–14 mm in fruit, with an occasional pyramidal prickle. Calyx white to green, campanulate/cupulate, (4–)6–10 mm long, stellate-tomentose on tube externally, with acicular prickles on lobes and veins of tube; lobes often unequal, narrowly triangular to ovate, becoming strongly recurved between corolla lobes, (3–)4–7 × 2.5–5 mm, acute to apiculate; adherent basally in fruit, becoming reflexed, 4–10 × 3–8 mm, glabrescent, often with small acicular prickles. Corolla blue to purple, occasionally white, often with a prominent internal veining, deeply stellate, 2–3.4 cm diameter, tube 1–2 mm long, glabrous; lobes obovate, ovate or lanceolate, 5–14 × 2.5–5 mm, acute, often with inrolled margins, stellate-tomentose externally and on veins, margins and lobe apices internally, strongly reflexed after anthesis. Stamens usually equal; filaments free for 0.5–1.8 mm, glabrous; anthers bright yellow to orange, poricidal, 5–7(–8) × 0.9–2 mm basally, curved to connivent. Ovary 0.9–3 × 0.8–2.5 mm, stellate-pubescent, densely pilose around junction with style, 2–4-locular; style often S-shaped, 9–14 mm long in hermaphrodite flowers, stellate-pubescent in lower part, exserted up to 7 mm, 1.8–5.8 mm long in 'male' flowers with the long median rays of the stellate hairs often forming a sheath around the stylar base and upper ovary; stigma clavate, occasionally bi-lobed or sinuate, 1–1.8 × 0.5–1.2 mm in hermaphrodite flowers, and 0.4–0.8 × 0.3–0.6 mm in 'male' flowers. Berries rough, mottled or striped green and cream when immature, finally yellow, dull, globose to spheroid with thick (0.3–2 mm) tough pericarp, 1.2–2.8 × 1.5–3.5 mm, ± glabrous. Seeds > 100 per berry, smooth, light to dark brown, obovoid to orbicular, occasionally reniform, 1.8–3.2 × 1.6–2.5 mm, not flattened, shallowly punctuate/verrucate; sclerotic granules usually absent. Fig. 26/1–8, p. 217.

KENYA. Meru District: Gariba, 8 km N of Isiolo, 19 Jan. 1962, *Stewart* 632!; Masai District: Selangai Game Post, 16 Dec. 1969, *Kibue* 103!; Machakos District: Nairobi–Mombasa Road near Athi, 28 Aug. 1959, *Verdcourt* 2356!
TANZANIA. Arusha District: Arusha National Park, 5 May 1965, *Richards* 20351!; Masai District: Masailand, track through Lisingita area, 9 Jan. 1969, *Richards* 23691!; Mbulu District: Katesh, 2 May 1962, *Polhill & Paulo* 2283!
DISTR. **K** 1, 4, 6; **T** 2; Somalia

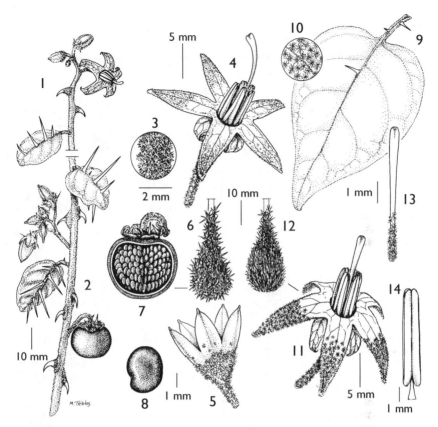

FIG. 26. *SOLANUM ARUNDO* — **1**, apical shoot; **2**, lower stem with young inflorescence and mature berry; **3**, stem indument; **4**, flower; **5**, calyx; **6**, ovary; **7**, fruit section; **8**, seed. *S. DENNEKENSE* — **9**, leaf; **10**, leaf indument; **11**, flower; **12**, ovary; **13**, stigma and style; **14**, stamen. 1 & 3 from *Tweedie* 1868; 2 from *Kibue* 103; 4–6 from *Hepper & Jaeger* 7262; 7–8 from *Bally* 12496; 9–10 from *Kirima et al.* 048167; 11–14 from *Newbould* 5859. Drawn by Margaret Tebbs.

HAB. *Acacia* or *Acacia-Commiphora* bushland, *Acacia-Balanites* wooded grassland, scattered tree grassland, thickets, grassland, semidesert, abandoned cultivations and disturbed or heavily overgrazed land, on lava rocks, ant-hills, bare slopes and riverine banks; 300–2150 m

SYN. *S. diplacanthum* Dammer in E.J. 48: 245 (1912); Bitter in F.R. 16: 145 (1923); T.T.C.L.: 580 (1949). Type: Tanzania, *Fischer* 133 (B†, holo.)
   S. helleri Standley in Smithsonian Misc. Collect. 68, 5: 15 (1917). Type: Kenya, Northern Frontier District: Northern Guaso Nyiro, *Heller* s.n. (US, holo., photo!)

NOTE. Bitter (in F.R. 16; 142, 1923) thought that his new section *Ischyracanthum* was composed of a small group of shrubs, ± restricted to Somalia, which stood between the sections *Oliganthes* and *Andromonoecum*. He considered that the prickle morphology and small leaves in the three species *S. ogadense* Bitter, *S. dennekense* Dammer and *S. diplacanthum* Dammer were so distinct from all other African Solanums that their separation into a new section was necessary. Whalen (1984) considered this to be an unusual species complex without any clear relatives either in Africa or elsewhere. Jaeger (1985) thought the andromonoecious flowers and yellow fruits were suggestive of section *Melongena* species, though the tough pericarp was a unique character not found in any other African Solanums. Many authors have reviewed andromonoecy in *Solanum*, including Whalen & Costich (in D'Arcy (ed), Solanaceae Biology & Systematics: 284–302 (1986)). Although this term strictly refers to both hermaphrodite and male flowers occurring on the same plant – the two sect. *Ischyracanthum* species dealt with here could more accurately be described as heterostylous – all flowers have well-developed anthers and all the ovaries examined had similar sized and shaped ovules. The styles however differ in length, with those from the more distal areas of the plant being sometimes (but not always) short and enclosed within the anther cone, while the majority are long and exserted. The protologues of two of these species and their descriptions in floras (e.g. Friis 2006) only give the characteristics of the hermaphrodite flowers. Bitter (1923) did however mention male flowers with 'reduced gynoecia' in the description of his new section but while he gave the stylar length of 'male' flowers in the protologue of *S. ogadense,* he only gave the stylar dimensions of hermaphrodite flowers in the descriptions of *S. dennekense* and *S. diplacanthum.*

The inflorescence stucture too appears complex in these species. Although Mattei described them as being forked and many-flowered in his protologue of *S. arundo,* they are often described as being few-flowered. Bitter (1923), when characterising his section *Ischyracanthum,* described the inflorescences as being "few (4–7–11)-flowered", later citing those in *S. dennekense* and *S. diplacanthum* as being 4–7- and 4–5-flowered respectively. Moreover, they were also described as being supposely simple in the latter two species but 'once-forked' in *S. ogadense* (Bitter 1923). However, from the available herbarium material, condensed unopened buds and pedicels scars on the rachides indicate that both simple and complex branched cymes or racemes can occur on the same plant in these species. Simple inflorescences (2–6(–10) flowered) often occur terminally, while forked or multiply branched inflorescences tend to be lateral and up to around 16-flowered, though there can be up to 30+ with the inflorescence being up to 3.8 cm in diameter. It is possible that the reference to inflorescences being 1–2(–3) flowered by some authors (e.g. Friis, 2006 for *S. arundo*) may be due to including only fully open flowers. In addition, a basal pedicel often occurs, either at or very near (within 2 mm) the base of the vestigial or short peduncle. Specimens exhibit few mature berries with often only one berry apparently maturing on each infructescence. This is probably due to the andromonoecy characterising both *S. dennekense* and *S. arundo.*

The powdery stellate indumentum of these species is also characteristic; it often sloughs off when the plants are handled, while the dense hair sheath clasping the stylar bases can usually be detached as a unit. Though small leaves characterise these species, their dimensions did not seem to be a consistent delimiting feature, and indeed seemed to be linked to habitat location.

Bitter (1923) thought that *S. arundo* was possibly similar to or synonymous with one of the three species that he recognised, but considered Mattei's protologue to be insufficient for a valid decision. In the absence of being able to examine type material he thought that the literal translation of Mattei's berry description inferred that the enlarged fruiting calyx was accrescent and that this species could be more closely allied to *S. dubium* in the section *Monodolichopus* Bitter than to those in his new section. However, it is likely that Mattei's description merely referred to the fruiting calyx lobes being persistent and adherent to the berry bases rather than accrescent and completely enclosing the berry.

Bukenya & Hall (1988) thought that plants of *S. arundo* cultivated in Legon Botanical Garden, Ghana had probably been introduced from northern Kenya, and that it was not naturalised in Ghana. They also noted that this species is concentrated in the drier parts of East Tropical Africa and that it has also been recorded from the west coast of India. The seeds of Ethiopian specimens seem to be larger than those found on East African specimens whereas the flowers of Kenyan plants are reported to be larger than those in Somalia.

Many herbarium specimens of *S. arundo* lack leaves – though they often include leaf fragments; they seem to become brittle on drying. Plants of this species can form pure stands or large colonies and are considered invasive in Somalia. They tend to form impenetrable barriers, and are used for hedging in **T** 5. The berries are eaten by camels and goats in Somalia, though they are reported to be bitter and not eaten in **K** 1 where, however, elephants are reportedly fond of the plant. Verdcourt & Trump (Common Poisonous Plants for East Africa, 1969) noted that there was a record of *S. arundo* being used to procure a miscarriage; though this tragically resulted in the woman dying 48 hours later, there was no mention of the plant part used. However, the species is also listed as a medicinal plant in tropical Africa in PROTA 11: 308 (2002).

The placement of Dammer's *S. diplacanthum* is difficult in the absence of extant type material. Most of the leaf characters cited in the protologue are analogous to those found in *S. arundo,* except that Dammer described the margins as entire or rarely smoothly repand – features more typical of *S. dennekense.* Later, however Bitter (1923) described the margins as usually sinuate-repand with 2–3 lobes. Since he presumably made his description from *Fischer*'s holotype in Berlin, this species is considered to be synonymous with *S. arundo.* All other features cited in the protologue agree with those typifying *S. arundo.*

Standleys' *S. helleri* is undoubtedly synonymous with *S. arundo*; most features described in the protologue are identical with the exception of the inflorescence structure which was described as solitary, paired or few-flowered racemose cymes. However, the type specimen seems to be part of a mature lower stem fragment; younger stems apices would probably have borne branched and several-flowered immature inflorescences.

64. **Solanum dennekense** *Dammer* in E.J. 38: 57 (1905); C.H. Wright in F.T.A. 4, 2: 573 (1906); Bitter in F.R. 16: 144 (1923); Polhill, *Solanum* in E & NE Africa: 33 (ined., 1961); E.P.A.: 865 (1963); Whalen in Gentes Herb. 12: 265 (1984); Jaeger, Syst. stud. *Solanum* in Africa: 419 (1985, ined.); K.S.T.L.: 580 (1994); U.K.W.F., 2nd ed.: 243 (1994); Friis in Fl. Somalia 3: 215 (2006) & in Fl. Eth. 5: 135 (2006) & in Fl. Somalia 3: 215 (2006). Type: Ethiopia, Arussi-Galla, Dennek, *Ellenbeck* 1965 (?B†, holo.)

Shrubs 1–5 m high, rarely herbaceous, armed; young stems whitish or occasionally yellowish to 6 mm diameter, with dense stellate-tomentose pubescence, hairs as for *S. arundo*, glabrescent; simple glandular hairs sometimes also present throughout; main stems and branches with scattered recurved sharply pointed prickles as in *S. arundo*, 6.5–10 × 5.5–8 mm basally. Leaves alternate or opposite, light green above, whitish to light green below, rough, broadly ovate to obovate, 2.8–4.2(–7.5) × 2.3–3.5(–5.6) cm, bases cordate (cuneate), margins entire to sinuate, definite lobes absent, apices obtuse to acute; veining prominent; both surfaces stellate-tomentose, the hairs to 1 mm diameter with up to 14 equal rays 0.2–0.5 mm long, those below usually denser and enmeshed; usually with a few (to 4) straight acicular prickles 4–9 × 0.4–1 mm on midribs; petioles yellowish, 0.3–1.5 cm long, with occasional straight acicular prickles. Inflorescences terminal to lateral, simple or forked to multiply-branched, lax, racemose cymes. Flowers few to many(–16), sometimes aromatic, 5-(–6)-merous; axes yellowish and densely stellate-tomentose, hairs as on stems; peduncles erect, 0.8–1.7 cm long in flower, short and woody in fruit; pedicels erect to recurved and 0.6–1.3 cm in flower, 0.9–1.5 cm in fruit and strongly recurved, thickened beneath calyx. Calyx yellowish, campanulate, 6–11 mm long, stellate-tomentose externally, glabrous internally with distinct venation, with straight acicular prickles to 2.5 × 0.2 mm usually on tube veins; lobes often unequal, narrowly triangular to -ovate, becoming strongly recurved between corolla lobes, 2.5–6 × 1.5–4 mm, obtuse to acute; adherent basally in fruit then becoming strongly reflexed, 4–8 × 3–7 mm. Corolla blue to purple, sometimes with white margins, often with a yellow basal star, stellate, 2.2–3.6 cm diameter, tube 2–3.5 mm long, glabrous; lobes ligulate to lanceolate, 7–15 × 1.6–6 mm, acute, densely stellate-tomentose externally, internally with stellate hairs on median veins, margins and towards lobe apices, lobes strongly reflexed after anthesis. Stamens usually equal; filaments free for 0.6–1.5 mm, glabrous; anthers bright yellow to orange, poricidal, (5–)6–8.3 × 1–1.6 mm, usually connivent. Ovary brownish, 1.5–3 × 1.5–3.2 mm, stellate-pubescent with long silky median rays adherent to ovary and stylar base, usually bilocular sometimes 4-locular; style usually straight, stellate-pubescent in lower part, 8–12 mm in hermaphrodite flowers, sometimes curved and exserted up to 5 mm, 3–6 mm enclosed within the staminal tube in 'male' flowers, stellate-pubescent in lower part, often with long median rays clasping the stylar base; stigma clavate, usually bi- occasionally tri-lobed, 1–2.2 × 0.6–0.8 mm. Fruit rough, yellow, globose to ovoid, 1–3 × 1.6–3.2 mm, with thick (0.3–1.5 mm) tough pericarp, sparsely to moderately stellate-tomentose. Seeds > 100 per berry, smooth, orange or brown, 2–3 × 1.4–2.5 mm, obovoid to orbicular, occasionally reniform, not flattened, shallowly punctuate/verrucate; sclerotic granules absent. Fig. 26/9–14, p. 217.

KENYA. **Northern Frontier District: Saku, Lebaaogut Bridge, 10 km from Marsabit Town towards Karare, 5 July 2006, *Kirika et al.* SFLP 04/17/06!**; Masai District: Narok, Aitong Area, 19 Apr. 1961, *Glover et al.* 724! & Olemelepo, Mara Game Reserve, 23 Jan. 1972, *Taiti* 1883!

TANZANIA. Musoma District: Mugango, 1 June 1969, *Tanner* 4284!; Masai District: near Loliondo, 24 Feb. 1954, *A.C. Brooks* 86!; Maswa/Kwimba/Mwanza District: Simiyu, Serengeti Plains, Sep. 1961, *Newbould* 5859!
DISTR. **K** 1, 6; **T** 1, 2; Ethiopia, N Somalia (see Note)
HAB. Grassland, often near old Masai manyattas/enkang, *Terminalia* wooded grassland, woodland, *Acacia-Bauhinia* bushland, *Acacia-Commiphora-Grewia* bushland; 1200–1850 m

SYN. *S. ogadense* Bitter in F.R. 16: 143 (1923); Lanza in Chiovenda *et al.*, Miss. Biol. Borana Bot.: 200 (1939); E.P.A.: 874 (1963). Type: Somalia, Ogaden Steppe, *Keller* s.n. (Z, syn.); Milmil, *Ruspoli & Riva* 342 (239) (FT, syn., photo.!)
     S. gillettii Hutch & E.A. Bruce in K.B. 1941: 162 (1941); E.P.A.: 867 (1963). Type: Ethiopia, Harar-Gobelli road, *Gillett* 5255 (K!, holo.)

NOTE. Herbarium specimens of this species are generally poor – with one collector noting that the leaves immediately crinkle on picking. Another noted that the dense whitish indumentum often sloughs off when touched. In Somalia and Ethiopia this species is often found near cultivated ground and the edges of fields or in groups in grassy places. The plants are browsed by goats and donkeys in **K** 6, but their frequency in old Masai homestead areas is probably due to their use as protective hedging for cattle against lions and other predators. The flowers are reportedly visited by large bees in Somalia and as being aromatic in **T** 1.
     Friis in Fl. Eth. reported the occurrence of this species in Yemen – but no specimens were seen from this area; he also cited an 'unlocalised' specimen of this taxon which was probably from eastern Eritrea.
     The photograph of Ruspoli & Riva's syntype clearly illustrates sparse recurved prickles and the entire broadly ovate leaves with prominently cordate bases characteristic of *S. dennekense*.
     The holotype of *S. gilletti* is a good specimen which exhibits all of the morphological features that characterise *S. dennekense*.

65. **Solanum coagulans** *Forssk.*, Fl. Aegypt.-Arab.: 47 (1775); Blundell, Wild Fl. E. Afr.: 190, t. 841 (1992); U.K.W.F., 2ⁿᵈ ed.: 243 (1994); Friis in Fl. Eth. 5: 144, fig. 158.16.4–6 (2006) & in Fl. Somalia 3: 218 (2006). Type: Yemen, *Forsskål* 1744 (C!, lecto. designated here [IDC microfiche 2200, 101: III.1,2])

Erect herb with a woody base, 0.1–0.7 m, densely armed; young stems densely stellate-pubescent, trichomes porrect, translucent, sessile or stalked, stalks to 0.1 mm, rays (7–)8, 0.15–0.35 mm, midpoints reduced or ± same length as rays; prickles straight, 2–8 mm long, less than 0.5 mm wide at base, uniform pale yellow lighter than the stems. Leaf blades drying concolorous to slightly discolorous, ovate to narrowly lanceolate (narrowly elliptic), uniform yellowish green, (1.5)5–14 × (0.8–)1.5–6.5 cm, base attenuate to truncate or broadly cordate, usually oblique, margin lobed, the lobes 3–6(–11) on each side, up to 1.2 cm long, shallow and wide, rarely obovate, apically rounded, extending up to $^1/_2$ ($-^3/_4$) of the distance to the midvein; apex rounded to acute; stellate-pubescent on both sides, trichomes on abaxial surface porrect, sessile or stalked, stalks to 0.2 mm, rays 6–10, (0.1–)0.2–0.45 mm, midpoints reduced, trichomes of adaxial surface smaller; primary veins 4–8(–12) pairs; petiole 1–2 cm, $^1/_8$–$^1/_2$ of the leaf length. Inflorescences not branched, 2–3 cm long, with 1–6(–10) flowers; peduncle 0–5 mm long; rachis 0–0.8(–3.8) cm; peduncle and rachis unarmed or with reduced prickles; pedicels 0.7–1.5 cm long, in fruit 1–2(–2.8) cm long, usually unarmed. Flowers perfect, 5-merous. Calyx lobes 4–6.5 mm long, narrowly deltate, apically acute to narrowly acuminate or caudate, usually armed. Corolla white to mauve, 1.5–2.3 cm in diameter, lobed for ± $^2/_3$ of its length; lobes deltate, 4.5–8 × 4–7 mm. Stamens unequal, free portion of 4 short filaments 0.5–1.5 mm long, free portion of long filament 1.8–2.5 mm long; anthers 4.5–6.5 mm. Ovary densely stellate-pubescent; style curved, 6–9 mm long. Berries 2–8 per infructescence, spherical, 0.7–1.3 cm in diameter; fruiting calyx with numerous prickles. Seeds 2.3–2.8 × 1.7–2.5 mm, almost black. Fig. 27, p. 221.

UGANDA. Karamoja District: Kangole, 22 May 1940, *A.S. Thomas* 3482!

Fig. 27. *SOLANUM COAGULANS* — flowering and fruiting habit. Reproduced with permission from Gentes Herbarum 12: 265, 1984

KENYA. Turkana District: Lorenkipi [Lorengipe], 30 Jan. 1965, *Newbould* 6868!; Meru District: Meru Game Reserve, Ura, 18 June 1963, *Mathenge* 160!; Tana River District: Garissa, 26 Dec. 1942, *Bally* 2199!

TANZANIA. Musoma District: Seronera, 24 Mar. 1961, *Greenway* 9893!; Masai District: road from Longido to Engare Naibor, 25 Mar. 1970, *Richards* 25667!; Lushoto District: Kivingo, 2 Jan. 1930, *Greenway* 2010!

DISTR. **U** 1; **K** 1–4, 6–7; **T** 1–3; Sudan, Eritrea, Ethiopia, Somalia; Egypt, Saudi Arabia, Yemen

HAB. Common weed of cultivated land, grazed ground, roadsides, coastal plains, and wooded grassland; usually on sand, silt or loam; 200–1500 m

SYN. *S. dubium* Fresen. in Mus. Senckenb. 1: 166 (1833); T.T.C.L.: 584 (1949). Type: Arabia, *anonymous collector* in ?Herb. Fresenius (FR, holo.)

    S. dubium Fresen. var. *brevipetiolatum* Dunal in DC., Prodr. 13(1): 333 (1852). Type: Saudi Arabia, Jiddah, *Schimper* I.837 (G-DC!, lecto., designated here [best duplicate distribution; sheet cited in protologue]; BM!, CAS, E, LE!, M, MPU!, NY, P!, W, US, isolecto.)

    S. dubium Fresen. var. *cisterninum* Dunal in DC., Prodr. 13(1): 333 (1852). Type: Saudi Arabia, Jiddah, *Schimper* I.903 (G-DC!, holo.; A, E, GOET!, LE!, M, MPU, P!, W, iso.)

    S. thruppii C.H. Wright in K.B. 1894: 129 (1894). Type: Ethiopia, Hahi, Mar., *James & Thrupp* s.n. (K000413997! lecto., designated here [best material])

S. *ellenbeckii* Dammer in E.J. 38: 58 (1905). Type: Somalia, Wooqooyi Galbeed and Togdheer region, Hensa, *Ellenbeck* 247 (B†, holo.)

S. *dubium* Fresen. var. *denseaculeatum* Bitter in F.R. Beih. 16: 301 (1923). Type: Sudan, Sennar, Khartoum, *Kotschy* 344 (W0000611!, lecto., designated here [best duplicate distribution; best reproductive material]; BM!, GH, GOET!, K!, LE!, M, MO!, NY, P!, W, isolecto.)

S. *dubium* Fresen. var. *subinerme* Bitter in F.R. Beih. 16: 301 (1923). Type: Sudan, Kordofan, near Abu-Gerad, *Kotschy* 39 (W0000619!, lecto., designated here [best material]; BM!, K!, LE!, M, P!, W, isolecto.)

S. *dubium* Fresen. var. *dolichoplocalyx* Bitter in F.R. Beih. 16: 302 (1923). Type: Ethiopia, Ogaden, Mil Mil, *Riva in herb. Ruspoli* 877 (265) (FT, holo.)

S. *depressum* Bitter in F.R. Beih. 16: 303 (1923). Type: Ethiopia, Vallata del Ueb Karanle, *Riva in herb. Ruspoli* 1147 (496) (FT, holo.)

S. *ellenbeckii* Dammer var. *oligoplum* Bitter in F.R. Beih. 16: 305 (1923). Type: Somalia, Gedo and Bay region, Homare sul Gonare, *Riva in herb. Ruspoli* 1231 (792) (FT, holo.)

NOTES. *Solanum coagulans* is a common weed of northeast Africa and the Arabian Peninsula, immediately recognisable in fruit by its densely spiny accrescent calyx covering most of the pericarp. The lowermost stamen is 1.5–2 mm longer than the others, although this is easily overlooked in pressed specimens. Numerous thin pale yellow prickles also distinguish this species.

This species has been frequently referred to as *Solanum dubium* or *Solanum thruppii*. The epithet "*coagulans*" has been historically applied to members of *Solanum incanum* sensu stricto that have similarly lobed yellowish leaves and are sympatric with S. *coagulans* in northern Africa. *Solanum incanum* is a larger woody plant with equal stamens, non-accrescent calyx, thicker prickles, and larger fruits.

Somewhat similar densely spiny accrescent calyces are found in New World S. *sisymbriifolium* (which is invasive in South Africa and has been recorded from the Kenyan coast, p. 226) and S. *rostratum*, and the Madagascan endemics S. *toliaraea* D'Arcy & Rakot. and S. *mahoriense* D'Arcy & Rakot.

66. **Solanum melastomoides** *C.H. Wright* in K.B. 1894: 128 (1894); K.T.S.L.: 582 (1994); Friis in Fl. Eth. 5: 144, fig. 158.16.1–3 (2006) & in Fl. Somalia 3: 218 (2006). Type: Somalia, Bwobi, March 1885, *James & Thrupp* s.n. (K! [K000701275], lecto., designated here [best material]; K!, isolecto.)

Erect herb or shrub to 1(–3) m, armed; young stems densely stellate-pubescent, trichomes porrect, translucent, sessile or stalked, stalks to 0.1 mm, rays 8–9, 0.1–0.25 mm, midpoints reduced; prickles 5–10 mm long, 0.5–6 mm wide at base, straight or slightly curved. Leaf blades drying concolorous to slightly discolorous, dull green, sometimes reddish, ovate to elliptic, (1.5–)2–4(–5) × 1.3–2.2(–3.2) cm, base attenuate to cuneate, usually equal, margin shallowly sinuate (entire), lobes 1–2(–3) on each side, shallow or semicircular, rarely obovate, apically rounded, up to 0.8 cm long, extending up to $^1/_4$ (–$^3/_4$) of the distance to the midvein; apex rounded; stellate-pubescent on both sides, trichomes on abaxial surface porrect, sessile or stalked, stalks to 0.1 mm, rays 4–10, 0.1–0.25 mm, midpoints reduced, trichomes of adaxial surface smaller; primary veins (1–)3–6 pairs; petiole 0.6–1(–1.5) cm, $^1/_4$–$^1/_2$ of the leaf length. Inflorescences not branched or branched once, 3–4.5 cm long, with 2–6(–8) flowers; peduncle 0–13 mm long; rachis 0.1–2 cm; peduncle and rachis usually unarmed; pedicels 0.8–1.5 cm long, in fruit 1.2–1.7 cm long, unarmed. Flowers largely perfect, 5-merous. Calyx lobes ovate to widely deltate, 2–3(–4.5) mm long, obtuse to acuminate, unarmed. Corolla mauve, noticeably zygomorphic, 2–2.8 cm in diameter, lobed for $^1/_4$–$^1/_2$ of its length; lobes broadly deltate, 4–7 × 7–12 mm. Stamens unequal, free portion of the short filaments 0.7–1 mm long, free portion of the long filament 4.7–7.2 mm long; anthers 4.5–7 mm. Ovary stellate-pubescent; style 10–16 mm long, curved. Berries 1–3 per infructescence, spherical, 1.2–1.8 cm in diameter; fruiting calyx unarmed. Seeds 2.5–3 × 2–2.4 mm, almost black.

KENYA. Northern Frontier District: Dandu, 7 Apr. 1952, *Gillett* 12716!
DISTR. **K** 1; Ethiopia, Somalia
HAB. Open bushland or rocky places, on limestone or red sand; ± 750 m

NOTES. Easy to identify with one stamen filament 4–6 mm longer than the others, and fruit filled with black seeds; distinguished from *S. coagulans* by its unarmed non-accrescent calyx. Only one collection from the FTEA area is known.

The sheet cited in the protologue of *S. melastomoides* (Wright, 1894) contains two James & Thrupp collections. The syntype from Harradigit has uncharacteristically large leaves, inflorescences, and flowers, and a greater cover of prickles; the syntype from Bwobi is chosen as the lectotype because its smaller size and the presence of darker curved spines make it more representative of *S. melastomoides*. In the F.T.A. Wright (1906) cited a second specimen of *S. melastomoides*, collected by Miss Edith Cole from Dooloo, and Bitter (1923) has repeated this. Jaeger (1985) correctly identifies the type as *James & Thrupp* s.n. but does not recognise the fact that it constitutes two gatherings. Lester (unpublished manuscript) erroneously adds *Cole* s.n. to the type citation of *S. melastomoides*.

67. **Solanum somalense** *Franch.* in Révoil, Faune et Fl. Çomalis: 47 (1882); C.H. Wright in F.T.A. 4, 2: 212 (1906); Chiovenda in Miss. Stef.-Paoli 1: 129 (1916); Bitter in E.J. 54: 504 (1917); Polhill, *Solanum* in E & NE Africa: 17 (ined., 1961); E.P.A.: 878 (1963); Whalen in Gentes Herb. 12(4): 273 (1984); Jaeger, Syst. studies *Solanum* in Africa: 361 (1985, ined.); Friis in Fl. Eth. 5: 128, fig. 158.10.1–6 (2006) & in Fl. Somalia 3: 210 (2006). Type: N Somalia, Çomalis, *Revoil* 83 (P, holo. photo!; P, iso. photo!)

Shrub or woody herb to 2.5 m high, occasionally scrambling, bark reddish brown with prominent lenticels; stems grey to yellow when young from dense stellate pubescence of short hairs composed of 8 equal ± rays 0.1 mm with the central erect eglandular ray of the same size, often mixed with stalked stellate and short glandular hairs, glabrescent. Leaves usually alternate, coriaceous, yellowish to light green, ovate to elliptic, (1.8–)2.4–7 × (0.8–)1.5–6 cm, bases cuneate (cordate), decurrent and often oblique, margins entire, apices obtuse to acute; dark green and stellate-pubescent on both surfaces when mature, hairs as on stems but larger, denser on margins, veins and midribs of lower surfaces where stalked brown glands usually also present; petioles 0.5–1.6(–2.5) cm long, sometimes curved downwards. Inflorescences terminal often becoming lateral, simple, forked (sometimes several times), or simple and forked lax cymes 3–7 cm broad, 3–16(–22+)-flowered. Flowers zygomorphic; peduncles 0.4–2 cm long becoming stout and woody; pedicels erect and 0.7–1.8 cm long in flower but curved beneath calyces, often recurved and 0.7–2 cm long in fruit; axes pubescent, often with stellate hairs and stalked glands inter-mixed, sometimes with only scattered hairs at maturity. Calyx cupulate, 3–4.5(–6) mm long, stellate-pubescent intermixed with stalked glands externally, often with only scattered hairs in fruit; lobes narrowly triangular to linear, (1.5–)2–4.5 × 1–2.8 mm; adherent becoming strongly reflexed to pedicels in fruit when 3–7 × 1.6–3.5 mm. Corolla purple occasionally white, rotate/stellate to stellate, 2–3.2 cm diameter; tube to 1.2–2 mm long; lobes broadly ovate to lanceolate with prominent median veins, 6–11 × 2.2–6 mm, margins sometimes inrolled, stellate-pubescent externally with hairs denser towards lobe apices, sparsely pubescent to glabrous internally, spreading to strongly reflexed after anthesis. Stamens unequal; filaments free for 0.5–1(–1.3) mm, glabrous; anthers yellow to orange/brown sometimes with purple apices, four 4.8–7.5 × 0.5–1.2 mm, the fifth 7–10 × 0.8–1.6 mm with an incurved apex, usually splayed exposing the style. Ovary 1–2 × 1.1–1.8 mm, glabrous, bilocular; style sigmoidal, curved apically over the longest anther, 0.9–1.3 mm long, widening to 0.6–1.1 mm beneath the stigma, ± glabrous, exserted 3.5–6 mm; stigma capitate to bilobed, 0.6–0.9 mm diameter. Berries smooth, glossy, orange to yellow, globose to ovoid, 0.7–1.4 cm broad and 0.8–1.5 cm long, pericarp coriaceous and sometimes translucent, occasionally with scattered stellate hairs. Seeds 11–30 per berry, yellow to light brown, reniform, ovoid or orbicular, 3–4.4 × 3–4 mm, flat, foveolate; sclerotic granules absent. Fig. 23/12–17, p. 188.

Kenya. Northern Frontier District: Dandu, 16 Mar. 1952, *Gillett* 12549! & Wajir, Catholic
	Mission Compound, 27 Apr. 1978, *Gilbert & Thulin* 1132! & Wajir, Jan. 1955, *Hemming* 474!
Distr. **K** 1, 7; Ethiopia, Djibouti and Somalia
Hab. *Commiphora* woodland, *Acacia-Commiphora* bushland, *Commiphora-Lannea-Boswellia* bushed
	grassland, thicket edges, roadsides, steep lava slopes; 60–1100 m

Syn. *S. withaniifolium* Dammer in E.J. 38: 58 (1905); C.H. Wright in F.T.A. 4, 2: 572 (1906). Type:
		Somalia, Dabab, *Ellenbeck* 166 (B†, holo.)
	S. somalense Franch. var. *withaniifolium* (Dammer) Bitter in E.J. 54: 506 (1917)
	S. anisantherum Dammer in E.J. 38: 187 (1906). Type: Kenya, Kwale District: between Teita
		and Vanga [Wanga], *Fischer* 404 (?B†, holo.)
	S. somalense Franch. var. *anisantherum* (Dammer) Bitter, in Dammer in E.J. 54: 505 (1917)
	S. somalense Franch. var. *parvifrons* Bitter in E.J. 54: 506 (1917). Type: Somalia, Somadu,
		Ellenbeck 255 (?B†, holo.)

Note. The synonymy of *S. anisantherum* with *S. somalense* is based on Bitter's protologue which
described subulate calyx lobes, the longer fifth anther and the apically curved styles.
	Bitter's two varieties of *S. somalense* are largely based on minor variations in inflorescence
sizes, floral and vegetative dimensions and are unlikely to be of specific significance in this
species. Although Polhill (1961) thought that Bitter's smaller-dimensioned var. *parvifrons* was
a valid variety, the variability exhibited by this species across its distributional range suggests
that such formal recognition would be unwise. Both varieties are therefore considered to be
synonyms of *S. somalense*.
	Polhill (1961) considered section *Anisantherum* to be monotypic and more naturally allied
to section *Torvaria* (Dunal) Bitter in the subgenus *Leptostemonum* rather than to the subgenus
Eusolanum in which Bitter (1917) had placed it. However, as Jaeger (1985) pointed out, Bitter
was uncertain which of these two subgenera the species belonged to but finally favoured the
latter largely because of the absence of prickles. Seithe (in E.J. 81: 296, 1962) later moved this
section into her subgenus *Stellatipilum* in which she had grouped all stellate-haired species
and which incorporated the subgenus *Leptostemonum*. The section *Anisantherum* is composed
of two disjunct species, the NE African *S. somalense* and the predominantly Indian species *S.
pubescens* Willd. However, Jaeger thought the provisional records of two Yemeni specimens of
the latter species could signify an earlier distributional range continuous from India to NE
Africa. Whalen (1984) was uncertain of the placement of this species group within
Leptostemonum.
	Jaeger (1985) considered that the biological significance of the unequal anthers in *S.
somalense* was unknown when it was first described, though this phenomenon is also seen in
other sections of the genus such as *Androceras* and *Nycterium*. Wright (1906) seemed to
confuse *S. somalense* with *S. carense* Dunal (a member of the section *Anthoresis*); though his
morphological description matches that of *S. somalense*, three of the Somalian specimens
that he cited for *S. carense* (page 220 – *James & Thrupp* s.n.; *Edith Cole* s.n. and *Appleton* s.n.)
also belong to the former species.
	Herbarium specimens indicate that relatively few fruits develop from each multi-flowered
inflorescence. A few Ethiopian specimens (e.g. *Burger* 3001) exhibited prominent
multicellular-headed glandular hairs interdispersed with the dense stellate hairs. The
density of the striking stellate-haired pubescence varies considerably with that on the
Somalian specimens generally being more dense than the often glabrescent forms found in
Ethiopia. *Bally & Smith* described the flowers of their Kenyan specimen (14565) as having
the distinct odour of nutmeg. There are conflicting reports on the edibility of the fruit of
this species; they are occasionally eaten in Somalia, but *Gillett* (3901) reported the berries as
being poisonous to man. However, sheep, cows, goats and stock camels often browse and
graze on the plants in both Somalia and Ethiopia, sometimes heavily, though *McKinnon*
(8/4) noted that this could result in non-fatal 'tympany'. Medicinally the burnt, ground
fruits are used on wounds in Somalia, where children also make spinning tops by pushing
thorns into their berries.
	The Somalian species *S. melastomoides* Wright (see species 66) is superficially similar to *S.
somalense*. The pubescence and vegetative characters are similar, as are the long arching styles
and the unequal stamens. However in *S. melastomoides* the curvature of the fifth stamen is due
to the filament being much longer than the other four, and not to the anther length itself as
in *S. somalense*.

68. *Solanum capsicoides* All., Auct. Syn. Meth. Stirp. Hort. Reg. Taurensis: 64 (1773) & in Mélanges Philos. Mathém.: 64 (1774); Chiovenda in Ann. Bot. 10: 19 (1912) & in Bot. Miss. Con. Kenya: 88 (1935), as *S. capiscoides* Guatteri; Polhill, *Solanum* in E & NE Africa: 37 (ined., 1961), as *S. capiscoides* Guatteri; Nee in Biol. & Tax. Solanaceae: 574 (1979); Purdie, Symon & Haegi, Fl. Australia 29: 118 (1982); Troupin, Fl. Rwanda 3: 378 (1985); Jaeger, Syst. stud. *Solanum* in Africa: 477 (1985, ined.); Bukenya & Hall in Bothalia 18: 84 (1988); RHS Gard. Dict. 4: 317 (1997); Gonçalves in F.Z. 8, 4: 69 (2005); Friis in Fl. Eth. 5: 145 (2006). Type: cultivated at Turin, *Allioni* s.n. (TO, holo., photo!; WIS, photo!)*

Syn. *S. ciliatum* Lam., Encycl., 2; 21 (1799); Heine in F.W.T.A. 2ⁿᵈ ed., 2: 334 (1963). Type: cultivated in Paris, *Lamarck* s.n. (P-LAM!, holo.)
   S. ciliare Willd., Enum. Hort. Berol: 237 (1809). Type: cultivated at Berlin of unknown origin, *Willdenow* s.n. (B-W, holo., photo!)
   S. aculeatissimum sensu Chiovenda in Racc. Bot. Mss. Consol. Kenya: 88 (1935), *non* Jacq.

Note. This species is native to coastal Brazil and is now a successful weed in many tropical and subtropical areas; it only occurs sporadically in Africa where it has been intoduced as an ornamental in Ghana, Sierra Leone, Ethiopia, Zimbabwe and South Africa (Natal). It prefers drier open lowland habitats, has apparently spread only slowly in Africa (Nee 1979) and has failed to become naturalised. The species is often known by the synonymous name *S. ciliatum* Lam. There has been much confusion between this species and *S. aculeatissimum* with Polhill (1961) suggesting that the latter was probably synonymous with *S. capsicoides*, and if so the latter name would have to take precedence. The two are most easily differentiated by their seed characteristics; those in *S. capsicoides* are larger (> 4 mm long) flat and winged, whereas those in *S. aculeatissimum* are smaller (< 3 mm long) rounded and without wings. It is probable that the Kenyan specimens listed as this species by Chiovenda (1912, 1935) were misidentified and belong to *S. aculeatissimum*. It is commonly known as the cock-roach berry, and the berries have been reported as poisonous to calves in Australia.

69. *Solanum atropurpureum* Schrank in Syll. Ratisb., 1: 200 (1824): T.T.C.L.: 580 (1949), as *atripurpureum*; Polhill, *Solanum* in E & NE Africa: 37 (ined., 1961), as *atripurpureum*. Type: Brazil, "unde D de Martius seminis misit"; specimen not cited

Commonly known as the Blue potato bush, this species is native to Brazil. Section *Acantophora*; it is a herb or subshrub growing to 2 m, and armed with dense slender spreading violet prickles up to 2.5 cm long. Leaves broadly ovate, up to 23 cm long and sinuate to deeply pinnatifid with stellate-pubescent lower surfaces. Flowers typically up to 1.9 cm diameter with calyces which are prickly externally; described as yellow in T.T.C.L.; this is unlikely as (apart from the tomato) no yellow-flowered Solanums occur in Africa. It is likely that the flowers were in fact purple, drying paler and somewhat yellowish. Polhill (1961) recorded that it had been cultivated at Amani and the plants cited in T.T.C.L. were found on the roadside near Amani; this area is notorious for the occurrence of oddities. As this part was going to press, Maria Vorontsova discovered that scans of specimens confirmed the presence of this taxon cultivated at Amani in the past.

70. *Solanum rantonnetii* Lesc. in Hérincq, Hort. Franc. sér. II, i : 197 (1859); Carr. in Rev. Hort.: 420 (1868) as *rantonetii*; Bitter in F. R. 12: 458 (1913) & in Abh. Nat. Ver. Bremen, 23: 152 (1914); U.O.P.Z.: 447 (1949) as *rantonnettii*; Polhill, *Solanum* in E & NE Africa: 12 (ined., 1961). Type: Grown from seed sent from La Plata (Argentina) to M. Rantonnet in Hort d'Heyeres; type specimen not cited; Plate 16 in Hérincq, L'Horticult. Franc. Ser.II, i : 197 (1859) could be selected as the lectotype.

---

* As this part was going to press, it was found that *Renvoize & Abdallah* 1925 from Tanzania, Mufindi (EA) is this species.

SYN. *Lycianthes rantonnetii* (Lesc.) Bitter in Abh. Nat. Ver. Bremen, 24: 332 (1920); Gonçalves in F.Z. 8 (4): 6 (2005)

NOTE. This species was recorded in U.O.P.Z. as being planted in many gardens in Zanzibar. Native to Paraguay and Argentina, it is an unarmed evergreen bush, shrub or climber which can grow up to 3 m. It is characterised by smooth almost glabrous stems, ovate entire to undulate large (to 10 cm long) leaves; clusters of dark blue to violet rotate flowers up to 3 cm diameter which have striking yellow throats and are often fragrant and mature yellow fruits. Belonging to the section *Lycianthes*, the calyces are characteristically composed of up to five small apical teeth alternating with up to five basal swellings. It is commonly known as the Paraguay Nightshade, and though it is a common ornamental in the F.Z. region no specimens of it were encountered in the herbarium during this revision, though Polhill (1961) recorded is as being cultivated in Kenya. This taxon is often described under *Lycianthes rantonnettii*.

71. *Solanum sisymbriifolium* Lam., Tabl. Encycl. 2: 25 (1794)

NOTE. A specimen from Kenya, Mombasa, Ostrich Farm, Ngutatu Baobab Sanctuary; 26 Aug. 2002, *Luke & Luke* 9041! has been tentatively identified by MSV as this taxon. It is a heavily armed herb or small shrub to 1.5 m tall. Leaves deeply pinnatifid to shallowly bipinnatifid, twice as long as wide, 5–15 × 2–7 cm. Flowers purple or white, in leaf-opposed or lateral inflorescences; corolla 2–3 cm in diameter. Fruit a (depressed-)globose bright red berry 12–16 mm across; fruiting calyx tightly investing the berry until ripe, then strongly reflexed, the lobes 15 × 5 mm.

    Native to dry regions from Ecuador to Argentina, but widely introduced in tropical and subtropical areas, and naturalized sporadically in Africa; in South Africa it is classified as a noxious weed (AGIS 2007). This taxon may be arriving in Kenya.

SPECIES OF UNCLEAR STATUS

Postscript: as this part was going to press, Maria Vorontsova discovered that scans of specimens made at the East African Herbarium indicated the presence of another *Solanum* taxon cultivated at Amani in the past:

72. *Solanum pectinatum* Dunal in DC., Prodr. 13(1): 250. 1852. Type: Peru, Maynas, *Poeppig* 2224 (G-DC, holo. [photos at GH, WIS]; F, iso.)

Erect, coarsely prickly perennial 1–2 m tall. The indument of simple, uniseriate hairs sets this species apart from other species of section *Lasiocarpa*, but habit, leaf morphology, inflorescence structure, floral and fruit morphology are all normal for the section.

Native to Costa Rica through Panama. Cultivated for its fruits, which have a sweet-tart flavor.

19. **WITHANIA**

Pauq., Diss. Belladone: 14 (1825), *nom. conserv.*; Dunal in DC., Prodr. 13(1): 453 (1852); Hunziker in Gen. Solanaceae: 264–270 (2001); Hepper in Solanaceae III: 211–227 (1991)

Hypnoticum C.F.W. Meissn., Gen. Comm.: 184 (1840)

Annual or perennial shrubs and herbs; hairs eglandular, usually branched, occasionally simple. Leaves alternate or in unequal pairs. Flowers occasionally solitary, usually in few- to many-flowered fascicles, axillary, usually hermaphrodite, occasionally dioecious; pedicels short to almost absent. Calyx usually actinomorphic, campanulate to urceolate-campanulate, with 5–6 lobes. Corolla usually actinomorphic, broadly campanulate, with 5–6 often recurved lobes. Stamens included or slightly exserted; filaments fused to lower part of corolla tube where enlarged and flattened, usually glabrous, slender where free; anthers equal, oblong, bilobed, basifixed by filaments inserted between thecae, sometimes convergent around stigma. Ovary superior, glabrous, bilocular, ovules numerous; disc annular, smooth, often crenulate, occasionally absent; style usually glabrous; stigma discoid-capitate, bilobed. Fruit a berry, mature pericarp thin and translucent, enclosed by enlarged and inflated accrescent chartaceous urceolate calyx, with the mouth wide open, narrow or almost completely closed. Seeds numerous, compressed, with or without sclerotic granules.

10–19 species; indigenous to the Old World and widely distributed throughout Africa and the Mediterranean to the warmer parts of Asia. Hepper (1991) considered *Withania* to be composed of ten species. Hunziker (cf. 2001) recently enlarged this genus by adding nine mesophytes formerly included in the genera *Mellissia* Hook. and *Physaliastrum* Makino, thereby extending the geographical range of this genus from the Canary Islands in the west, through Asia to China and Japan in the east. Symon (1991) also emphasised the closeness of *Mellissia* (a critically endangered endemic of St Helena) to *Withania*, but retained them as distinct genera. The genus *Withania* is close to *Physalis*, with which it is often confused, and like the latter also belongs to the subfamily Solanoideae. *Withania* is one of the two Solanaceous genera which D'Arcy (in Solanaceae 3: 105, 1991) considered to be truly Old World and which Symon (in Solanaceae 3: 146, 1991) considered to be a distinctive African Gondwanan element.

Withania somnifera (*L.*) *Dunal* in DC., Prodr. 13(1): 453 (1852); Engl., Hochgebirgsfl. Trop. Afr.: 374 (1892); E. & P. Pf.: 19 (1895); P.O.A. C: 351 (1895); C.H. Wright in Fl. Cap. 4(1): 107 (1904) & in F.T.A. 4, 2: 249 (1906); Z.A.E.: 283 (1914); F.P.N.A. 2: 204 (1947): W.F.K.: 89 (1948); T.T.C.L.: 592 (1949); K.T.S.: 538 (1961): F.F.N.R.: 377 (1962); E.P.A. 2: 857 (1963); Heine in F.W.T.A. 2nd ed., 2: 330 (1963): Troupin, Fl. Rwanda 3 : 383 (1985), as *Whithania*; Hepper in Solanaceae III: 223 (1991); U.K.W.F. 2nd ed: 244 (1994); K.T.S.L.: 583, ill., map (1994); Hepper in Fl. Egypt 6: 77 (1998); Thulin in Nord. J. Bot. 22(4): 388 (2002) & in Fl. Somalia 3: 203 (2006); Gonçalves in F.Z. 8(4): 56 (2005); Friis in Fl. Eth. 5: 154 (2006). Type: India, Tsierutti, *Linnaeus* 247.1 (LINN!, lecto., designated by Schönbeck-Temesy in Rechinger, Fl. Iranica, 100: 27 (1972), misprinted as LINN 241/1) [See also Jarvis, Order out of Chaos: 742 (2007)]

Annual or perennial woody herbs or shrubs, erect, spreading or decumbent, 0.3–3 m high, laxly to densely branched, sometimes strongly aromatic; main stems erect, terete, all parts (in FTEA area) tomentose to pilose with whitish-yellow branched hairs, denser on younger parts, woody parts glabrescent. Leaves usually membranaceous, ovate to lanceolate, (2.5–)4–11(–12.5) × (1.7–)3.5–7.8 cm, bases cuneate, margins usually entire, sometimes sinuate, apices obtuse to acute, densely tomentose when young with branched hairs, denser on veins, midribs and lower surfaces; petioles (0.5–)1–2.5(–4) cm but longer outside floral area. Inflorescences in compact fascicles of (3–)5–9 (–20+) flowers, axillary, epedunculate; pedicels erect to pendulous, 1–3 mm long in flower, 2–5.5 mm long in fruit, densely clustered around stem, densely tomentose; calyx cupulate to urceolate-campanulate, 2.5–5.5(–6.6) mm long overall, cup fused for 2–3.5 mm, usually actinomorphic with 5 narrowly triangular to ligulate recurved acute lobes 1–3.5 × 0.5–1.5 mm wide, densely tomentose externally. Corolla white, yellow, green, greenish-yellow or -white, usually hermaphrodite, broadly campanulate, (3–)4.5–6.5(–8) mm long overall with tube ± 1 mm long, usually actinomorphic with five broadly triangular recurved lobes 1.5–3(–4.5) × 1–2 mm wide, densely tomentose externally. Stamens usually exserted; filaments free for 1.3–2.5 mm; anthers yellow to orange, equal, oblong to ovoid, 0.7–1.3 mm long, apiculate. Ovary dark brown, ovoid, 1.1–1.5 × 0.7–1.5 mm, smooth; disc greenish to brown, 1–2.4 mm diameter; style 2–3(–4.5) mm long, often exserted; stigma 0.2–0.5 mm broad. Fruit a smooth red, orange or yellow globose berry, 5–8(–10) mm diameter, mature pericarp often thin and translucent, enclosed by

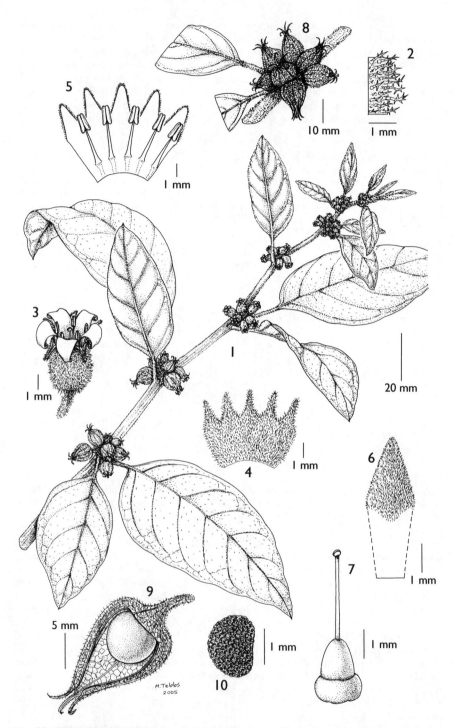

FIG. 28. *WITHANIA SOMNIFERA* — **1**, habit in flower and fruit; **2**, stem indument; **3**, flower; **4**, opened calyx; **5**, opened flower; **6**, outer corolla lobe surface; **7**, gynoecium; **8**, infructescence; **9**, berry within accrescent calyx; **10**, seed. 1 from *Faden* 74/943; 3–7 from *Paget-Wilkes* 1033; 2 & 8–10 from *Meyerhoff* 60M. Drawn by Margaret Tebbs.

enlarged chartaceous urceolate conspicuously veined usually opaque calyx 10–18(–22) × 8–12(–15) mm with the mouth almost completely closed and subtended by recurved calyx lobes 3–6(–9) × 0.5–2 mm. Seeds yellowish to brown, orbicular or discoid, sometimes reniform, 1.3–2.3 × 1.2–1.9 mm, foveolate; sclerotic granules usually absent. Fig. 28, p. 228.

UGANDA. Kigezi District: Ruzhumbura, Apr. 1939, *Purseglove* P668!; Karamoja District: Rupa, Matheniko County, July 1965, *Wilson* 1643!; Masaka District: Kabula, near Saza H.Q., 24 June 1950, *G.H.S. Wood* Y41!

KENYA. Northern Frontier District: Matakweni Hill, 3 km WSW of Wamba, 2–3 July 1974, *Faden & Faden* 74/943!; Naivasha District: outcrop above lake W of Lake Naivasha, 21 Jan. 1968, *E. Polhill* 79!; Teita District: Tsavo East National Park, 2 km W of Lugards Falls on Galana River, 9 Mar. 1977, *Hooper & Townsend* 1273!

TANZANIA. Musoma District: Chamuho, Ikizu, Lake Province, 2 Nov. 1953, *Tanner* 1697!; Morogoro District: Tungi, 29 Nov. 1961, *Semsei* 3457!; Iringa District: E part of Ruaha Park, 9 km NE of Msembe, 7 Aug. 1970, *Thulin & Mhoro* 636!

DISTR. U 1–4; K 1–7; T 1–7; widespread in Africa; also found throughout southern Europe and the Arabian Peninsula to India, Sri Lanka and China

HAB. *Acacia-Commiphora* bushland, grassland, wooded grassland, forests, river-banks, lakeshores, forest edges and clearings, a weed of shambas, bomas, old cultivations, fallow land, waste and disturbed places; may be locally common; 0–2300 m (– 2800 m fide U.K.W.F.)

CONSERVATION NOTES. Widespread; least concern (LC)

SYN. *Physalis somnifera* L., Sp. Pl.: 182 (1753); A. Rich., Tent. Fl. Abyss. 2: 95 (1850)
　　P. flexuosa L., Sp. Pl.: 182 (1753). Type: "Habitat in India", *Herb. Linnaeus* 247.2 (LINN!, lecto. designated by Deb in J. Econ. Bot. 1: 52 (1986))
　　Physaloides somnifera (L.) Moench, Meth. Pl.: 473 (1794); Hiern, Cat. Afr. Pl. Welw. 3: 752 (1898); A. Rich., Tent. Fl. Abyss. 2: 95 (1850)
　　Physalis tomentosa Thunb., Prodr. Fl. Cap.: 37 (1794). Type: none cited
　　P. somnifera L. var. *communis* Nees in Linnaea 6: 455 (1831). Type: numerous syntypes given including: Egypt, *Aucher Eloy* 600; Arabia, *Schimper* 167; Sudan, *Kotschy* 307 (BM!, isosyn.); Cape of Good Hope, *Drège* s.n. (G-DC, P), Herb. *Requien & Herb. Dunal* s.n., (?G, ?P, ?MPU, syn.)
　　P. somnifera L. var. *flexuosa* (L.) Nees in Linnaea 6: 454 (1831)
　　Withania somnifera (L.) Dunal var. *communis* (Nees) Dunal in DC., Prodr. 13(1): 454 (1852); P.O.A. C: 351 (1895)
　　W. somnifera (L.) Dunal var. *flexuosa* (L.) Dunal in DC., Prodr. 13(1): 454 (1852)
　　Hypnoticum somniferum (L.) Rodrig., *ined.* fide Dunal, DC., Prodr. 13(1): 453 (1852)
　　Withania somnifera (L.) Dunal var. *macrocalyx* Chiov. in Fl. Somalia: 237 (1929) & Fl. Somalia 2: 336 (1932), correcting the typographical error of *W. mucronata* Chiov. in Nuov. Giorn. Bot. Ital. 34: 845 (1927); E.P.A.: 858 (1963); Hepper in Solanaceae 3: 223 (1991). Type: Somalia, *Paoli* 24 (FT, holo.)
　　W. obtusifolia Täckholm in Svensk Bot. Tidskr. 26: 370 (1932) & Stud. Fl. Egypt, ed. 2: 476 (1974); Hepper in Solanaceae 3: 223 (1991). Type: Egypt, Gebel Elba, *Täckholm* s.n. (CAI, holo.)
　　W. microphysalis Suesseng. in Mitt. Bot. Staats. München 3: 93 (1951). Type: Namibia, *Rehm* s.n. (?M, holo., specimen locality not cited)
　　W. somnifera (L.) Dunal subsp. *obtusifolia* (Täckh.) Abedin *et al.* in Pakistan Journ. Bot. 23(2): 280 (1991)
　　W. chevalieri Gonç. in Garcia de Orta, Ser. Bot. 14(1): 149 (1999). Type: Cabo Verde Islands, Island of Sal, *Chevalier* 44296 (COI, holo: P, iso.)

NOTE. All the characters given in the protologue of *W. chevalieri* overlap with those found in *W. somnifera*, with the exception of the style being 0.5 mm shorter. Thulin too (2002) thought that Gonçalves' new species was a variable form of *W. somnifera* with smaller flowers and fruiting calyces.

　　The species is very variable morphologically and especially vegetatively. Hunziker (in Gen. Solanaceae: 264, 2001) described it as being polymorphic, while Hepper (1991) noted that many herbarium specimens were intermediate between *W. somnifera* and *W. obtusifolia*, though he maintained them as separate species. Abedin *et al.* (in Pakistan J. Bot. 23: 279, 1991) while conceding that the features distinguishing the latter two taxa overlapped, thought that the variability warranted subspecific recognition in the Saudi Arabian region. Brenan & Greenway (in T.T.C.L., 1949) described the occurrence of *W. aristata* (Ait.) Pauq. in forests above 1680 m in the Usambaras. Their description of the inflorescences as having solitary or clustered flowers with aristate calyx lobes could indicate the occurrence of this species at high altitudes. However, this species seems to be confined to the Canary Islands, and Hepper (1991) thought that this together with *W. frutescens* were unlike the other species in the genus. Since no specimens examined during this revision were encountered which matched the morphological characteristics of *W. aristata*, it has not been included in this account.

The typification of this species was discussed by Thulin (2002) who concluded that the Indian locality given on LINN 247.1 is correct despite the protologue including "Mexico" – where it has never been recorded – together with "Crete and Spain".

USES. The species has been used medicinally since ancient times; it was known to the ancient Egyptians with fruiting branches being included in the floral collar laid around the golden effigy of Tutankhamun. A large number of withanolides and alkaloids have been isolated from its leaves and roots, and there are numerous reports of its medicinal use throughout Africa. In East Africa root extracts are used to treat intestinal worms (**T** 1), stomach disorders (**T** 1, 5), thrush (**T** 1), pneumonia (**T** 1), gonorrhoea (**T** 2, **K** 1,6), irregular menstruation, coughs (**T** 2, 7), abscesses (**T** 2), childhood fevers and nightmares (**K** 6) and male sterility, as an emetic (**K** 2), a tonic (**K** 6), a diuretic (**K** 6), and an eye-wash (**K** 6, **T** 1). The plant is used for purificatory rites in **U** 2, crushed leaves are heated for use as an external pain killer in **K** 6 and to curdle milk by the Masai (**K** 6). There are conflicting reports of the edibility of this species which is considered to evoke bad luck especially if cut in Somalia, where, as in **T** 5, it is rarely eaten by stock though it is eaten by goats, cattle, donkeys and giraffe in Ethiopia. The flowers are reportedly visited by hunting wasps in Ethiopia.

INDEX TO SOLANACEAE

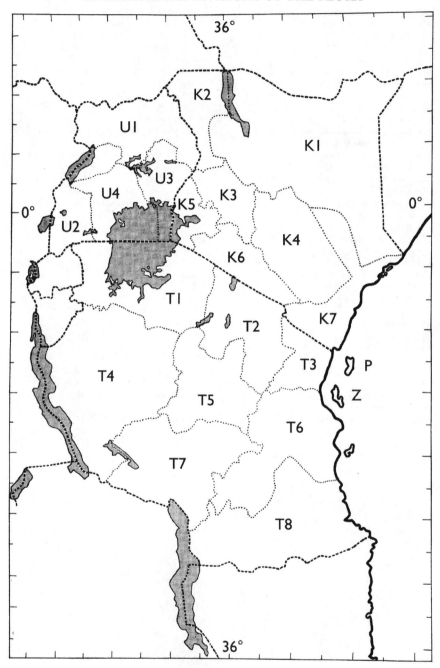

LIST OF ABBREVIATIONS

A.V.P. = O. Hedberg, Afroalpine Vascular Plants; **B.J.B.B.** = Bulletin du Jardin Botanique de l'Etat, Bruxelles; Bulletin du Jardin Botanique Nationale de Belgique; **B.S.B.B.** = Bulletin de la Société Royale de Botanique de Belgique; **C.F.A.** = Conspectus Florae Angolensis; **E.J.** = A. Engler, Botanische Jahrbücher für Systematik, Pflanzengeschichte und Pflanzengeographie; **E.M.** = A. Engler, Monographieen Afrikanischer Pflanzen-Familien und Gattungen; **E.P.** = A. Engler, Das Pflanzenreich; **E.P.A.** = G. Cufodontis, Enumeratio Plantarum Aethiopiae Spermatophyta; in B.J.B.B. 23, Suppl. (1953) et seq.; **E. & P. Pf.** = A. Engler & K. Prantl, Die Natürlichen Pflanzenfamilien; **F.A.C.** = Flore d'Afrique Centrale (*formerly* F.C.B.); **F.C.B.** = Flore du Congo Belge et du Ruanda-Urundi; Flore du Congo, du Rwanda et du Burundi; **F.E.E.** = Flora of Ethiopia & Eritrea; **F.D.-O.A.** = A. Peter, Flora von Deutsch-Ostafrika; **F.F.N.R.** = F. White, Forest Flora of Northern Rhodesia; **F.P.N.A.** = W. Robyns, Flore des Spermatophytes du Parc National Albert; **F.P.S.** = F.W. Andrews, Flowering Plants of the Anglo-Egyptian Sudan *or* Flowering Plants of the Sudan; **F.P.U.** = E. Lind & A. Tallantire, Some Common Flowering Plants of Uganda; **F.R.** = F. Fedde, Repertorium Speciorum Novarum Regni Vegetabilis; **F.S.A.** = Flora of Southern Africa; **F.T.A.** = Flora of Tropical Africa; **F.W.T.A.** = Flora of West Tropical Africa; **F.Z.** = Flora Zambesiaca; **G.F.P.** = J. Hutchinson, The Genera of Flowering Plants; **G.P.** = G. Bentham & J.D. Hooker, Genera Plantarum; **G.T.** = D.M. Napper, Grasses of Tanganyika; **I.G.U.** = K.W. Harker & D.M. Napper, An Illustrated Guide to the Grasses of Uganda; **I.T.U.** = W.J. Eggeling, Indigenous Trees of the Uganda Protectorate; **J.B.** = Journal of Botany; **J.L.S.** = Journal of the Linnean Society of London, Botany; **K.B.** = Kew Bulletin, *or* Bulletin of Miscellaneous Information, Kew; **K.T.S.** = I. Dale & P.J. Greenway, Kenya Trees and Shrubs; **K.T.S.L.** = H.J. Beentje, Kenya Trees, Shrubs and Lianas; **L.T.A.** = E.G. Baker, Leguminosae of Tropical Africa; **N.B.G.B.** = Notizblatt des Botanischen Gartens und Museums zu Berlin-Dahlem; **P.O.A.** = A. Engler, Die Pflanzenwelt Ost-Afrikas und der Nachbargebiete; **R.K.G.** = A.V. Bogdan, A Revised List of Kenya Grasses; **T.S.K.** = E. Battiscombe, Trees and Shrubs of Kenya Colony; **T.T.C.L.** = J.P.M. Brenan, Check-lists of the Forest Trees and Shrubs of the British Empire no. 5, part II, Tanganyika Territory; **U.K.W.F.** = A.D.Q. Agnew (or for ed. 2, A.D.Q. Agnew & S. Agnew), Upland Kenya Wild Flowers; **U.O.P.Z.** = R.O. Williams, Useful and Ornamental Plants in Zanzibar and Pemba; **V.E.** = A. Engler & O. Drude, Die Vegetation der Erde, IX, Pflanzenwelt Afrikas; **W.F.K.** = A.J. Jex-Blake, Some Wild Flowers of Kenya; **Z.A.E.** = Wissenschaftliche Ergebnisse der Deutschen Zentral-Afrika-Expedition 1907–1908, 2 (Botanik).

FAMILIES OF VASCULAR PLANTS REPRESENTED IN THE FLORA OF TROPICAL EAST AFRICA

The family system used in the Flora has diverged in some respects from that now in use at Kew and the herbaria in East Africa. The accepted family name of a synonym or alternative is indicated by the word "see". Included family names are referred to the one used in the Flora by "in" if in accordance with the current system, and "as" if not. Where two families are included in one fascicle the subsidiary family is referred to the main family by "with".

PUBLISHED PARTS

Foreword and preface
*Glossary
Index of Collecting Localities

Acanthaceae
  Part 1
  **Part 2
*Actiniopteridaceae
*Adiantaceae
Aizoaceae
Alangiaceae
Alismataceae
*Alliaceae
*Aloaceae
*Amaranthaceae
*Amaryllidaceae
*Anacardiaceae
*Ancistrocladaceae
Anisophylleaceae — as Rhizophoraceae
Annonaceae
*Anthericaceae
Apiaceae — see Umbelliferae
Apocynaceae
  *Part 1
  **Part 2
*Aponogetonaceae
Aquifoliaceae
*Araceae
Araliaceae
Arecaceae — see Palmae
*Aristolochiaceae
*Asclepiadaceae — see Apocynaceae
Asparagaceae
*Asphodelaceae
Aspleniaceae
Asteraceae — see Compositae
Avicenniaceae — as Verbenaceae
*Azollaceae

*Balanitaceae
*Balanophoraceae
*Balsaminaceae
Basellaceae
Begoniaceae
Berberidaceae
Bignoniaceae
Bischofiaceae — in Euphorbiaceae
Bixaceae
Blechnaceae
*Bombacaceae
*Boraginaceae
Brassicaceae — see Cruciferae
Brexiaceae
Buddlejaceae — as Loganiaceae
*Burmanniaceae
*Burseraceae
Butomaceae
Buxaceae

Cabombaceae
Cactaceae
Caesalpiniaceae — in Leguminosae
*Callitrichaceae
Campanulaceae
Canellaceae
Cannabaceae
Cannaceae — with Musaceae
Capparaceae
Caprifoliaceae
Caricaceae
Caryophyllaceae
*Casuarinaceae
Cecropiaceae — with Moraceae
*Celastraceae
*Ceratophyllaceae
Chenopodiaceae
Chrysobalanaceae — as Rosaceae

Editorial adviser, National Museums of Kenya: Quentin Luke
Adviser on Linnaean types: C. Jarvis

Parts of this Flora, unless otherwise indicated, are obtainable from:
Royal Botanic Gardens, Kew, Richmond, Surrey TW9 3AB, England. www.kew.org or www.kewbooks.com

* Only available through CRC Press at:
UK and Rest of World (except North and South America):
CRS Press/ITPS,
Cheriton House, North Way, Andover, Hants SP10 5BE.
e: uk.tandf@thomsonpublishingservices. co.uk

North and South America:
CRC Press,
2000NW Corporate Blvd, Boco Raton, FL 33431-9868, USA.
e: orders@crcpress.com

** Forthcoming titles in production
For availability and expected publication dates please check on our website, www.kew.books.com or email
publishing@kew.org

Information on current prices can be found at www.kewbooks.com or www.tandf.co.uk/books/

ROYAL BOTANIC GARDENS

First published in 2012 by
Royal Botanic Gardens, Kew
Richmond, Surrey, TW9 3AB, UK
www.kew.org

ISBN 978 1 84246 395 6

Distributed on behalf of the Royal Botanic Gardens, Kew in North America by the University of Chicago Press, 1427 East 60th Street, Chicago, IL 60637, USA.

British Library Cataloguing in Publication Data
A catalogue record for this book is available from the British Library

Design and typesetting by Margaret Newman,
Kew Publishing, Royal Botanic Gardens, Kew.

Printed in the the USA by The University of Chicago Press

Kew's mission is to inspire and deliver science-based plant conservation worldwide, enhancing the quality of life.

Kew receives half of its running costs from Government through the Department for Environment, Food and Rural Affairs (Defra). All other funding needed to support Kew's vital work comes from members, foundations, donors and commercial activities including book sales.